The Organic Chemistry of Palladium

Volume I METAL COMPLEXES

ORGANOMETALLIC CHEMISTRY

A Series of Monographs

EDITORS

P. M. MAITLIS
MCMASTER UNIVERSITY
HAMILTON, ONTARIO
CANADA

F. G. A. STONE
UNIVERSITY OF BRISTOL
BRISTOL, ENGLAND

ROBERT WEST
UNIVERSITY OF WISCONSIN
MADISON, WISCONSIN

The Organic Chemistry of Palladium

Peter M. Maitlis
McMaster University
Hamilton, Ontario, Canada

Volume I   *Metal Complexes*

1971   *Academic Press*   New York and London

CHEMISTRY

ACADEMIC PRESS, INC.
111 Fifth Avenue, New York, New York 10003

United Kingdom Edition published by
ACADEMIC PRESS, INC. (LONDON) LTD.
Berkeley Square House, London W1X 6BA

LIBRARY OF CONGRESS CATALOG CARD NUMBER: 77-162937

PRINTED IN THE UNITED STATES OF AMERICA

E quindi
uscimmo
a riveder
le stelle
　　　—*Dante*

To Marion, Niccola, Sally, and Emily

Contents

Preface

The last twenty years have seen the emergence of the new field of organo-transition metal chemistry. It has now become one of the most important and exciting areas of chemical endeavor. This stems not only from the use of organo-transition metal complexes as catalysts and models for catalytic reactions but also from their intrinsic importance to our understanding of structure and bonding.

Only a few years ago our knowledge of the field could be comfortably fitted into one volume; today the same space would hardly suffice to cover one of the more popular aspects of the subject, and any attempt to deal comprehensively with it therefore requires some subdivision. Useful lessons can be learned from a classification according to either the metal, the ligand, or the reaction, but each such arrangement has its drawbacks. In the light of current knowledge a treatment based on the metal is perhaps the most rewarding, especially if this is accompanied by references to other, neighboring metals, and further subdivision by ligands and reactions.

The choice of palladium for the first work on the Transition Metals in The Organometallic Chemistry Series was dictated to a large degree by the number of organic reactions which palladium catalyzes. These include olefin oxidation, the oligomerization of olefins, dienes and acetylenes, carbonylation, coupling of arenes, vinylation, acetoxylation, isomerization, halogenation, and many others. In addition, the chemistry of the organometallic complexes of palladium and platinum is well explored and, in outline at least, well understood.

A complete treatment of even this limited topic covers a great deal of material and a wide variety of reactions. I have accordingly chosen to divide the monograph into two volumes, the first covering the metal complexes and their structures, bonding, and reactions, while the second deals with the catalytic processes and other reactions induced by palladium. Although I have treated the entire topic in a unified fashion, each volume is self-contained and may be read separately.

The catalytic reactions are of such variety that palladium promises to become as important in organic synthesis as the Grignard reagent or hydroboration, and may well be more versatile than either. In this connection, a most significant feature of palladium chemistry is the ease of reoxidation of Pd(0) to Pd(II). This has allowed the active Pd(II) to be regenerated *in situ* and has made the industrial synthesis of acetaldehyde from ethylene (Wacker process) under homogeneous conditions not only feasible but economically more attractive than any other route. This use of a rather rare and expensive metal in an industrial process only foreshadows other equally important developments.

Even when the commercial exploitation of a *homogeneously* catalyzed reaction appears to be unfavorable, the study of such processes can lead to the development of heterogeneous catalysts of high specificity for unusual reactions. The one-step palladium(II)-catalyzed homogeneous synthesis of vinyl acetate from ethylene is now, as a heterogeneous reaction, the most economical method for the large-scale production of vinyl acetate. While this is likely to be the pattern of much future industrial use of the platinum metals, only studies under homogeneous conditions can lead to a detailed understanding of the reactions involved and to rational new developments.

The aim of this two-volume work has been to collect the available data, both on the complexes and on the catalyzed reactions, and to fit them into a coherent pattern. A number of mechanisms are well established, but in many cases one can only speculate on reaction paths. In order to facilitate the development of rational hypotheses, I have included information on the inorganic chemistry of palladium and also on the chemistry, where it is relevant, of the neighboring elements. By doing so, I hope also to have defined more exactly the unique features of the metal. On present evidence we can conclude that the special features exhibited by Pd(II), for example, arise from its greater lability by comparison with Pt(II) and the lower affinity toward oxygen and nitrogen donor ligands by comparison with the even more labile Ni(II). Of particularly great interest is the question of whether, by a suitable choice of ligand, one metal cannot be "tuned" in such a way that it chemically resembles another. If this is possible, and present indications are that it is, then it should also be feasible to carry out catalytic reactions characteristic of palladium with other metals.

Although this work is designed specifically for the research worker in the field and for the organic chemist who wishes to make use of the wide variety of metal-catalyzed reactions now available to him, the inclusion of comparative studies and the introductory sections should also make it useful as a supplementary text for graduate courses.

Since I have tried to make these volumes of particular use to the organic chemist, I have emphasized the broader aspects of the mechanisms and have not included extensive tabulations of physical properties of complexes, which may all be found in the appropriate references. The newcomer to the field should, however, be warned that many of the properties of a given complex, such as the melting or decomposition points and even the color, may vary within quite wide limits.

Organic chemists, in particular, may be dismayed at my decision not to use rigorous IUPAC nomenclature for the complexes. In practice, this is cumbersome and tends to emphasize the metal, or other trivial ligands, at the expense of the organic ligand of interest, with the result that even a reader familiar with the field may take several minutes to grasp the structure. The simplifications used, which are essentially of such a nature as to emphasize the organic ligand [e.g., allylpalladium chloride dimer in place of di-μ-chlorobis(π-allyl)dipalladium], together with diagrammatic formulas will enable the text to be read more easily.

This work covers the literature on palladium comprehensively to 1970. In addition, a large number of references to work published in 1970 are included, and I hope that no significant advance in this area which appeared in press before 1971 has been omitted. Coverage of the chemistry of the neighboring elements and some aspects of the inorganic chemistry of palladium is, of necessity, more curtailed.

This work was started in 1968–1969 at Imperial College, London. I should like to express my appreciation to Professor G. Wilkinson and his colleagues for their generous hospitality and to the National Research Council of Canada for the award of a Senior Fellowship which made these volumes possible.

I should also like to thank all who so kindly read parts of the manuscript for their comments, in particular, Dr. R. F. Heck, Dr. P. Henry, Dr. J. Powell, Dr. J. F. Harrod, Dr. J. M. Davidson, and Professor G. C. Bond.

Peter M. Maitlis

Contents of Volume II

Catalytic Reactions

The Organic
Chemistry
of Palladium

Volume I   METAL COMPLEXES

Chapter I

The Inorganic Chemistry of Palladium Compared to That of the Neighboring Elements

A. INTRODUCTION

1. Historical

The early history of palladium is most unusual. The collaboration between William Hyde Wollaston and Smithson Tennant which began in 1800 resulted in a great improvement in the refining and fabrication of platinum. In the course of his careful study of native platinum from South America Wollaston discovered palladium in 1803 and rhodium in 1804. Wollaston announced his discovery in an anonymous leaflet in which he also offered the metal for sale. He did not disclose his identity as the discoverer of the metal until 2 years later.[1] His reason for doing this was probably to claim priority for the discovery while keeping secret his method for the commercial manufacture of malleable platinum.

Wollaston obtained the metal from native platinum by dissolving it in aqua regia; gold was precipitated from the solution with ferrous sulfate and platinum [as the hexachloroplatinate(IV)] on addition of ammonium chloride. On addition of iron to the filtrate, palladium was obtained as a gray powder.

The name was derived from Pallas (Minerva), a name given to a small planet which had been discovered in 1802 by the astronomer Olbers, a friend of Wollaston.

2. Occurrence, Production, and Consumption

Palladium is a relatively rare metal, occurring in the earth's crust to the extent of about 8.6×10^{-13} parts. It is found together with other platinum metals, usually as the metal, but sometimes combined with antimony, sulfur, gold, or mercury. The major part of the world's supply comes from ores found in South Africa, Canada, and the U.S.S.R. The actual methods used today to obtain the metal in high purity depend on the source and are closely guarded commercial secrets.

Total world production of the platinum metals amounted to 2.95×10^6 troy ounces in 1966. Consumption of all the platinum metals by industry in the United States amounted to 1.675×10^6 ounces; of this 894,000 ounces were Pd; 690,000, Pt; 70,000, Rh; 11,000, Ir; 8300, Ru; and 1800, Os. The main uses for palladium were: electrical (531,000), chemical (221,000), dental (67,000), jewelry (32,000), and petroleum industry (29,000).[2]

The U.S.S.R. is the largest producer of platinum metals and in 1966 the United States imported 985,000 ounces, largely from Russia.[2] In contrast, consumption of Pd in the United States in 1940 was 51,400 ounces of which about 45% was used in dental alloys and about 42% in the electrical industry.[3]

The current (January 1971) price of palladium metal is around $1.30 per gram.

3. Physical Properties and Electronic Structure

Massive metallic palladium is a white metal, as hard as platinum and extremely ductile. It forms ductile alloys with many metals with great ease, and can be deposited on metals. Since it also has high corrosion resistance, these properties account for its uses in electrical contacts, especially where long life and reliability are important such as in microcircuits and telephone systems. It is also used in dental alloys and for jewelry.[4] Its melting (approximately 1550°) and boiling points (approximately 2870°)[4] are lower than those of platinum (1775° and approximately 4050°, respectively).

Palladium is one of the "4d-elements" of the transition metals and is usually classed, together with iron, cobalt, nickel, ruthenium, rhodium, osmium, iridium, and platinum, as a group VIII metal. Its atomic weight is currently accepted to be 106.4 and its atomic number is 46. Six isotopes occur naturally— ^{102}Pd (1.0%), ^{104}Pd (11%), ^{105}Pd (22.2%), ^{106}Pd (27.3%), ^{108}Pd (26.7%), and ^{110}Pd (11.8%). Isotopes of mass 98, 99, 100, 101, 103, 107, 109, 111, 112, 113, 114, and 115 are radioactive, that of mass 107 being the longest-lived with a half-life of 5×10^6 years. Of the naturally occurring isotopes only ^{105}Pd has a nuclear magnetic moment ($I = \frac{5}{2}$); however, coupling to ^1H and other nuclei has not been observed in palladium compounds.

The metal is most closely related to nickel and platinum, the three of them making up the nickel triad of group VIII, to its neighbor rhodium (atomic number 45), and to some extent to its neighbor on the other side, silver (atomic number 47). The electronic ground state of the metal is $1s^2$, $2s^2$, $2p^6$, $3s^2$, $3p^6$, $3d^{10}$, $4s^2$, $4p^6$, $4d^{10}$ and has the 4d shell completely filled. This contrasts with Ni and Pt in which the valence shells are $3d^8 4s^2$ and $5d^9 6s^1$, respectively. At the end of the 4d series the 4d energy level lies below the 5s even in the zero-valent state. For Ni and Pt the $(n + 1)$s level lies below the nd level in the atoms, but the energy levels are reversed in the ions. For this reason, as with the other transition metals in their compounds in positive oxidation states, the electrons which have been removed first by oxidation are usually taken to have been the $(n + 1)$s electrons. Thus all three elements in the +2 state have the electron configuration $n d^8$. Rhodium has a valence shell of $4d^5 5s^1$; again loss of one electron to Rh(I) gives the configuration $4d^8$. Silver has the valence shell of $4d^{10} 5s^1$; one electron is easily lost to give Ag(I), $4d^{10}$. Owing to the positive charge now drawing the 4d energy levels down so much, further loss of electrons, which must come from the 4d levels, is difficult, and the chemistry of Ag in the +2 and +3 states is very limited. The elements following (Cd, In, Sn, etc.) can only lose those valence shell electrons which are in excess of the number required for a complete $4d^{10}$ shell. The high positive charge on the ions Cd^{2+}, In^{3+}, and Sn^{4+} makes it impossible for even the strongest oxidizers to remove any electrons from the now very low energy, filled 4d shell.

The electron configurations of these metals in positive oxidation states are straightforward; however, a large number of compounds are known where the metal is formally in a zero-oxidation state. The compounds involving Ni, Pd, and Pt are invariably diamagnetic and from their properties the metals are regarded as all having a $n d^{10}$ configuration. They are, therefore, not in the same state as the free metal atom; formally energy must be supplied to bring them up to the zero-oxidation state.

Palladium metal crystallizes in a cubic close-packed lattice, as do all the other transition metals at the end of the d series (Co, Ni, Cu, Rh, Pd, Ag, Ir, Pt, and Au).[5] The metal–metal distance in the solid is 2.751 Å. This compares with values of 2.492 Å for Ni, 2.775 Å for Pt, 2.690 Å for Rh, and 2.889 Å for Ag. The effect of the lanthanide contraction is clearly seen in the similarity of the metal–metal distances for Pd and Pt. This feature is also evident in the bond lengths of Pd–X and Pt–X bonds, which are very nearly the same for a given atom, X, in a similar environment (see Section B,4,e).

4. Chemical Properties of the Metal

Palladium is one of the noble metals and is therefore quite difficult to oxidize —more so than nickel, but less so than platinum. Reliable comparative oxidation potentials do not appear to be available. The rate of reaction with suitable oxidizing agents depends on the state of division of the metal and some finely divided types, e.g., palladium sponge [from hydrogen reduction of $(NH_4)_2$ $PdCl_4$], palladium black [obtained by hydroxylammonium chloride reduction of Pd(II) salts in aqueous solution], or colloidal palladium [obtained by hydrazine reduction of palladium(II) salts in basic solution in the presence of a stabilizer such as albumen] have quite high reactivity. In particular, this is the case in the heterogeneous catalysis of various reactions, e.g., $C_2H_4 + O_2 \rightarrow$ CH_3COOH (Pd-black) or $H_2 + I_2 \rightarrow 2HI$ (colloidal Pd). A very important property of palladium is its ability to absorb hydrogen up to a point represented by the stoichiometry Pd_2H. At higher temperatures ($>120°$) it also becomes permeable to hydrogen, a phenomenon which allows the separation of hydrogen from other gases. Hydrogen when released by palladium appears to be activated and is highly reactive; for example, it reacts in the cold with Cl_2 or I_2, or at $200°$ with NO to give NH_3 and H_2O. This is probably part of the reason for the usefulness of the metal in catalytic hydrogenation of olefins (see Chapter V, Volume II).

At atmospheric pressure between $788°$ and $830°$ Pd reacts in air to form, to some extent at least, the oxide PdO; above this temperature the oxide decomposes again to the metal. This process is complete at $920°$.[6] Somewhat similar results are obtained for the other noble metals; platinum, for example, showing a gain in weight above $538°$, followed by progressive loss in weight above $607°$ when heated in oxygen in a thermobalance. This oxidation occurs largely on the surface.[6]

Neither ozone nor water have any effect on palladium, and even halogens only react on heating. Fluorine was reported by Moissan to react about $400°$ to give red PdF_2.[7] This appears to be either very impure material (probably still containing metal, since the pure compound has more recently been reported to be pale violet[8]) or to be impure "PdF_3."[†][9] Palladium reacts readily with chlorine above $300°$ to give $PdCl_2$. Interestingly, although palladium has a well-established chemistry of the IV oxidation state, no evidence for the direct formation of PdF_4 and $PdCl_4$ (unknown) by these reactions has been reported.

The more usual method of oxidizing palladium is in solution. In contrast to platinum, it dissolves in concentrated nitric acid, as well as, like platinum, in aqua regia ($HCl–HNO_3$). The product from the former is probably a Pd(IV) hydroxonitrate[11]; from the latter, ammonium chloride precipitates $(NH_4)_2$

† "PdF_3" = $Pd^{2+}PdF_6{}^{2-}$.[10]

$PdCl_6$. Salts of the $PdCl_6{}^{2-}$ ion are well known, but they disproportionate in solution to Cl_2 and $PdCl_4{}^{2-}$. The metal also dissolves in aqueous cyanide in the presence of an oxidizing agent (see Section B, 4, b).

B. THE CHEMISTRY OF PALLADIUM COMPLEXES WITH INORGANIC LIGANDS

1. General

Palladium has a well-established chemistry for some four oxidation states: O, I, II, and IV.† A few compounds of Pd(III) are also known, but no structural data is available and it is very possible that some or all of these will be shown *not* to have the metal in a true +III state. Again, most of the compounds of palladium which are formally in the +I oxidation state are found to be diamagnetic. Since Pd(I) has a d^9 configuration, this implies that the odd electrons of two (or possibly, in some cases, four or six) Pd(I) atoms are coupled together in metal–metal bonds. In common with most other metals which form predominantly covalent compounds in lower oxidation states, palladium shows quite a high tendency to form cluster compounds with metal–metal bonds. In Pd(II) these clusters are weak and very long metal–metal interactions are found, but in the lower oxidation states, the evidence, though fragmentary at the time of writing, indicates these species to be common and held together strongly by short Pd–Pd bonds.

Here again palladium is intermediate between nickel and platinum, but is perhaps closer to nickel on present indications. Nickel has well-established oxidation states of 0, I, II, III, and IV. Ni(IV) is, in general, less prevalent than Pd(IV) and only exists where special stabilization occurs, e.g., in the carboranyl complex $[(B_9C_2H_{11})_2Ni(IV)]^0$,[13] $(diars)_2NiCl_2{}^{2+}$,‡ and $NiF_6{}^{2-}$. The nickel analog of $PdCl_6{}^{2-}$ does not appear to exist. On the other hand, Ni(III), in complexes such as $(Et_3P)_2NiBr_3$,[14] $[(diars)_2NiCl_2]^+Cl^-$,[15] and

---

† The formal oxidation states (or oxidation numbers) are designated by Roman numerals. They are defined as the formal charge remaining on the metal when all the ligands in their closed-shell electron configurations are removed.[12] For most simple inorganic ligands, halide$^-$, SCN^-, $NO_3{}^-$, H_2O, Ph_3P, NH_3, CO, etc., the situation is unambiguous. For organic ligands, nitrosyls, H, and some others the situation is less straightforward and the conventions used will be explained in the appropriate place. It must not, however, be assumed that the complexes are ionic and that the metal bears a large positive charge. Ionic bonding is very rare in the chemistry of palladium and most interactions are covalent, if somewhat polarized

‡ See Glossary.

$(Me_2PhP)_2NiBr_3$,[16] is much better documented than Pd(III). Furthermore, Ni(I) is known to exist in at least some paramagnetic monomeric compounds without metal–metal bonds, e.g., $(Ph_3P)_3NiX$ (X = halogen).[17, 18] There is also more chemistry of Ni(0) [$Ni(CO)_4$ and its many substitution products and analogs] than there is at present of Pd(0).

The tendency for the greater stabilization of higher oxidation states lower down the periodic table is exemplified in the chemistry of platinum. In contrast to the other two metals, Pt(IV) is a common and relatively stable state, and with fluoride as ligand even Pt(V) and Pt(VI) become accessible. The +II state is again the most common oxidation state and an extensive chemistry of Pt(0) is now developing. There do not appear to be any true well-defined compounds of Pt(III) and even the existence of Pt(I), except in cluster compounds, is not certain.

It is also instructive to compare these metals with the Co, Rh, and Ir subgroups. Thus, Pd(II) (d^8) corresponds to Rh(I). This oxidation state is well known for Rh and Ir though perhaps somewhat less so for Co, and many properties typical of Pd(II), such as the square planar stereochemistry and the high reactivity in catalytic systems, are also observed here. Rh(II) and Ir(II) (d^7) are again rare but not unknown, especially in diamagnetic metal–metal bonded species, such as $[Rh(acetate)_2 \cdot H_2O]_2$, while Co(II) is, of course, quite common and paramagnetic. The contrasts are more obvious in the higher oxidation states where the d^6 configuration M(III) [corresponding to Pd(IV)] is the most generally stable, with the exception of some aquo and halo complexes of Co(II). Furthermore, the +IV, +V, and +VI oxidation states for Rh, and especially for Ir, are also known.

Comparisons with the Fe, Ru, and Os subgroup are perhaps not so fruitful, except for the M(0) state which again shows many parallels to the chemistry of the d^8 states of the other elements.

Copper, silver, and gold are so disparate themselves that meaningful comparisons with Ni, Pd, and Pt are hard to make. The M(I) (d^{10}) state is common to all, but, apart from a few cases where Ag(I) and Cu(I) adopt similar stereochemistries to trigonal Pt(0) [as in $(Ph_3P)_3Pt$], the resemblances end there. The ease of oxidation of d^{10} to d^8, so marked with all the metals of the nickel triad, is virtually absent for Cu and Ag, although gold does share this property. Au(III) has many similarities to Pt(II).

2. The Zero-Oxidation State

The best known derivative of these metals in the zero-oxidation state is $Ni(CO)_4$; it is readily formed both from nickel metal and from nickel(II) salts on treatment with carbon monoxide. Its properties and those of the other carbonyls have been extensively reviewed by Calderazzo, Ercoli, and Natta.[19] Neither palladium nor platinum appear to form the analogous simple binary carbonyls, despite many attempts to prepare them. A polynuclear $[Pt(CO)_2]_n$ has been briefly described by Chatt and co-workers,[20] and the preparation of $[Pt(CO)_2]_5$ has been claimed by Matveev et al. in a Russian patent.[21] Krück and Baur[22] have reported the synthesis of $(Ph_3P)_2Pt(CO)_2$ from $(Ph_3P)_2Pt$ $(PF_3)_2$ and CO, but $Pt(CO)_4$ could not be isolated by analogous reactions. This problem is considered in Chapter II, this volume.

All these metals, however, do form complexes of the type L_nM, where L is a neutral ligand, such as a t-phosphine, t-phosphite, t-arsine, isocyanide, bipyridyl, etc., and M = Ni, Pd, and Pt.† Mixed complexes, such as $(Ph_3P)_2Pd$-$(Ph_3Sb)_2$,[23] and phosphine platinum carbonyls, $(Ph_3P)_3PtCO$,[24–26] have also been reported. Other types known are the anionic cyanides and alkynyls, $K_4M(CN)_4$ (M = Ni, Pd, but not Pt) and $K_4Ni(C\equiv CR)_4$, $K_2M(C\equiv CR)_2$ (M = Pd, Pt). Last, if nitric oxide is considered to bond to a metal as NO^+ (after having transferred an electron to the metal from a π^*-antibonding molecular orbital) then a number of compounds of the type XMNO are also formally in the zero-oxidation state. With the exception of the cyclopentadienyl nitrosyls, C_5H_5MNO (see Chapter VI, this volume), the structures of most of these are, however, not clear yet.

Before considering the zero-valent metal complexes in more detail some general points about them should be noted. They are all rather labile and also easily oxidized, usually to the (II) state. The highest coordination number in the solid is probably 4, although Chatt et al.[27] have suggested that one complex may be 5-coordinate. The exact state of most of them in solution is rather unclear since they are nearly all thermodynamically unstable with respect to loss of ligand: $L_nM \rightleftarrows L + L_{n-1}M$ and even, $L_{n-1}M \rightleftarrows L + L_{n-2}M$. The limiting case is probably L_2M for most except bidentate chelating ligands and some strongly electronegative phosphines.[28]

a. Stabilization of Low Oxidation States by Back-Bonding

Since a metal in the zero-oxidation state already has a full complement of electrons, any additional electrons which a Lewis base ligand can donate will not be accepted unless there is some mechanism for dissipating the excess negative charge built up on the metal. For this reason simple Lewis bases, such

---

† Many olefins and acetylenes also form complexes of the type $L_2M(un)$ [and some $(olefin)_nM$] in which the metal can formally be regarded as being in the (0) state. These are more fully discussed in Chapter III, this volume.

as ammonia or water, do not usually form stable metal(0) complexes. Only those ligands which can, to some degree, also remove this excess charge are, therefore, expected to form such complexes and this is borne out in practice.

The mechanism by which this occurs can be described in a number of ways. The simplest way, pictorially, is to view the removal of charge from the filled metal d orbitals as occurring through overlap with vacant ligand π^* or d orbitals, by the formation of π-type bonds.

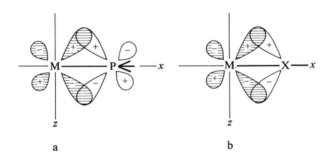

FIG. I-1. Pictorial representation of metal–ligand π bonding. (a) For a t-phosphine or phosphite, vacant d orbitals on phosphorus overlap with metal d_{xz} orbital. (b) For a ligand X with vacant π^* orbitals (CO, bipy) which can overlap with the metal d_{xz} orbital.

In Fig. I-1b is shown the second alternative, where the ligand X has vacant π^* orbitals of correct symmetry. This type of interaction is presumed to occur with ligands such as carbon monoxide, isonitriles, bipyridyl, and many others. Apart from the requirement that the orbitals on metal and ligand must be of the same symmetry, overlap must be significant and the energies of the two orbitals must be comparable for bonding to occur.

The process whereby a bond of σ symmetry is formed between the appropriate σ orbital on the ligand and a vacant orbital on the metal (in the cases discussed here, probably sp³ for tetrahedral, sp² for trigonal, and sp for linear stereochemistry), and a bond of π symmetry between appropriate orbitals on metal and ligand can be represented as M ← L. This implies a forward donation of charge via the σ bond from L to M and a back-bonding from M to L. This type of bonding is termed synergic because the two effects strengthen each other.

A similar situation is depicted in Fig. I-1a, which represents the bond between a t-phosphine or t-phosphite and a metal. The phosphorus can utilize its vacant d orbitals to π bond to the metal. Although it is generally agreed that there is, in complexes in which phosphorus is the donor atom, some degree of π bonding, the exact amount is a subject for discussion and has not yet been satisfactorily resolved. A simple picture would lead one to suppose that electron-withdrawing substituents on the phosphorus (F, OPh, etc.) would increase the back-bonding from the metal and stabilize low oxidation state (electron-rich) complexes. However, Tolman[29] has shown that in $R_3PNi(CO)_3$ the A_1 carbonyl stretching frequency is inversely related to the σ basicity of the phosphine or phosphite. Furthermore, in the complexes $(R_3P)_4Ni$, the stability of the complex toward exchange with $R_3'P$ is determined more by steric than by electronic factors; for example, the complex $(Me_3P)_4Ni$ is surprisingly stable.

Another factor is that in a series of complexes of a given ligand, the relative amounts of σ and π bonding may vary from one metal to another. Meier $et\ al.$[30] have explained the differences in reactivity of $M[P(OPh)_3]_4{}^0$ complexes in terms of nickel forming strong π bonds and weak σ bonds to the phosphite, whereas the opposite is true for platinum. In these terms then, palladium occupies an intermediate position with respect to both π and σ bonding and forms the least stable complexes of this type.

b. Preparation of Ni(0), Pd(0), and Pt(0) Complexes

i. *Phosphine Complexes.* The most usual methods involve reduction of an M(II) compound; however, in a few cases the metal will react directly with the ligand. Thus PF_3[31] and the chelating ligand o-phenylenebis(diethylphosphine)[32] react with palladium metal (and nickel, but *not* directly with platinum) to give tetrakis(trifluorophosphine)palladium(0) and bis[o-phenylenebis-(diethylphosphine)]palladium(0), (I-1), respectively.

(I-1)

A wide variety of reducing agents have been used and many of these reactions parallel those used to form the metal carbonyls. For example, Krück and Baur[22, 33] used copper metal as a reducing agent and halogen acceptor; excess ligand has also been used.

$$MCl_2 + Cu + PF_3 \xrightarrow{100°/pressure} M(PF_3)_4 \quad (M = Pd, Pt)$$

$$PdO + 5Ph_3P \xrightarrow{EtOH/\Delta} (Ph_3P)_4Pd + Ph_3PO$$

The most commonly used methods involve reduction of phosphine metal halide complexes in the presence of excess phosphine. For palladium this has been carried out using hydrazine[34] or excess sodium borohydride[27] as reducer:

$$2(Ph_3P)_2PdCl_2 + 4Ph_3P + 5N_2H_4 \rightarrow 2(Ph_3P)_4Pd + 4N_2H_5Cl + N_2$$

$$L_2MX_2 + BH_4^- \rightarrow L_2M \quad (L = \text{ } \text{AsMe}_2 \text{ } \text{AsMe}_2, \text{ } \text{PMe}_2 \text{ } \text{PMe}_2 \text{ } ; \text{ } M = Ni, Pd)$$

The reaction of the phosphine platinum halides with hydrazine has been studied more carefully and turns out to be very complex. The products depend on the nature of the phosphine and on the exact conditions. Thus, bis(trialkylphosphine)platinum halides yielded the hydrido complexes, not the zerovalent complexes.[35]

$$cis\text{-}(alkyl_3P)_2PtCl_2 + N_2H_4 \rightarrow trans\text{-}(alkyl_3P)_2PtHCl$$

At 80° cis-$(Ph_3P)_2PtCl_2$ in ethanol gave $trans$-$(Ph_3P)_2PtHCl$[36, 37]; some of the hydroxide, $(Ph_3P)_2Pt(OH)Cl$, was also formed.[36] Using slightly different conditions, Malatesta and Cariello[38] isolated tris(triphenylphosphine)-platinum(0) (I-2).

$$4(Ph_3P)_2PtI_2 + 3N_2H_4 \rightarrow (Ph_3P)_3Pt + [(Ph_3P)_2Pt_2(N_2H_4)_2I_4] + N_2 + 4HI$$

(I-2)

The formation of (I-2) and analogous complexes was favored by higher reaction temperatures and an excess of ligand.[39, 40]

Intermediates in this reaction have been isolated and a reaction scheme similar to the following may be envisaged[35, 41]:

$$cis\text{-}(R_3P)_2PtCl_2 + N_2H_4 \rightarrow [(R_3P)_2PtN_2H_4Cl]^+Cl^- \longrightarrow$$

$$cis\text{-}(R_3P)_2PtHCl \xrightarrow{N_2H_4/R_3P} (R_3P)_{3,4}Pt + N_2H_5Cl$$

The reactions also proceed without hydrazine, in alcohol, especially under basic conditions, to give the hydridochlorides[35]:

$$cis\text{-}(R''_3P)_2PtCl_2 + KOH + RR'CHOH \rightarrow trans\text{-}(R''_3P)_2PtHCl + RR'CO + KCl + H_2O$$
$$(R = R' = Me; R = Me, R' = H)$$

and a mechanism involving transfer of hydride from the α carbon on the alcohol to the metal in a metal alkoxide intermediate has been proposed.

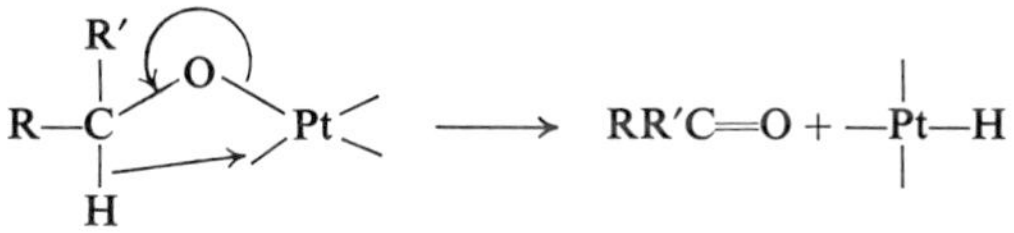

An interesting commentary on this reaction is provided by the work of Sacco et al.,[42] who showed that the following reactions occurred with phosphines and rhodium(III):

$$3L + RhCl_3 \cdot hydrate \rightarrow L_3RhCl_3 \xrightarrow{\text{EtOH/H}_2O} L_3RhHCl_2 \xrightarrow[-HCl]{\text{EtOH}/\Delta} L_3Rh^ICl$$
$$\text{(I-3)} \qquad\qquad \text{(I-4)} \qquad\qquad \text{(I-5)}$$

When L = EtPh$_2$P, (I-3) was the product below 25°. On heating this complex in aqueous alcohol, the hydridochloride (I-4) was formed; the hydride here appeared to come from the water. With L = Ph$_3$P, the trichloro complex (I-3) was not isolated, and the first isolable product was (I-4). On further heating, reductive dehydrohalogenation to give the rhodium(I) complex (I-5) occurred. This latter complex was not obtained using L = EtPh$_2$P. When the reactions were carried out using Et$_2$PhP, the trichloro complex (I-3) was the product; even prolonged heating in ethanol only gave traces of (I-4). The Lewis basicity of the phosphine is, therefore, very important in determining the extent and nature of the reaction.

In the reactions with platinum only the cis-(R$_3$P)$_2$PtCl$_2$ is useful as a starting material, the trans compound reacts much too slowly. This is presumably a consequence of the high trans effect of the phosphine and the kinetic inertness of Pt(II)[43]; the effect of the former is to labilize the chloride only in the cis complex. Palladium(II) complexes are, in contrast, much more labile than their Pt(II) counterparts and also isomerize very easily, hence this difficulty does not arise. For Ni(0) complexes the reducing agents of choice are sodium borohydride, sodium naphthalenide,[27] or trialkylaluminums. These latter reagents are also useful for preparing Pd(0) complexes, but cannot be used for the analogous Pt(0) complexes owing to the higher thermal stability of the ethyl-Pt(II) complexes, which are presumed to be intermediates.

$$Pd(acac)_2\dagger + EtOAlEt_2 + PPh_3 \rightarrow (Ph_3P)_4Pd^0 \ ^{44}$$

$$Pd(acac)_2 + Et_3Al + PPh_3 \xrightarrow{Et_2O} (Ph_3P)_4Pd^0 \ ^{45}$$

Yields of 90% or over of very pure material are obtained in the latter reaction and it appears to be the method of choice for synthesizing $(Ph_3P)_4Pd^0$.

Zero-valent phosphite complexes have been prepared by Meier et al.[30] from the dihalide and a base in methanol, e.g.,

$$(RO)_3P + MCl_2 + base + MeOH \rightarrow [(RO)_3P]_4M$$

An interesting variation on these reactions is the reduction of organometallics; no reducing agent other than the phosphine is necessary and the reactions appear to go cleanly under mild conditions,

$$cis\text{-}(Et_3P)_2PdMe_2 + PPh_3 \xrightarrow{\varDelta} (Ph_3P)_4Pd \ ^{49}$$

This last reaction is obviously related to the formation of $(Ph_3P)_4Pd$ from $Pd(acac)_2$, PPh_3, and Et_3Al. Other alkyl complexes undergo these reactions and Johnson et al. have prepared the zero-valent triphenylphosphine, arsine, and stibine complexes from (I-6).[50]

$$Pd(acac) + Ph_3E \rightarrow C_8H_{11}CH(COMe)_2 + (Ph_3E)_4Pd \quad (E = P, As, Sb)$$

(I-6)

Other π-allyl complexes behave similarly,

$$\left[\text{(—PdCl)} \right]_2 + PPh_3 \xrightarrow{Me_2CO/H_2O} [CH_2\!=\!CHCH_2PPh_3]Cl + (Ph_3P)_4Pd \ ^{51}$$

$$\text{(—M—)} + R_3P \rightarrow (R_3P)_4M \ (+ CH_2\!=\!CHCH_2CH_2CH\!=\!CH_2 ?) \ ^{52, 53}$$

$$(R = Ph, Et, M = Ni; R = Ph, M = Pd)$$

† See Glossary.

but platinum gives a bis(σ-allyl) complex,

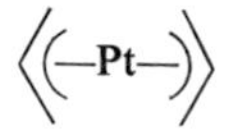

$$\langle\!\!\langle -Pt- \rangle\!\!\rangle + Ph_3P \rightarrow (Ph_3P)_2Pt(CH_2CH\!\!=\!\!CH_2)_2 \; [53]$$

The other product from reaction of bis(π-allyl)palladium has not been identified, but is presumably 1,5-hexadiene. Intermediates of the type (I-7) (M = Ni, R = Me, R' = Et; M = Pd, R = H, R' = Ph) have been reported,[54] but without details.

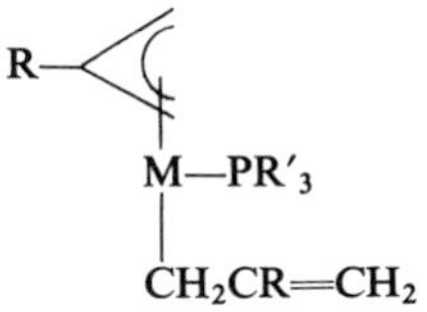

$$R\!\!-\!\!\langle$$
$$M\!\!-\!\!PR'_3$$
$$|$$
$$CH_2CR\!\!=\!\!CH_2$$
(I-7)

The nickel complexes (I-8) (R = Ph, F, OPh, etc.) are conveniently prepared from dicyclopentadienylnickel, which is commercially available.[55-58]

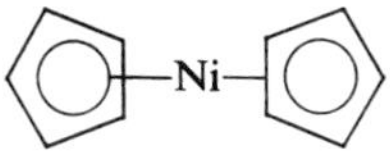

$$\langle\!\!\bigcirc\!\!\rangle\!\!-\!\!Ni\!\!-\!\!\langle\!\!\bigcirc\!\!\rangle + R_3P \rightarrow (R_3P)_4Ni$$
(I-8)

A good alternative is by reaction of the phosphine or phosphite with bis(1,5-cyclooctadiene)nickel(0). Bipyridyl, o-phenanthroline, and diphos† have also been used to give the bis chelate Ni(0) complexes.[57]

The ligands L in $L_nM(0)$ are labile and can frequently be exchanged for other neutral ligands. The prototype reactions are the substitutions of CO in nickel carbonyl by other ligands; these have been reviewed by Manuel[59] and

$$(RNC)_2Pd + 3(R'O)_3P \rightarrow [(R'O)_3P]_3PdCNR + RNC$$
$$(R = p\text{-}CH_3C_6H_4, R' = p\text{-}ClC_6H_4; R = p\text{-}CH_3OC_6H_4, R' = Ph)$$

$$(RNC)_2Pd + R'_3P \rightarrow (R'_3P)_4Pd$$
$$(R = p\text{-}CH_3C_6H_4, R' = PhO; R = p\text{-}CH_3OC_6H_4, R' = Ph; R = Ph, R' = PhO)$$

$$(p\text{-}CH_3C_6H_4NC)_2Pd + 3R'_3P \rightarrow (R'_3P)_3Pd + 2p\text{-}CH_3C_6H_4NC$$
$$(R' = p\text{-}ClC_6H_4, p\text{-}CH_3C_6H_4)$$

† See Glossary.

Calderazzo et al.[19] Reactions of this type for palladium were discovered by Malatesta.[34,60,61]

The replacement of olefins in complexes of type (I-9) also affords a route to mixed ligand complexes.[62]

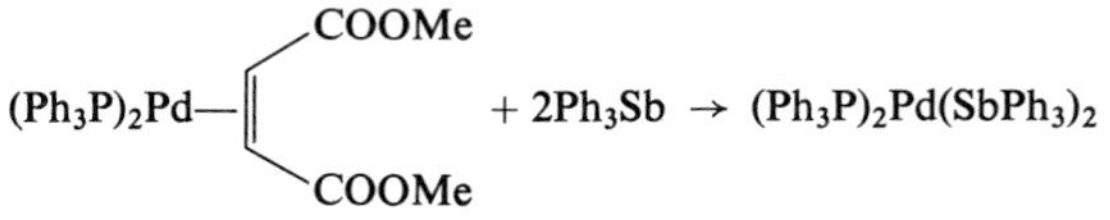

$$(Ph_3P)_2Pd \mathbin{-} \Big\langle \begin{matrix} COOMe \\ \\ COOMe \end{matrix} \quad + 2Ph_3Sb \rightarrow (Ph_3P)_2Pd(SbPh_3)_2$$

$$\text{(I-9)} \hspace{5cm} \text{(I-10)}$$

The complex (I-10) is formed very readily and is also produced by the following reactions[23,63]:

$$(Ph_3P)_4Pd + 2Ph_3Sb \rightarrow \text{(I-10)} \leftarrow (Ph_3P)_2PdCS_2 + 2Ph_3Sb$$

Reactions of this type are also known for nickel[29,30,64,65] and platinum.[30,38] The kinetics of some of these reactions have also been measured.[30]

Another type of reaction has also been reported by Krück and Höfler[66]:

$$Ni(PF_3)_4 + NaOMe \xrightarrow{\text{MeOH}/\Delta} Ni[P(OMe)_3]_4$$

Few metal(0) complexes stabilized by phosphines bearing alkyl groups have been mentioned in the literature. Mukhedkar et al. have prepared $(Ph_2MeP)_4$-Pd[67] and Tolman has reported $(Me_3P)_4Ni$; the latter is very reactive, but surprisingly stable toward substitution by other phosphines.[29]

ii. *Cyano and Isonitrile Complexes.* The first complex of Pd(0) described was the tetracyanometallate(0) (I-12) (M = Pd), which was obtained by the reduction of the tetracyanometallate(II) (I-11) (M = Pd) using potassium in liquid ammonia.[68,69]

$$K_2M(CN)_4 + 2K \xrightarrow{\text{NH}_3(l)} K_4M(CN)_4$$

$$\text{(I-11)} \hspace{4cm} \text{(I-12)}$$

This reaction also gave an isolable nickel complex (I-12) (M = Ni)[68,69]; but only the very sensitive $[Ph_4P]_2Pt(CN)_2$ could be obtained.[70] The palladium complex was significantly more reactive than the nickel.[69] A potentiometric study of the mechanism by Watt et al.[71] has shown that there are two steps in the reaction of (I-11) (M = Ni) with potassium in liquid ammonia; the first is a fast reduction to a Ni(I) complex, $K_4Ni_2(CN)_6$, the second is the much slower reduction to the zero-valent complex.

The similarity between CN^- and $RC{\equiv}C^-$ led Nast and co-workers to prepare the alkynyl analogs. Although the nickel complex was normal and

4-coordinate,[72] those of Pd and Pt were unusual in that they were apparently 2-coordinate[73-75] (see Chapter II, this volume).

The isocyanides RNC are also isoelectronic with the acetylides, RC_2^-, and Malatesta showed that RNC would also stabilize both Ni(0) and Pd(0); Pt(0) complexes of this type are apparently unknown. Two routes have been described to prepare these compounds, one involving reduction of $(RNC)_2PdI_2$ in strongly alkaline solution by an excess of RNC, and the other involving organopalladium complexes.

Malatesta has suggested that the mechanism for his reaction is as follows[61]:

$$(RNC)_2PdI_2 + 2KOH + 2RNC \rightarrow [(RNC)_4Pd](OH)_2 + 2KI$$

$$[(RNC)_4Pd](OH)_2 + RNC \rightarrow (RNC)_4Pd + RNCO + H_2O$$

$$n(RNC)_4Pd \rightarrow [(RNC)_2Pd]_n + 2nRNC$$

The isolated product was the polymer $[(RNC)_2Pd]_n$, $(R = Ph, p\text{-}MeC_6H_4, p\text{-}MeOC_6H_4)$. The other complexes, again bis(isocyanide)palladium complexes, were prepared by Fischer and Werner by the reaction of (π-cyclohexenyl)(π-cyclopentadienyl)palladium with RNC.[46,47]

$$RNC + \text{[structure]} \longrightarrow (RNC)_2Pd \quad (R = Pr^i, C_6H_{11})$$

The nickel complexes, $Ni(CNR)_4$, are four-coordinate and monomeric and presumably tetrahedral.[76] In addition, Otsuka et al. have described $Ni(CNR)_2$ and a variety of other nickel and palladium complexes with isonitrile ligands.[77,78]

c. Properties of Ni(0), Pd(0), and Pt(0) Complexes

Few crystal structure determinations of zero-valent complexes of Ni, Pd, and Pt have yet been reported, and of those the majority also contain CO or an organic ligand. Two types of coordination are observed, tetrahedral, as in $(Ph_3P)_3PtCO$,[24-26] and planar trigonal as in $(Ph_3P)_3Pt$.[79] These, together with linear 2-coordination are the stereochemistries expected for d^{10} complexes, and all the known M(0) complexes of the nickel triad are diamagnetic.

Nickel carbonyl $[Ni(CO)_4]$ is, of course, tetrahedral, and a number of other complexes ML_4 $[M = Ni, Pd, and Pt; L = PF_3, P(OEt)_3]$ have also been shown to be tetrahedral by spectroscopic methods.[64,80,81,81a]

A general characteristic of these complexes is the ease with which dissociation to 3- and 2-coordinate species occurs in solution.

$$ML_4 \rightleftharpoons ML_3 + L$$

$$ML_3 \rightleftharpoons ML_2 + L$$

The extent to which this dissociation occurs is found to depend very greatly on the ligand L. When L = CO, PF_3, or $P(OR)_3$, appreciable dissociation does not occur, but from a study of the kinetics of exchange and substitution of $Ni(CO)_4$ and $M[P(OR)_3]_4$ Basolo and his co-workers showed that these reactions occur by a dissociative mechanism.[30, 82, 83] The size of L is also very important.[29]

The effect is more dramatically seen in the triphenylphosphine complexes, $(Ph_3P)_4M$, which all gave very low molecular weights in nonpolar solvents.[17, 38, 40] The observed molecular weight for $(Ph_3P)_4Pt$ could not be explained simply on the basis of dissociation to PPh_3 and $(Ph_3P)_3Pt$ alone and further equilibria

$$(Ph_3P)_3Pt \rightleftharpoons (Ph_3P)_2Pt + PPh_3$$

$$(Ph_3P)_2Pt \rightleftharpoons Ph_3PPt + PPh_3$$

were postulated.

A number of complexes including $(Ph_3P)_2Pt$, $[(Ph_3P)_2Pt]_3$, $[Ph_3PPt]_4$, $[(Ph_3P)_5Pt_3]$, and $[(Ph_3P)_4Pt_3]$ have been briefly reported, but structural details are not yet available.[84–87,87a] An insoluble complex $[(Ph_3P)_2Pd]_x$ has also been mentioned.[88, 89]

$$(Ph_3P)_4Pd \xrightarrow{\quad CH_2=CHCl \text{ or } CH_2=C=CH_2 \quad} [(Ph_3P)_2Pd]_x$$

Species such as LNi, where L is usually a triarylphosphite, have been implicated as intermediates in many of the cyclooligomerization reactions described by Wilke and his collaborators.[90, 91]

The best characterized of these is tris(triphenylphosphine)platinum (I-13), the X-ray structure of which has been determined by Albano et al.[79] This

(I-13)   (I-14)

showed that the three phosphines were trigonally arranged about the platinum, with the metal very slightly (0.1 Å) above the plane defined by the three P atoms. The P–Pt distances [2.25–2.28(1)† Å] were shorter than the sum of the

† Figures in brackets are e.s.d.'s, see glossary.

covalent radii, 2.42 Å, and indicated a P–Pt bond order greater than 1 owing to d_π–d_π bonding.

By comparison, the carbonyl (I-14), which exists in two forms, is very nearly tetrahedral and has significantly longer P–Pt bonds [2.333–2.369(10) Å], reflecting a decrease in d_π–d_π bonding here.[25, 26] The palladium analog of (I-14) has also been reported.[92]

The complexes $[(p\text{-}XC_6H_4)_3P]_3Pd$ (X = H, Me, Cl) have also been described.[34, 93]

Kinetic evidence for the intermediacy of a species $(Ph_3P)_2Pt$ of significant stability has been put forward by Allen and Cook[94] from the rates of replacement of acetylenes (ac) in

$$(Ph_3P)_2Ptac' \rightleftharpoons (Ph_3P)_2Ptac' + ac$$

Birk, Halpern, and Pickard[95] have examined the kinetics of reactions of $(Ph_3P)_2PtC_2H_4$ with alkyl halides, phenylacetylene, and triphenylphosphine and of the autoxidation of $(Ph_3P)_3Pt$. The kinetics observed could best be interpreted in terms of the fast establishment of the equilibrium

$$(Ph_3P)_2PtC_2H_4 \rightleftharpoons (Ph_3P)_2Pt + C_2H_4 \qquad K_{eqm} = (3.0 \pm 1.5) \times 10^{-3}$$

followed by slower subsequent reactions. From their electronic spectra in solution, $(Ph_3P)_4Pt$ was shown to be identical to $(Ph_3P)_3Pt$, indicating that the former was completely dissociated. However, they also found that K_{eqm} for

$$(Ph_3P)_3Pt \underset{+PPh_3}{\overset{-PPh_3}{\rightleftharpoons}} (Ph_3P)_2Pt$$

was around 10^{-4} in benzene. Hence, neither the $(Ph_3P)_2Pt$ nor $(PPh_3)_4Pt$ was present to any significant extent. Heimbach[17] has also stated that $(Ph_3P)_3Ni$ has a normal molecular weight in solution.

Basolo and co-workers have determined some rates of substitution and exchange of phosphite complexes by NMR methods.[30, 83] For the reaction of L_4Ni with cyclohexylisocyanide, the relative rates for L are $P(OPh)_3 > P(OEt)_3 > P(OMe)_3 > P(OCH_2)_3CPr$. From these results and from the exchange

$$M[P(OEt)_3]_4 + P(OEt)_3 \rightleftharpoons M[P(OEt)_3]_4 + P(OEt)_3$$

(M = Pd, Pt), the enthalpies of activation for $M[P(OEt)_3]_4 \rightarrow M[P(OEt)_3]_3 + P(OEt)_3$ in aromatic solvents were estimated at 26 (Ni), 22 (Pd), and 27 (Pt) kcal·mole^{-1}. By comparison, ΔH^+ for $NiL_4 \rightarrow NiL_3 + L$ in hexane was 32 [L = $P(OEt)_3$] and 24 (L = CO) kcal·mole^{-1}. A similar order of stability for the trifluorophosphine complexes $M(PF_3)_4$ can be inferred from their decomposition temperatures Ni, $>155°$; Pd, $> -20°$; and Pt, $>90°$.

Little is known about the relative stabilities of the triphenylphosphine complexes; however, toward air (oxygen) the relative reactivities for $(Ph_3P)_4M$ are $Ni > Pd > Pt$. In contrast, $M(PX_3)_4$, where $X = F$ or OR, are not easily autoxidized. This is not surprising since oxidation by an external agent must compete here with the electron-withdrawing ligands already present.

d. Reactions of Ni(0), Pd(0), and Pt(0) Complexes (Oxidative Addition)

By far the most interesting and important feature of these M(0) complexes is that many are coordinatively unsaturated and readily undergo oxidative addition. A frequently noted adjunct of stability and inertness in the chemistry of low oxidation states, especially for the organometallics, is that the effective atomic number (E.A.N.) formalism is obeyed.† On this basis L_4M (where L is a 2-electron donor, R_3P, CO, etc., and M = Ni, Pd, or Pt) is coordinatively saturated. However, $(Ph_3P)_3Pt$ is not, and even complexes like $Ni(CO)_4$ undergo substitution reactions in which the rate-determining step is now recognized to be dissociation.[82, 96]

$$Ni(CO)_4 \rightarrow Ni(CO)_3 + CO$$

Furthermore, these M(0) complexes undergo oxidative addition with a wide variety of reagents to give products which are frequently best represented as in the (II) oxidation state. These reactions are very analogous to those known for some time already for Rh(I) and Ir(I),[97-99] and which involve addition of molecules X–Y to square planar d^8 complexes to give octahedral d^6 complexes in which the X–Y bond is usually broken and X and Y act as monodentate ligands. A wide variety of molecules X–Y will add, including H–H, R–halide, H–halide, RSO_2–Cl, RCC–H,[100, 101]

$$trans\text{-}(Ph_3P)_2IrCOCl + XY \rightarrow (Ph_3P)_2IrCO(Cl)X(Y)$$

In some cases (with O_2 and SO_2), complexes are obtained whose geometry is better described as 5-coordinate, whereas with acetylenes and olefins the adducts are difficult to describe in simple terms as true oxidations.

† This simply states that the number of electrons associated with the metal in the oxidation state of the compound under consideration plus the number of electrons assumed to be donated by the ligands equals the atomic number of the next rare gas. For example, in $Cr(CO)_6$, Cr is in the (0) state and has a valence-shell electron configuration of $[Ar] + 6$; if each CO is assumed to donate 2 electrons, the E.A.N. of Cr will be $[Ar] + 18$, and equivalent to Kr. Similar considerations apply to $Ni(CO)_4$, ferrocene $[(C_5H_5)_2Fe]$, if the Fe is in the (II) state and each C_5H_5 ligand supplies 6 electrons, tricarbonylcyclobutadieneiron $[C_4H_4Fe(CO)_3$, $C_4H_4 = 4$ electron donor, $Fe(0) = [Ar] + 8$ and $3 \times CO = 6$ electrons], and many others.

The ease with which oxidative addition occurs depends on the metal, the ligands, the nature of X and Y, and the solvent. For the reaction $d^8 \to d^6$, the lower the formal oxidation state of the metal and the heavier the metal, in a given triad, the more easily does this occur. Hence the orders of reactivity are Os(0) > Ir(I) > Pt(II); Os(0) > Ru(0) > Fe(0); Ir(I) > Rh(I) > Co(I); Pt(II) > Pd(II) > Ni(II), etc.

Ligands which are good σ donors facilitate addition; thus, while the iridium complex (I-15a) readily adds hydrogen to give (I-16a), the rhodium

$$\left[\begin{array}{c} \text{PR}_2 \quad \text{PR}_2 \\ \text{M} \\ \text{PR}_2 \quad \text{PR}_2 \end{array}\right]^+ + \text{H}_2 \longrightarrow \left[\begin{array}{c} \text{PR}_2 \quad \overset{\text{H H}}{\underset{\text{R}_2\text{P}}{\text{M}}} \quad \text{PR}_2 \\ \text{PR}_2 \end{array}\right]^+$$

(I-15)   (I-16)

(I-15a, I-16a) M = Ir, R = Ph
(I-15b, I-16b) M = Rh, R = Ph
(I-15c, I-16c) M = Rh, R = Me

complex (I-15b) does not. On replacement of the phosphine phenyls by methyls the rhodium complex (I-15c) now does add hydrogen to give (I-16c).[102, 103]

The stereochemistry of addition to trans-$(Ph_3P)_2IrCOCl$ and related complexes varies, and examples of both cis and trans addition have been found.[101, 104, 105]

Chock and Halpern[106] have investigated the reactions of $(Ph_3P)_2IrCOX$ with H_2, O_2, and MeI. The activation parameters and solvent dependence for the latter reagent are similar to those found for the Menschutkin reaction $[R_3N + RX \to R_4N^+X^-]$ and have been interpreted to favor a highly polar transition state of the type

$$[L_4Ir^{\delta+} \ldots R \ldots I^{\delta-}]$$

rather than

$$\left[L_4Ir \overset{R}{\underset{I}{\cdots}}\right]$$

Systems more similar to those of the Pd(0) and Pt(0) reactions discussed below are provided by the reactions of $(Ph_3P)_3RhCl$ studied by Wilkinson and his co-workers.[107] This also adds many molecules oxidatively, but with loss of one Ph_3P in a primary dissociation step (but see Ref. 108).

$$(Ph_3P)_3RhCl \rightleftarrows (Ph_3P)_2RhCl + Ph_3P$$

$$(Ph_3P)_2RhCl + XY \rightleftarrows (Ph_3P)_2RhCl(X)Y$$

(I-17)

where X = H, Y = H; X = H, Y = Cl; X = Me, Y = I, etc. The complex (I-17) is formally 5-coordinate, but frequently is solvated.

The iridium complex (I-18) is stable toward dissociation, but does react with hydrogen and methyl iodide; the tendency to undergo oxidation is so great that it will react with its own ligand, to give (I-19) in which oxidative addition of a phenyl hydrogen to the metal has occurred.[109, 110] The cobalt(I) complex

$$(Ph_3P)_3IrCl \xrightarrow{\Delta}$$

(I-18)   (I-19)

$(Ph_3P)_3CoH \cdot N_2$ probably exchanges H for D via similar intermediates[111] and the interesting equilibrium for Ru(0) $\rightleftarrows$ Ru(II) (another $d^8 \rightarrow d^6$ redox system) has been reported by Chatt and Davidson.[112]

The ease with which such reactions occur suggests that similar processes may be occurring in some of the Ph_3P–Pt(0) systems (see also Blake and Nyman[112a]).

In recent years a large number of addition reactions have been observed with $(Ph_3P)_4M$. In this case the oxidative addition of X–Y to L_nM^0 (d^{10}) leads to the square planar L_2MXY complexes where M is usually in the (II) state (d^8). Most of the work has involved platinum, but a number of reactions involving Pd and Ni complexes have also been reported. For example, (I-13) reacted with inorganic acids HX according to[113]

$$(Ph_3P)_3Pt \underset{KOH}{\overset{HX}{\rightleftarrows}} [(Ph_3P)_3PtH]^+X^- \underset{+PPh_3}{\overset{-PPh_3}{\rightleftarrows}} (Ph_3P)_2PtHX$$

(I-13)   (I-20)   (I-21)

When the anion X^- had a low tendency to coordinate with Pt(II), complexes (I-20) (X = ClO_4, BF_4, HSO_4) could be isolated, whereas when X had higher nucleophilicity toward Pt(II) the covalent trans complexes (I-21) were obtained (X = CN, NCS). When X = Cl, the product depended on the solvent, polar solvents favored formation of the ionic (I-20) (X = Cl), nonpolar solvents the covalent (I-21) (X = Cl). $(Ph_3P)_4M$ (M = Ni, Pd) did not give hydrides in reaction with acids under these conditions, instead the dihalo complexes were isolated and H_2 was evolved. A mechanism has been proposed for this.

$$M(PPh_3)_4 \rightarrow M(PPh_3)_3 \underset{-HX}{\overset{+HX}{\rightleftharpoons}} (Ph_3P)_2MHX \underset{-HX}{\overset{HX}{\rightleftharpoons}}$$

$$(Ph_3P)_2MH_2X_2 \rightarrow H_2 + (Ph_3P)_2MX_2$$

$$(M = Ni, Pd; X = halide)$$

Very recently, however, Kudo et al.[114] reported that both $(Ph_3P)_3PdCO$ and $(Ph_3P)_4Pd$ reacted with HCl at $-50°$ in ether to give the hydride, e.g.,

$$(Ph_3P)_3PdCO + HCl \rightarrow trans\text{-}(Ph_3P)_2Pd(H)Cl$$

However, nickel(0) complexes which do not dissociate readily in solution $\{(diphos)_2Ni, [(RO)_3P]_4Ni\}$ undergo oxidative addition to give the five-coordinate hydrido cations, e.g.,[115, 116]

$$[(EtO)_3P]_4Ni + HX \rightarrow \{[(EtO)_3P]_4NiH\}^+ X^-   (X = Cl, CF_3COO, \tfrac{1}{2}SO_4)$$

A number of organic halides react with $(Ph_3P)_4M$ to give the trans complexes (**I-22**),[117−121]

$$(Ph_3P)_4M + RX \longrightarrow \begin{array}{c} Ph_3P \diagdown \diagup R \\ M \\ X \diagup \diagdown PPh_3 \end{array}$$

$$(\text{I-22})$$

M = Pd, R = Me, X = I	M = Pt, R = Me, X = Br
R = MeCO, X = Cl	R = MeCO, X = Cl
R = EtOCO, X = Cl	R = PhC≡C, X = Br
R = Ph, X = I	R = PhCH=CH, X = Br
R = CF_3, C_2F_5, C_3F_7, X = I	R = CCl_3, X = Cl
	R = PhCO, X = Cl
	R = CF_3, C_2F_5, C_3F_7, X = I
	R = C_6F_5CO, X = Cl

and with $(diphos)_2M$ to give the cis complexes $diphosMR_fI$.[118]

$$(diphos)_2M + R_fI \rightarrow diphosMI_2 + diphosMR_f \cdot I$$
$$M = Pd, R_f = CF_3$$
$$M = Pt, R_f = CF_3, C_2F_5$$

A wide variety of polar inorganic molecules add oxidatively to $(Ph_3P)_{3,4}M$; some reactions of the platinum complex are summarized in Chart I-1.[122−132]

$$(Ph_3P)_{3,4}Pt + A\text{—}X \rightarrow (Ph_3P)_2PtX(A)$$

$X = Cl$; $A = Ph_2B$,[122] Ph_3Sn,[123] Ph_3PAu,[123] $HgCl$ [123]

$X = I$; $A = I$,[124] $(Ph_3P)_3Cu$,[123] $(Ph_3P)_2PtI$,[123] $(Ph_3P)_2NiI$ [123]

$X = A = SiCl_3$[122]

$X = H$; $A = (p\text{-}FC_6H_4)_3Si$,[125] SeH,[126] SH[126]

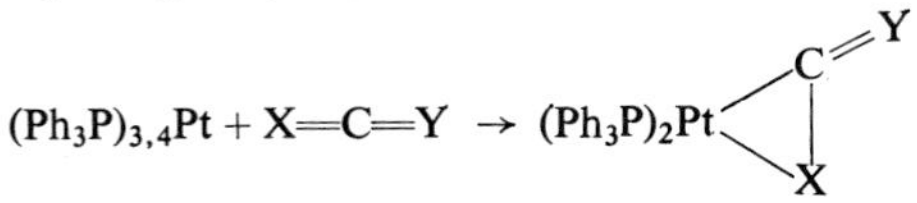

$$X = Y = S \text{ [120]}$$
$$X = S, Y = O \text{ [120]}$$
$$X = S, Y = RN \text{ [120]}$$
$$X = O, Y = (CF_3)_2 \text{ [120, 12]}$$
$$X = S, Y = (CF_3)_2 \text{ [120]}$$

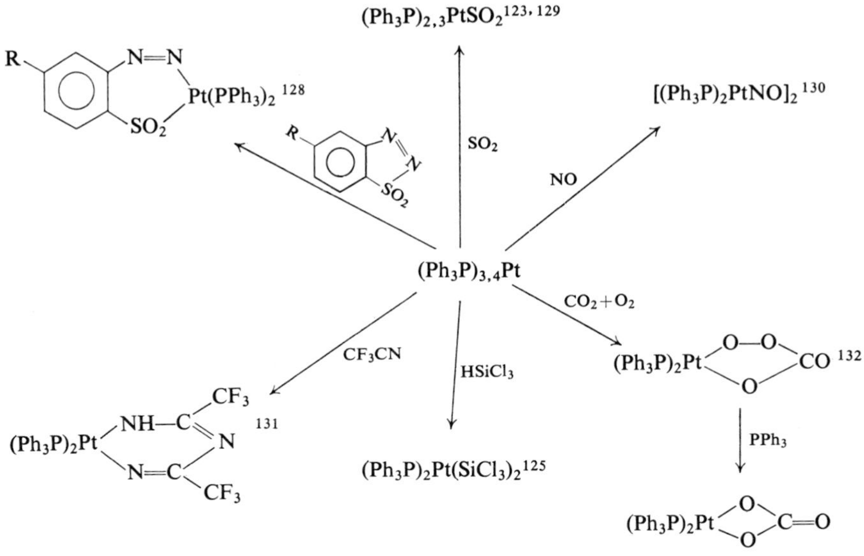

CHART I-1. Some oxidative addition reactions of $(Ph_3P)_{3,4}Pt$.

Some oxidative addition reactions of $(Ph_3P)_4Pd$ are collected in Chart I-2; $(Ph_2MeP)_4Pd$ also reacts with a number of fluorocarbon halides $(C_3F_7I, CF_2\text{=}CFCl)$ to give the trans adducts, $(Ph_2MeP)_2Pd(R_f)X$.[67]

Some polyhalo compounds $(CBrCl_3, C_2Cl_6)$ react with $(Ph_3P)_4Pt$ to give $(Ph_3P)_2Pt(hal)_2$.[121] Another interesting variant is the reaction with perchloro and perfluorovinyl compounds; olefin π complexes, e.g., (I-23), are sometimes isolated as intermediates.[88, 95, 133, 134]

No such intermediates were isolated in the reactions of $(Ph_3P)_4Pd$ with vinylic chlorides, but the insertion of the metal into the C–Cl bond was again stereospecific.[88]

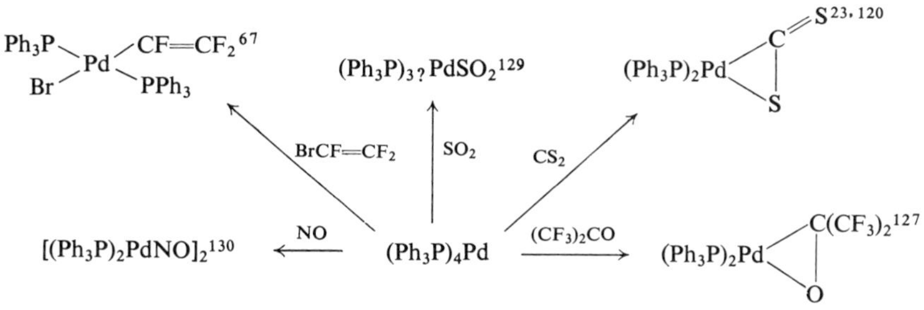

CHART I-2. Some oxidative addition reactions of $(Ph_3P)_4Pd$.

The mechanism of the rearrangement of the platinum complex **(I-23)** to **(I-24)** has been investigated[121, 135] and has been shown to be either intramolecular or to involve an intermediate carbonium ion. The solvent dependence of

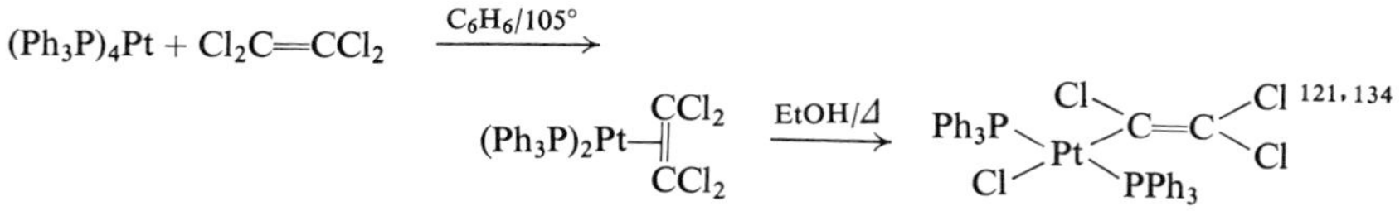

(I-23)   **(I-24)**

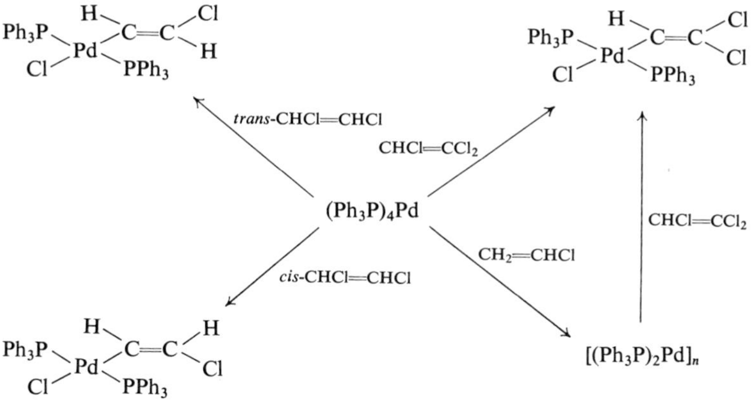

the rates of isomerization of **(I-23)** in alcohols is similar to that observed for solvolysis of alkyl halides and supports a carbonium ion intermediate. In benzene, however, another mechanism, involving primary loss of one PPh_3 from **(I-23)**, appears to play a role. No mechanistic studies on the direct reaction of $(Ph_3P)_4M$ with simple organic halides have yet been reported. However, the kinetics of the reaction of $(Ph_3P)_2PtC_2H_4$ with MeI, $PhCH_2Br$, and

ICH_2CH_2I have been interpreted in terms of the rapid establishment of the equilibrium:

$$(Ph_3P)_2Pt(C_2H_4) \rightleftharpoons (Ph_3P)_2Pt + C_2H_4$$

followed by reaction of $(Ph_3P)_2Pt$ with X–R to give the observed products.[95] A mixture of $(Ph_3P)_4Pt$ and an organic halide has been used as an initiator for free radical polymerization.[136] Allyl and methallyl halides also react with $(Ph_3P)_4Pd$ to give σ-allyl complexes,[51, 117] and with $(Ph_3P)_4Pt$ to give ionic π-allyl complexes.[120, 137, 138] These are discussed in Chapter V.

The complexes $(Ph_3P)_{3,4}M$ all react with oxygen to give the peroxo complexes (**I-25**). Triphenylphosphine is also formed[95] (but see Ref. 139). X-ray structures of the benzene and toluene solvates of the platinum complex show that it has the stoichiometry $(Ph_3P)_2PtO_2$ and has an approximately planar arrangement of the two P atoms, the Pt, and the two oxygens.[140] The nickel and

$$(Ph_3P)_{3,4}M + O_2 \longrightarrow \begin{matrix} Ph_3P \\ Ph_3P \end{matrix} \!\! M \!\! \begin{matrix} O \\ O \end{matrix} + 1(2)PPh_3 \quad (M = Ni, Pd, Pt)$$

(**I-25**)

platinum peroxo complexes (**I-25**) are yellow, whereas the palladium one is green.[141] All are very reactive and thermally not very stable. They decompose to the metal and triphenylphosphine oxide, the nickel complex already when warmed above −35°.† Both the palladium and nickel complexes liberate oxygen on treatment with bromine at low temperatures. These complexes are

$$O_2 + (Ph_3P)_2MBr_2 \xleftarrow[\text{(M = Ni, Pd)}]{Br_2} (Ph_3P)_2MO_2 \xrightarrow{\Delta} 2Ph_3PO + M$$

(**I-25**)

very active oxidizing agents whose utility is only just being investigated. Carbon dioxide is oxidized to carbonato[141] complexes (via peroxycarbonato complexes[132]); SO_2 to sulfato complexes; NO_2 to bis(nitrato) complexes[142, 143]; NO to bis(nitrito) complexes[143]; and acetone to a peroxo complex.[144] This has been ascribed to the oxygen in (**I-25**) being in an approximation to the $^3\Sigma_u^-$ excited state, in valence bond terms, $(Ph_3P)_2Pt^{\delta+}-O-O^{\delta-}$, allowing facile nucleophilic addition to $\text{\textbackslash}C{=}O$ and similar systems.[144]

The oxidations also occur catalytically; for example, triphenylphosphine is oxidized to Ph_3PO in the presence of (**I-25**). The palladium complex is the most convenient for this purpose and conversions of 500 moles of PPh_3 per mole of (**I-25**) (M = Pd) have been achieved at 50°.[145]

† For $(R_3P)_2NiO_2$ the thermal stabilities are R = cyclohexyl > piperidyl > phenyl.

The mechanism has been investigated by Halpern and co-workers[95] who found the reaction to occur in two stages.

$$(Ph_3P)_3Pt + O_2 \rightarrow (Ph_3P)_2PtO_2 + Ph_3P$$

This process is unaffected by low concentrations of PPh_3, but in the presence of an excess a slow reaction occurring by the following path is postulated.

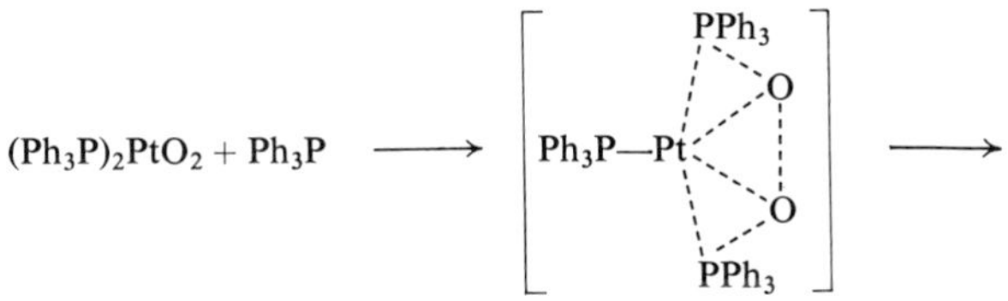

$$(Ph_3P)_2PtO_2 + Ph_3P \longrightarrow \left[\begin{array}{c} PPh_3 \\ O \\ Ph_3P-Pt \\ O \\ PPh_3 \end{array} \right] \longrightarrow$$

$$Ph_3PPt(OPPh_3)_2 \xrightarrow{PPh_3} (Ph_3P)_3Pt + 2Ph_3PO$$

Ph_2S, Ph_3Sb, $(EtO)_3P$, and CO were reported not to be catalytically oxidized and the presence of some of them inhibited the oxidation of PPh_3.[63] However, n-Bu_3P was oxidized to n-Bu_3PO and $C_6H_{11}NC$ to the isocyanate $C_6H_{11}NCO$.[63] The rates of oxidation apparently vary with the metal and the ligand; for Pt the order has been reported as $Ph_3P > C_6H_{11}NC > Bu_3P$, while exactly the reverse holds true for palladium.[146]

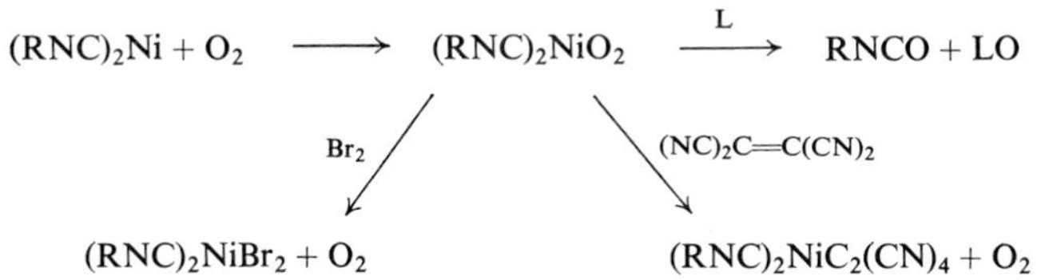

$$(RNC)_2Ni + O_2 \longrightarrow (RNC)_2NiO_2 \xrightarrow{L} RNCO + LO$$

with branches: Br_2 path giving $(RNC)_2NiBr_2 + O_2$ and $(NC)_2C=C(CN)_2$ path giving $(RNC)_2NiC_2(CN)_4 + O_2$.

Otsuka et al.[77,78,147] have prepared the peroxyisonitrile complexes $(RNC)_2MO_2$ ($R = Bu^t$ or C_6H_{11}, $M = Ni$; $R = Bu^t$, $M = Pd$) by oxygenation

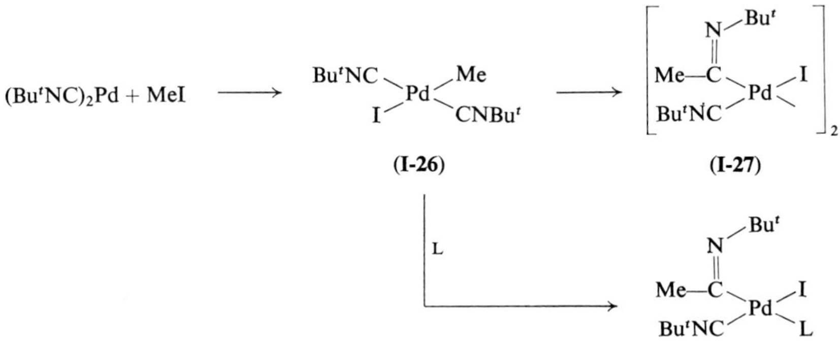

$$(Bu^tNC)_2Pd + MeI \longrightarrow \begin{array}{c} Bu^tNC \\ Pd \\ I \end{array} \begin{array}{c} Me \\ CNBu^t \end{array} \longrightarrow \left[\begin{array}{c} N-Bu^t \\ Me-C \\ Pd-I \\ Bu^tNC \end{array} \right]_2$$

(I-26)   (I-27)

with L leading to:

$$\begin{array}{c} N-Bu^t \\ Me-C \\ Pd-I \\ Bu^tNC \quad L \end{array}$$

(I-28)

of $(RNC)_2M$ [or $(RNC)_4Ni$] at low temperatures. These complexes were polymeric; on warming the nickel complex in the presence of a ligand, L (PPh_3, $R'NC$), oxidation of the ligand occurred.

The oxygen in the nickel complex was replaced by tetracyanoethylene, bromine, and boron trifluoride. Olefins were not oxidized by this catalyst.

The palladium isonitrile complex, $(Bu^tNC)_2Pd$, underwent an oxidative addition with methyl iodide, to give (I-26), which isomerized above 11° to give (I-27) or (I-28).[78]

e. *Conclusions*

In summary, zero-valent metal complexes of the types L_nM (L = *t*-phosphine, *t*-phosphite, PF_3, Ph_3As, Ph_3Sb, RNC, etc., and $n = 2, 3$, or 4) are well known for Ni, Pd, and Pt. The structures are largely unknown, but it is likely that some at least will turn out to be cluster compounds. The reported low reactivity of $[Ph_3PPt]_4$ agrees with this.[85, 87]

One characteristic property is their tendency to oxidatively add polar (and some nonpolar) molecules to give square planar complexes in the (II) oxidation state. Ligands with small electronegative substituents stabilize these zero-valent complexes and complexes bearing ligands with larger, less electron-withdrawing ligands, undergo oxidative addition more easily. Relative stabilities and reactivities for complexes of the three metals are less easy to judge on present evidence, but it appears that at least two series exist: one with ligands like PF_3, RNC where it appears that the Ni(0) complexes are the more stable, and another, with ligands such as Ph_3P where it appears that, toward air, for example, the relative reactivities are Ni > Pd > Pt.

Cross-relations with the neighboring elements are not very illuminating; the closest formal relationships are to the square planar d^8 complexes of Rh(I) and Ir(I). On the other side, Cu(I), Ag(I), and Au(I) are isoelectronic with Ni(0), Pd(0), and Pt(0). Similarities are seen in the stereochemistries which these metals in the (I) oxidation state adopt and in the ready oxidation of Au(I) (d^{10}) to square planar Au(III) (d^8).

3. The M(I) Oxidation State

The chemistry of this oxidation state is not, at the moment, very extensive. Indications, however, are that it may well become important especially with regard to our understanding of catalytic processes.

a. *Ni(I)*

The +I state is most clearly defined for nickel, where the first complex, $K_4Ni_2(CN)_6$, was isolated by Bellucci and Corelli in 1913.[148] The complex has been shown to be diamagnetic and binuclear with a Ni–Ni bond.[149,149a] Watt et al.[71] have shown that the potassium in liquid ammonia reduction of $K_2Ni(CN)_4$ to $K_4Ni_2(CN)_6$ is very fast, whereas subsequent reduction to $K_4Ni(CN)_4$ is very slow. $K_4Ni_2(CN)_6$ reacts with dimethyl acetylene dicarboxylate to give a complex, (**I-29**), in which the acetylene has inserted into the metal–metal bond, and a trimer, $C_6(COOMe)_6$.[150] This is analogous to

$$K_4Ni_2(CN)_6 + MeOOCC\equiv CCOOMe \rightarrow K_4 \begin{bmatrix} (NC)_3Ni\diagdown \diagup COOMe \\ C \\ \| \\ MeOOC \diagup C \diagdown Ni(CN)_3 \end{bmatrix}$$

(**I-29**)

some of the reactions of $Co_2(CN)_{10}{}^{4-}$.

Some mononuclear Ni(I) complexes were discovered by Heimbach[17] and also by Porri et al.[18] The following methods have been used to prepare them.

$$(Ph_3P)_4Ni + \tfrac{1}{2}X_2 \xrightarrow{\ C_6H_6\ } Ph_3P + (Ph_3P)_3NiX$$

$$(Ph_3P)_4Ni + (Ph_3P)_2NiX_2 \xrightarrow{\ Et_2O\ } 2(Ph_3P)_3NiX$$

$$\left[\diagdown \!\!\!\diagdown\!\!-NiBr \right]_2 + 6PPh_3 \xrightarrow{\ 110°,\ Et_2O\ } 2(Ph_3P)_3NiBr$$

(X = halogen)

The magnetic moment was found to be 2.2 B.M., in agreement with the value expected for one unpaired electron (d^9) together with spin-orbit coupling for a tetrahedral molecule. The ESR spectra all showed signals characteristic for the presence of one unpaired electron and there seems to be no evidence for Ni–Ni interactions.[17] Preliminary data from a crystal structure determination agreed with the conclusion that the molecule is pyramidal.[18] Both papers reported evidence for dissociation in solution and Heimbach[17] has claimed the isolation of $(Ph_3P)_2NiX$ in 100% yield by the following method.

$$(Ph_3P)_2NiC_2H_4 + (Ph_3P)_2NiX_2 \xrightarrow{\ Et_2O\ } C_2H_4 + 2(Ph_3P)_2NiX$$

On reaction with triphenylphosphite, disproportionation to Ni(0) and Ni(II) occurred.[17]

$$(Ph_3P)_3NiCl + (PhO)_3P \rightarrow [(PhO)_3P]_2NiCl_2 + [(PhO)_3P]_4Ni$$

Corain et al.[151] have reported the preparation of some Ni(I) cyanide complexes (I-30). The magnetic moments of the complexes (I-30) were in the range

$$[Ph_2P(CH_2)_nPPh_2]Ni(NCS)_2 + KCN \xrightarrow{\text{EtOH(aq.)}}$$

$$[Ph_2P(CH_2)_nPPh_2]_{1.5}Ni(CN)_2 \xrightarrow{\text{NaBH}_4/\text{EtOH}} [Ph_2P(CH_2)_nPPh_2]_{1.5}NiCN$$

$$\text{(I-30)} \quad (n = 3, 4)$$

2.0–2.3 B.M., but these authors preferred a square planar dimeric structure.

Some intriguing olefin Ni(I) complexes, e.g., (I-31), have been briefly described by Porri et al.[152] and Dubini and Montino.[153] The complex (I-31)

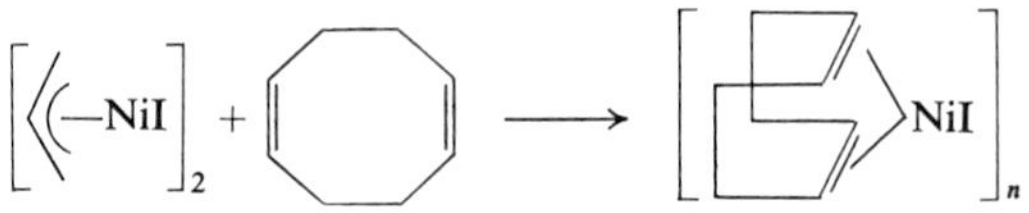

(I-31)

was apparently a very effective polymerization catalyst for dienes and acetylenes.

b. *Pd(I) and Pt(I)*

Mononuclear complexes of these types have not been reported for Pd and Pt. This is not surprising since both these metals have quite a high tendency to form diamagnetic complexes involving M–M bonds. Apart from the complexes already cited, prepared by Nyholm and co-workers,[123] [(Et₃P)₂PtGePh₃]₂ has been reported by Glockling and Hooton[154] and "(Ph₃P)₃Pt₂S(CO)₂"[120] has now been shown[155] to have the structure (I-32).

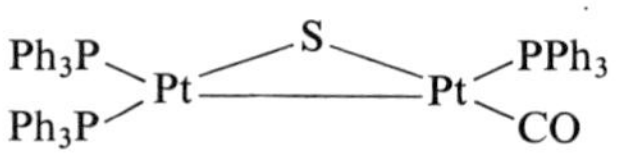

(I-32)

Oranskaya and Mikhailova[156] studied the vapor pressure of PdCl₂ between 883° and 1030°K and found no evidence for PdCl₁. The best evidence for the existence of Pd(I) comes from two crystal structure determinations by Allegra of the compounds [(C₆H₆Pd)₂(Al₂Cl₇)₂] and [(C₆H₆Pd)₂(AlCl₄)₂] formed by the reaction of PdCl₂ with benzene in the presence of Al and AlCl₃.[157, 158] These are discussed further in Chapter VI,C, this volume, but the essential features are linear ClPdPdCl groups sandwiched between two benzene rings. The Pd–Pd distance [2.58(1) Å] is much shorter than in the metal.

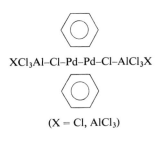

XCl$_3$Al–Cl–Pd–Pd–Cl–AlCl$_3$X

(X = Cl, AlCl$_3$)

Davidson and Triggs[159] observed that palladium acetate reacts with benzene in acetic acid–perchloric acid to give biphenyl and a magenta Pd(I) species, stable in acetic acid, which could be isolated as explosive purple crystals of formula $[C_6H_6PdH_2O \cdot ClO_4]_n$ on addition of acetic anhydride. Other reducing agents (1-hexene, phenylboronic acid) are effective in this reaction if benzene is also present. On addition of chloride ions to the complex, disproportionation to Pd metal and $PdCl_4^{2-}$ occurred. The NMR spectrum of the complex did not show any proton resonances and other properties were consistent with the presence of a paramagnetic d^9 species, perhaps in equilibrium with a diamagnetic binuclear complex.

Moiseev and his co-workers have reported evidence which may imply the presence of a lower oxidation state of Pd [possibly Pd(I)] in the isomerization of 1-butene by $PdCl_2$ under aqueous acidic conditions.[160,161] The isomerization of 1-hexene by α(?)-$PdCl_2$ in the absence of solvent has also been interpreted by these workers as due to the Pd(I) species[162] (see Chapter III,B, Volume II).

Klanberg and Muetterties[163] have reported the formation of the cluster compound (I-33) which is formally a Pd(I) complex; a structure based on a tetrahedron of Pd atoms with a Ph$_3$P and a PH$_3$ attached to each and chlorines bridging all four faces has been proposed and seems very plausible.

$$(Ph_3P)_2PdCl_2 + PH_3 \rightarrow [(Ph_3P)(PH_3)PdCl]_4$$
(I-33)

Some other compounds that may well contain Pd(I), but about which no structural information is available, are the various carbonyl halides (see Chapter II,B, this volume).

c. *Comparisons with Other d^9 Molecules*

Analogs to the M(I) state here are the M(0) state of Co, Rh, and Ir and the M(II) state for Cu, Ag, and Au. The M(0) state for the cobalt subgroup is usually diamagnetic and polynuclear (with metal–metal bonds), but evidence for $\dot{C}o(CO)_4$ in equilibrium with $Co_2(CO)_8$ in the vapor phase has been obtained.[164] Chatt *et al.*[32] have also reported a bis(phosphine)Co(0) complex. Complexes of Ag(II) and Au(II) are known, although they are rare; they appear to have square planar structures. In contrast, Cu(II) is the most common and stable oxidation state (at least under aqueous conditions) for copper. For most ligands Cu(II) adopts a very distorted octahedral coordination (frequently to square pyramidal 5-coordination or even, in the limiting case, to square planar), but complexes in which it adopts a distorted tetrahedral coordination are also known (e.g., Cs_2CuBr_4). With more covalent ligands, the oxidation state Cu(I) is preferred, but some intermediate ligands, especially carboxylate, cause the formation of bridged dimers. Appreciable interaction there *can* lead to complex coupling of the unpaired electrons. Although the information on hand at the moment is limited, it is possible that similar patterns of behavior may emerge for the M(I) oxidation states of Ni, Pd, and Pt.

4. The M(II) Oxidation State

For nickel and palladium this represents by far the most important state in most of their normal chemistry and for platinum it is at least as important as the (IV) state.

a. *Electronic Structures for the d^8 State*

In the absence of ligands which make special demands on the metal, Ni(II) is usually paramagnetic and octahedrally coordinated, whereas Pd(II) and Pt(II) are diamagnetic and square planar. The (II) oxidation state has the d^8 configuration. For a field of six ligands of octahedral symmetry, the degeneracy of the d orbitals in the free ion (in the valence state) is lifted and two orbitals ($d_{x^2-y^2}$, d_{z^2}) are raised in energy by $\frac{3}{5}\Delta_0$, the remaining three becoming stabilized by an equivalent amount, $\frac{2}{5}\Delta_0$, where Δ_0 is the crystal field splitting energy for an octahedral field (see Fig. I-2).† For simplicity the effects of π bonding to the ligands are ignored here.

The eight d electrons fill up these orbitals as shown, the $d_{x^2-y^2}$ and d_{z^2} orbitals each then being only singly occupied. As a result Ni(II) in octahedral complexes is paramagnetic to the extent required for two unpaired electrons plus a smaller contribution from orbital angular momentum.

In the presence of ligands which have special steric or electronic requirements nickel(II) can also adopt a tetrahedral coordination, and again is para-

---

† Readers who are unfamiliar with the basis of crystal and ligand field theory are referred to the excellent discussion of this in simple terms by Cotton and Wilkinson.[165]

magnetic with two unpaired electrons. An alternative for very bulky ligands or for covalent ligands which can π bond to the metal is for a tetragonal distortion to take place. This implies the removal† of two trans ligands; as a consequence of the lowering of the symmetry to D_{4h} the degeneracy of the d orbitals is further lifted. Two ($d_{x^2-y^2}$ and d_{xy}) are raised in energy. The remaining three (d_{xz} and d_{yz}, which are degenerate, and d_{z^2}) are lowered. The usual order quoted for Ni(II) is $d_{x^2-y^2} > d_{xy} > d_{z^2} > d_{xz} = d_{yz}$, but for Pd(II) and Pt(II) the d_{z^2} level, as might be expected, has been shown to be lowest from the electronic spectra and the levels are now $d_{x^2-y^2} > d_{xy} > d_{xz} = d_{yz} > d_{z^2}$.[166, 167] In all d^8 systems, however, the three lowest lying orbitals will, in any case, be occupied.

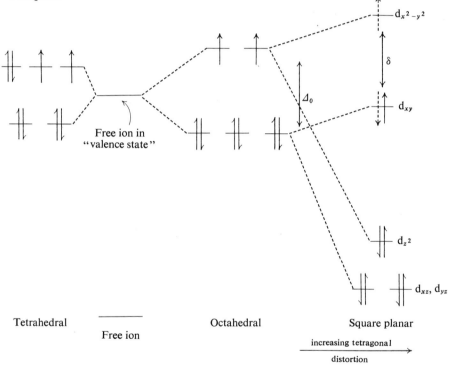

Tetrahedral   Octahedral   Square planar

Free ion

increasing tetragonal

distortion

FIG. I-2. Crystal field diagrams for Ni(II).

† This is meant only in a purely formal sense; in practice one merely observes that in square planar coordination four ligands are close and interact strongly with the metal, while the fifth and sixth are far and interact only weakly at best. The case considered is when both ligands are removed to the same extent, but this need not always be so. If one trans ligand is "removed" further than the other, the result will be a 5-coordinate square pyramidal structure. These are well known and similar crystal field splitting diagrams can easily be derived for them.

The crucial point therefore is the energy difference, δ, between the two highest ($d_{x^2-y^2}$ and d_{xy}) orbitals. If this is less than the energy required to pair the remaining two electrons in the lower orbital, it will be energetically more favorable to occupy these two orbitals singly, giving a paramagnetic species. If δ is greater than the pairing energy, then the two electrons will pair in the d_{xy} orbital leaving the $d_{x^2-y^2}$ orbital vacant and giving a diamagnetic species. The pairing energy is usually considered to remain fairly constant and the observed variations are explained in terms of some ligands giving rise to a large splitting (strong field ligands), whereas others give rise only to a small one (weak field ligands). Furthermore the splitting is normally found to increase from 3d to 4d and to be highest for the 5d metals in a given group. In practice it is found that although Ni(II) does form some tetragonally distorted high spin (paramagnetic, two unpaired electrons) complexes,† only one is known for Pd(II), PdF_2, and none for Pt(II).

Typical colors for Pd(II) complexes are in the range yellow-brown-red.

b. *Oxidation Potentials*

Table I-1 abstracts some data on reaction potentials from the review by Goldberg and Hepler.‡ [169] For comparison, E^0 for $M = M^{2+} + 2e^-$ is -0.25 V

TABLE I-1

Oxidation Potentials[a]

		E^0 (V)	
X		M = Pd	M = Pt
$M_{(m)} + 4X^- = M^{II}X_4^{2-}$ (aq) $+ 2e^-$ (in acidic solution)			
Cl		-0.59	-0.75
Br		-0.49	-0.67
I		-0.18	-0.40
CN		$+0.4$	~0.09
$M^{II}X_4^{2-}$ (aq) $+ 2X^- = M^{IV}X_6^{2-}$ (aq) $+ 2e^-$ (in acidic solution)			
Cl		-1.26	-0.77
Br		-0.99	-0.64
I		-0.48	-0.39

[a] Taken from Goldberg and Hepler.[169]

† Orioli and Sacconi[168] have noted that in square pyramidal Ni(II) complexes the low spin complexes show a marked elongation of the Ni–apical ligand bond length. This, as might be expected, is not observed for the high spin square pyramidal complexes.

‡ These are all algebraic oxidation half-reaction potentials defined with reference to $Ag = Ag^+(aq) + e^-$, $E^0 = -0.8$ V.

for Ni, -0.98 V for Pd,[169a] and estimated to be about -1.2 V for Pt. This leads to the general conclusion that the order of oxidizability of the metals is Ni$\gg$ Pd > Pt. This is also borne out by qualitative observations; for example, nickel dissolves in dilute mineral acids, palladium in concentrated nitric acid, whereas platinum only dissolves in aqua regia.

On the other hand, oxidation of M(II) to M(IV) is usually more difficult for Pd than for Pt. Comparative potentials are not known for Ni but from the paucity of Ni(IV) compounds reported it can safely be inferred that the oxidation to Ni(IV) is much more difficult. In this connection it is also of interest that the oxidation potentials for Pd and Pt depend to a large extent on the anion; in the presence of halides the order of ease of oxidation decreases in the order, I > Br > Cl. It also appears to be *easier* to oxidize Pd to $Pd^{IV}I_6^{2-}$ than Pt to $Pt^{IV}I_6^{2-}$, whereas for MCl_6^{2-} and MBr_6^{2-} the reverse holds true. This effect may also explain the recent report that the material prepared from a solution of palladium in nitric acid, commonly known as palladium(II) nitrate, is in fact a palladium(IV) hydroxonitrate.[11] Analogously anhydrous $Pd(NO_3)_4$ turns out *not* to be a strong oxidizer.[11]

c. *Stereochemistry of the (II) Oxidation State*

The octahedral coordination so common for Ni(II) is virtually unknown for Pd(II) and Pt(II) and square planar coordination is the general rule. The few cases of octahedral stereochemistry which are known have two trans ligands quite far removed from the metal. Just how significant the resulting interactions are is not clear. For example, the crystal structure of bis(dithiobenzoato)palladium (**I-34**) shows that four S atoms lie in the coordination plane of the metal (at 2.32–2.34, e.s.d. 0.003 Å) and two sulfurs from neighboring molecules complete a tetragonally distorted octahedron at 3.32 and 3.46 Å from the metal.[170] Again, in bis[o-phenylenebis(dimethylarsine)]-diiodopalladium, (**I-35**), the two iodine atoms complete a distorted octahedron about the metal. The Pd–I distances of 3.52 Å compare with a calculated sum of covalent radii of 2.65 Å,[171] and with an experimental value of 2.60 Å found in α-PdI_2.[172] In α-PdI_2 as well, two iodine atoms on adjacent chains complete a distorted octahedron at 3.52 Å.

Five-coordination is well established for all three metals. The tendency to go 5-coordinate depends on the ligand and the metal. It is highest for nickel and least for platinum. For example,[173] (TAS)NiBr$_2$, (**I-36**), exists as a tetragonally distorted square pyramid, whereas (TAS)PdBr$_2$ is ionic and square planar with only one bromine in the coordination sphere of the metal.

Many of the 5-coordinate Ni(II) and Pd(II) complexes investigated appear to have a coordination best described as distorted square pyramidal,[168, 174-179] whereas the two 5-coordinate Pt(II) structures so far determined are better described as trigonal bipyramidal.[180, 181] Some attempts have been made to define characteristics which favor the formation of 5-coordinate complexes. Westland[182] has proposed that the tendency decreases as $Sb > As > P > N$,

(I-34)

(I-35)

(I-36)

when these atoms are present in donor ligands. The evidence is incomplete, but stibines are known to stabilize unusual coordination numbers,[183] arsines and phosphines certainly both give rise to 5-coordination while amines appear not to. Collman[184] has made the point that "soft" ligands (see below) are good for stabilizing 5-coordination for Ir(I) and similar conclusions are no doubt true for Pd(II) as well; however, steric effects must also be important.

Equilibria of the type

$$[(diars)_2Pd]^{2+} + X^- \rightleftharpoons [(diars)_2PdX]^+$$

have been studied by Harris and Nyholm[171, 185] and Ettore et al.[173, 186] Harris and Nyholm showed that $(diars)_2PdI_2$ (I-35) acted as a 1:1 electrolyte in nitrobenzene, and that titration of $[(diars)_2Pd](ClO_4)_2$ (a 1:2 electrolyte) gave a sharp end-point after the addition of one equivalent of halide ion. They were able to isolate salts of the type $[(diars)_2PdX]^+ClO_4^-$ (1:1 electrolytes). In solution these may have a molecule of solvent occupying the sixth coordination site. These authors observed that the stability of complexes of the type $[(diars)_2PdX]^+$ decreased in the order $X = I > Br > Cl$.[185] Ettore et al.[186] have extended this range for different ligands X to give the complete series: $I^- > SeCN^- > SCN^- > S_2O_3^{2-} \approx Br^- > Cl^- > N_3^- > (NH_2)_2CS$. From measurements of the conductivities of $(diars)_2MCl_2$ in water, Harris, Nyholm, and Phillips[187] concluded that the greatest tendency for association between Cl^- and the metal was for Ni; Pd and Pt had approximately the same tendency to associate. This was interpreted in terms of a larger transfer of charge from the arsine to the metal for Ni and hence a higher effective positive charge on nickel than on Pd or Pt. Harris, Livingstone, and Reece[188] concluded from their study of the shifts in the visible spectra of PdX_4^{2-} in various solvents and in the presence of excess halide ion that species such as PdX_5^{3-} and PdX_6^{4-} probably existed under the latter conditions (X = Cl, Br). However, $PtCl_4^{2-}$ showed the same spectrum in the solid, in 1 M $HClO_4$ and in 10 M HCl and was hence assumed not to coordinate further halide ions. For the system

$$Pd(CN)_4^{2-} + CN^- \rightleftharpoons Pd(CN)_5^{3-}$$

Watt et al.[189] have derived a value for K_{eqm} of about 100.

It has long been recognized that different metals form complexes of differing orders of stability with ligands. This was discussed in terms of class "a" and "b" character by Chatt and Ahrland (see the review[190]). Class "a" metals formed complexes such that the order of stability of halo complexes, for example, was $F^- > Cl^- > Br^- > I^-$, whereas for class "b" complexes the reverse was true. Similar criteria applied to the group V and VI ligands, and for class "b" metals typical orders of stability are $F^- < Cl^- < Br^- < I^-$; $O \ll S \sim Se \sim Te$; $N \ll P > As > Sb > Bi$.

This classification has been modified and extended, largely by Pearson, who suggested the terms "hard" and "soft" to describe both the Lewis acids and Lewis bases,† on the basis of their polarizability. Thus "hard" metal ions (class "a") are small, highly charged, and bind best to the least polarizable (hardest) ligand atom of a group, forming more polar bonds. "Soft" metal ions (class "b") are large, have a low oxidation state, easily distorted outer electrons, and bind best to the "softer" (more easily polarizable) ligand atom of a group, forming more covalent and less polar bonds.

Soft bases include the heavier ligand atoms, CO, olefins, aromatic hydrocarbons, and CH_3^-, and typical hard ligand atoms are O, N, and Cl. Typical hard metals are H^+, Li^+, Na^+, Be^{2+}, Al^{3+}, and Co^{3+}, and typical soft metals are metals in the zero-oxidation state, Rh^I, Ir^I, Pd^{II}, Pt^{II}, Cu^I, Ag^I, and Au^I. Borderline metal ions include Ni^{II}, Fe^{II}, Co^{II}, Cu^{II}, Rh^{III}, and Ir^{III}.

These considerations allow a little further understanding of why Ni(II), for example, forms more stable aquo complexes than Pd(II), and why Ni(II) in general has a somewhat higher affinity for the first-row ligand atoms. This is also reflected in the rather higher electronegativity for nickel (1.75) than for palladium (1.35) and platinum (1.44).[165]

---

† This is described in the book by Basolo and Pearson.[43]

d. *Kinetics of Substitution of Square Planar Complexes*

The kinetics of replacement of a ligand, X, in a square planar complex by another ligand follow a two-term rate law:

$$MA_3X^{n+} + Y^- \xrightarrow{H_2O} MA_3Y^{n+} + X^-$$

$$Rate = k_1[MA_3X^{n+}] + k_2[MA_3X^{n+}][Y^-]$$

The first term has been shown to be due to the aquation reaction in which the rate-determining step is replacement of X^- by H_2O; the coordinated water is then replaced by Y^- in a second, fast, step:

$$MA_3X^{n+} + H_2O \xrightarrow{slow} MA_3H_2O^{(n+1)+} + X^-$$

$$MA_3H_2O^{(n+1)+} + Y^- \xrightarrow{fast} MA_3Y^{n+} + H_2O$$

Both terms therefore imply a bimolecular reaction and this is reasonable when it is recalled that square planar complexes have two vacant coordination sites and that 5-coordinated complexes are fairly common. It is therefore not unreasonable to postulate the existence of a 5-coordinate intermediate or transition state in these substitution reactions. Most of the work on substitution in square planar complexes has been carried out for Pt(II) since these are more inert and react more slowly thus making them more amenable to kinetic study.[43] Recent work on analogous reactions of Pd(II) complexes has confirmed that they too in general react by similar paths.[191, 192] The major exceptions are the substitutions of X in [Et$_4$dienMX]$^+$, where the ligand Et$_4$dien† effectively shields the vacant fifth and sixth coordination sites from attack by the entering ligand. For example, this complex (M = Pt) reacts very much more slowly than [dienPtX]$^+$,† and by a unimolecular path. The same applies to the palladium complex, [Et$_4$dienPdX]$^+$, which is more reactive than the platinum analog and undergoes substitution by paths which are independent of the concentration of the incoming ligand.

Few comparative studies of the relative rates of substitution of identical complexes of Ni(II), Pd(II), and Pt(II) have been carried out. One, by Basolo *et al.*, has shown that for

$$\text{trans-}(Et_3P)_2M(o\text{-tolyl})Cl + py \rightarrow \text{trans-}(Et_3P)_2M(o\text{-tolyl})py^+ + Cl^-$$

the relative rates are 5×10^6 (Ni); 1×10^5 (Pd); and 1 (Pt).[193] These authors have interpreted this result in terms of the greater ease with which Ni(II) and Pd(II) can achieve 5-coordination by comparison with Pt(II). Smith and

† See Glossary.

Sawyer[194] have estimated that Pd–O exchange must be at least 10^3 as fast as Pt–O exchange in $[(HOOC \cdot CH_2)_2NCH_2COO]_2M$.

e. Trans Effect, Trans Influence, and Bond Lengths

The occurrence of a trans effect was first noticed by Chernyaev and co-workers in their syntheses of square planar Pt(II) complexes. They generalized the concept that some ligands preferentially labilize ligands trans to themselves. For example, the following stereospecific syntheses of *cis*- and *trans*-$(NH_3)_2PtCl_2$ are rationalized by assuming Cl^- to have a higher trans effect than NH_3.

$$PtCl_4{}^{2-} + NH_3 \rightarrow NH_3PtCl_3{}^- \xrightarrow{\ NH_3\ } cis\text{-}(NH_3)_2PtCl_2$$

$$Pt(NH_3)_4{}^{2+} + Cl^- \rightarrow (NH_3)_3PtCl^+ \xrightarrow{\ Cl^-\ } trans\text{-}(NH_3)_2PtCl_2$$

The effect has been defined as a kinetic phenomenon by Basolo and Pearson[43] (and Langford and Gray[195]) as "the effect of a coordinated group upon the rate of substitution reactions of ligands opposite to it in a metal complex." The explanation for the kinetic trans effect usually given today is that a stabilization of the 5-coordinate transition state is achieved by ligands which have a high trans effect. Since many of these are π-bonding ligands (e.g., CO, R_3P, and C_2H_4) Langford and Gray have proposed that they are particularly effective as equatorial ligands (L') in a trigonal bipyramidal transition state, e.g.,[195]

An approximate order of decreasing trans effect is CO, CN^-, $C_2H_4 > R_3P$, $H^- > Me^-$, $SC(NH_2)_2 > Ph^-$, $NO_2{}^-$, I^-, $SCN^- > Br^-$, $Cl^- > py$, NH_3, OH^-, H_2O. Therefore the situation is more complex than this simple picture suggests; some of the highest trans effects are noticed for ligands such as H, CH_3, and other alkyls, where stabilization of the transition state by π bonding to these ligands cannot occur. Furthermore, with the greater availability of accurate X-ray bond length data it has become increasingly apparent that although most metal–ligand distances are normal† in complexes which do not have any

† That is, equal to the sum of the covalent radii.

TABLE I-2

Bond Lengths of Palladium Complexes from X-Ray Determinations

Bond	Nature of linkage	Range (Å)	Average value (Å)	No. of determinations	References
Pd–C	Pd–Ph	1.998	1.998	1	201
	Pd–CH	2.16	2.16	1	201a
	Pd–CS$_2$	2.00	2.00	1	23
	Pd–π-allyl	2.02–2.22	2.12	24	202–225
	Pd–π-olefin	2.10–2.28	2.19	3	209, 226, 227
	Pd–π-cyclopentadienyl	2.23–2.28	2.25	1	214
	Pd–π-benzene	2.20–2.81	—	2	157, 158
Pd–H	In (Et$_3$P)$_2$PdHCl, not determined				228
Pd–N	Most complexes	2.00–2.09	2.05	9	201a, 229–236
	Pd(L)$_2^a$	1.93–1.99	1.96	2	237, 238
Pd–P	Most complexes	2.31–2.33^b	2.32	7	176, 201, 211, 218, 224, 228, 239
	Trans to C in (Ph$_3$P)$_2$PdCS$_2$	2.42	2.42	1	23
	PPh$_2$ bridging Pd and Fe	2.15	2.15	1	239
Pd–As	All complexes	2.38–2.50	2.44	2	171, 240
Pd–O	All complexes	1.98–2.12	2.07	8	204, 207, 210, 221, 225, 231, 235, 241
Pd–S	All complexes	2.23–2.34	2.30	7	170, 225, 241–244
Pd–F	PdF$_2$, Pd(PdF$_6$)^c	2.04–2.17	2.11	2	245, 246

Pd–Cl	Terminal Cl (most complexes)	2.25–2.33	2.30	12	198, 209, 226, 227, 229, 233, 234, 236, 242, 247–249
	Terminal Cl (trans to σ-Pd–C)	2.38	2.38	2	201, 211
	Terminal Cl (trans to Pd–H)	2.43	2.43	1	228
	Bridging Cl	2.32–2.75	2.42	18	157, 158, 202, 203, 205, 206, 208, 212, 213, 215–217, 220, 222, 239, 247, 248, 250
Pd–Br	Most complexes[d]	2.34–2.57[e]	2.47[e]	5	219, 240, 243, 251, 252
	Terminal Br (trans to σ Pd–C)	2.59	2.59	1	201a
Pd–I	Most square planar complexes[f]	2.59–2.64	2.61	4	172, 176, 253, 254
	Iodines in fifth and sixth coordination sites of tetragonally distorted octahedral complexes	3.29–3.52	3.42	4	171, 173, 176, 253
Pd–Pd	Pd–metal	2.751	—	1	5
	$[(C_6H_6)_2Pd_2](AlCl_3X)_2$	2.57	—	2	157, 158
	$(C_3H_5PdOAc)_2$	2.94	—	1	204
	Nonbonded Pd–Pd	3.32–3.56	3.46	7	—

[a] For bis(glyoximato)palladium(II)[238] and bis(dimethylglyoximato)palladium(II).[237]

[b] An older value of 2.19 Å for a Pd–P in a 5-coordinate complex[175] is not included.

[c] Older values of 1.89 and 1.80 for Pd–F in Cs_2PdF_6 and Rb_2PdF_6[5] are not included.

[d] An insufficient number of determinations have been reported for a clear distinction between terminal and bridging ranges to be made. However, Sales et al.[243] report Pd–Br (bridging) 2.429(4) Å and Pd–Br (terminal) 2.404(4) Å.

[e] This does not include one value of 2.93 Å for an apical Pd–Br in the 5-coordinate $(C_{14}H_{13}P)_3PdBr_2$.[175]

[f] Not enough determinations have been carried out to determine ranges for bridging and terminal iodines. On present evidence they appear to have the same bond lengths.

stereochemical anomalies, this is not the case when one ligand is H, alkyl, or aryl. Under these circumstances, the distance to the metal of the ligand trans to H, alkyl, or aryl is usually significantly greater than when other ligands are present. The effect is clearly seen in some complexes of Pt; both square planar and octahedral complexes show it.

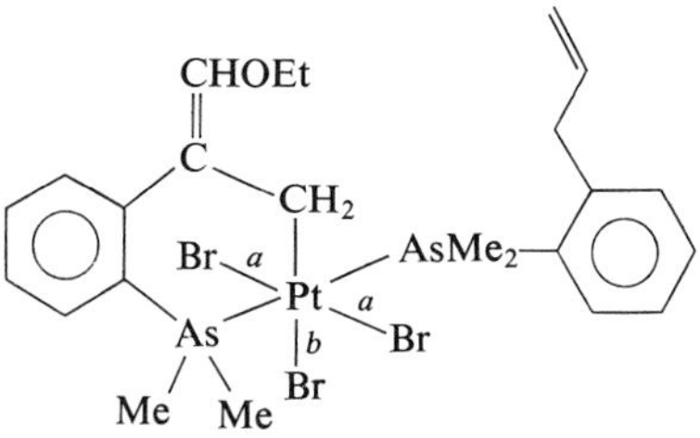

(I-37)

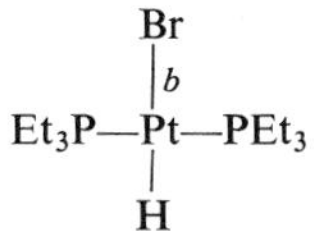

(I-38)

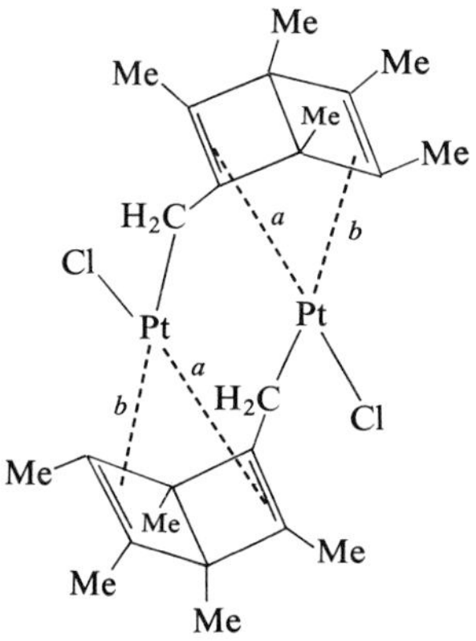

(I-39)

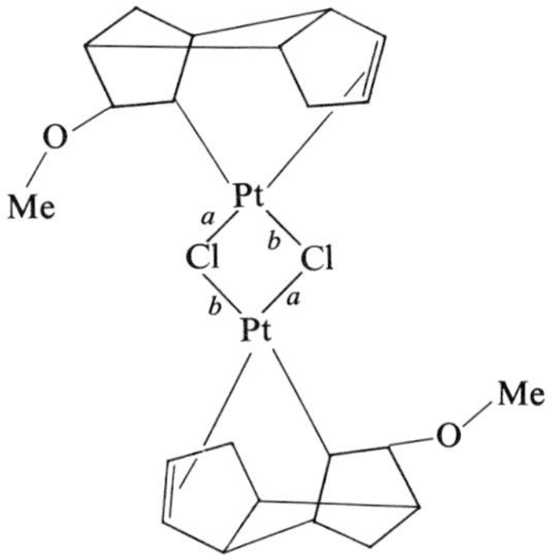

(I-40)

The length of the normal Pt–Br (terminal) bond can be defined from the two Pt–Br bonds, a, trans to each other in (I-37), as 2.47(1) Å.[196] By contrast, Pt–Br bonds trans to the CH_2- in (I-37), b, and trans to H in (I-38)[197] are 2.55(1) and 2.56(4) Å, respectively. This argument also presumes the radius of Pt(II) to be the same as that for Pt(IV); this is certainly the case for Pd(II) and Pd(IV)[198] and is also probable for Pt.

Similarly, the Pt–olefinic carbon bond lengths differ in (I-39).[199] The bond lengths, a, trans to Cl are 2.16(3) Å and those, b, trans to $-CH_2$, are 2.35(3) Å (average).

This also applies to the bridging chlorines in (I-40) where Pt–Cl, a, trans to an olefinic π bond, is 2.34 Å and that, b, trans to –CH is 2.51 Å.[200]

This effect is quite large and is apparently much greater than those which were previously reported to exist for bonds trans to π-bonding ligands such as olefins or CO. The magnitude of the latter has probably been exaggerated.

The few reports so far available for Pd bear this out; thus the value for Pd–Cl (terminal) for Cl trans to Ph (2.38 Å) is outside the range for other Pd–Cl (terminal), 2.25–2.33 Å[201,201a] (see also Ref. 201a). Some average bond lengths of palladium complexes are given in Table I-2.

This latter type of trans effect in the ground state has been termed *trans influence* and defined as "the ability of a ligand to weaken a bond trans to itself." Langford and Gray[195] have proposed an explanation for this in terms of the metal p orbitals giving best overlap with the valence orbitals of CH_3^- and H^-.† This reduces the amount of metal orbital available trans to this substituent and can therefore lead to a weakening (lengthening) of the trans bond. It does not, however, appear that the trans influence significantly strengthens (shortens) the metal–carbon bond. Furthermore, the trans influence is not always noticeable; for example, in the cyclooctenylnickel acetylacetonate (**I-41**), the difference in Ni–O distances (one trans to CH, the other to C=C) is at most 0.045 Å.[255]

(I-41)

A number of attempts have also been made to measure the extent of the trans influence by infrared and NMR spectra, particularly by Chatt and his co-workers.[256, 257] In *cis*-L_2PtX_2, ν_{PtX} varied approximately inversely as the trans *effect* of L, e.g., ν_{PtCl} decreased in the order $R_3N \gg R_3As > R_3Sb > R_3P$ and $R_2S > R_2Se \gg R_2Te$. This was taken to imply that the Pt–Cl bond strength was highest with amines and sulfides as ligands. Similar results were obtained from changes in ν_{PtH} in *trans*-$(Et_3P)_2PtHX$ with variation of X; this decreased in the order $NO_3^- > Cl^- > Br^- > I^- > NO_2^- > SCN^- > SnCl_3^- > CN^-$. *Trans*-$(Et_3P)_2PtHCN$ had the lowest ν_{PtH} and CN^- and $SnCl_3^-$ were also most effective at trans labilization.[35] Similarly, in the NMR spectra of a number of

---

† This applies both in the trigonal bipyramidal transition state for substitution and in the ground state. In the latter case the p orbitals are the only σ-bonding ones which have strong trans-directional properties. Hence if the trans-directing ligand overlaps very strongly with the p_x orbital, there will be less of this orbital available for bonding to the opposite ligand. In the transition state (trigonal bipyramidal) for substitution the leaving ligand will overlap even less with the p_x orbital and give the trans-directing group more of this orbital.

complexes of the type $(Et_3P)_2PtH(L)^+$, J_{PtH} has been shown by Church and Mays to vary with the trans ligand, L, in the order py > CO > CNR > CNAr $\approx$ PPh$_3$ > P(OPh)$_3$ > P(OMe)$_3$ > PEt$_3$.[258] These authors have also given a reasoned critique of some of the assumptions underlying these approaches and their shortcomings. Although there is again less data available for palladium, there can be little doubt that most of what has been shown to hold for Pt is also true for Pd.[259]

A feature which makes investigation of simple inorganic palladium complexes less easy than their platinum counterparts is their ready isomerization. In general, cis complexes of the type L_2MX_2 appear to be less common than the trans isomers for palladium. However, the medium in which measurements are made is all-important and some cis isomers are known, particularly in the solid state, e.g., $(NH_3)_2PdCl_2$[260] and $(Me_2SO)_2Pd(NO_3)_2$,[241] and also, of course, where L_2 is a chelating ligand such as diphos. In the early work of Chatt et al., the stereochemistries were usually determined by dipole moment measurements in nonpolar solvents such as benzene. Where the tendency for isomerization exists and the complexes are labile, there is thus an inherent defect in the method which favors the less polar trans isomers owing to solution entropy effects. Chatt and Wilkins[261] found that the tendency for cis isomers to exist in solution was only significant among complexes containing group V donor ligands. Here, for Sb the percentage of cis isomer was much less for Pd than Pt (Table I-3). More recent studies using infrared or NMR methods have

TABLE I-3

cis-$(R_3Sb)_2MCl_2$ (%) in Benzene Solution

M	R =				
	Me	Et	Pr^n	Bu^n	Ph
Pd	~40	6	3.8	3.7	14.6
Pt	~80	34.4	20.0	20.7	—

shown that the cis forms are more common than had been anticipated from this work. Coates and Parkin[262] showed that $(Me_3E)_2PdCl_2$ was cis in the solid for E = P and As, and other workers have observed the cis form in solution.[263,264] Goodfellow et al. concluded that the tendency for cis complexes to exist in solutions of $(Me_3E)_2MX_2$ decreased in the order M = Pt > Pd, E = As > P,

$X = I > Br > Cl^{265}$;† and Jenkins and Shaw[269] showed that $(Me_2PhP)_2PdCl_2$ was 67% cis in $CHCl_3$ and 100% cis in $MeOH–CS_2$. In contrast, however, $(Me_2PhP)_2PdBr_2$ was all trans, even in $MeOH–CS_2$.

f. Some Compounds of Pd(II)

i. PdF_2. This is an uncharacteristic complex of Pd(II) in that it is paramagnetic ($\mu_{eff} = 1.84$ B.M.). This is due to the presence of weak field ligands (F^-) in a slightly tetragonally distorted octahedron about the Pd (two fluorines at 2.171 Å, and four fluorines at 2.155 Å).[246] The magnetic moment is low for that expected for two unpaired electrons (expected $\mu_{eff} = 2.83$ B.M.) and antiferromagnetic interactions have been postulated to explain this. PdF_2 is isostructural with some other MF_2 complexes (M = Mn, Fe, Co, Ni, and Zn) and is the only simple Pd(II) complex where the tetragonal distortion is small enough for the seventh and eighth d electrons to be unpaired. The complex also has an uncharacteristic color for Pd(II) (violet) and is best formed by heating $Pd^{II}Pd^{IV}F_6$ (formerly known as PdF_3) with SeF_4.[8, 10] It is readily hydrolyzed by moist air to $PdO \cdot xH_2O$. A number of other related paramagnetic Pd(II) complexes, $Pd^{II}M^{IV}F_6$ (M = Pd, Pt, Sn, Ge), are known; their magnetic moments are all similar (2.72–2.98 B.M.) and all agree with the expected value for high-spin octahedral d^8.[270]

ii. $PdCl_2$. This is the most important compound of palladium and the usual starting material for the preparation of other complexes. It can be prepared by chlorination of palladium above 300° or from the solution of palladium in aqua regia (which contains $PdCl_6^{-2}$) by dilution and heating to dryness, when Cl_2 is lost. A recent patent by INCO (Mond) Ltd. describes its preparation as follows,[271]:

$$Pd(sponge) + Cl_2 \xrightarrow{\text{MeOH}} H_2PdCl_6 + H_2PdCl_4 + HCl \xrightarrow{\text{Pd}}$$
$$37:13$$

$$H_2PdCl_4 \xrightarrow{\text{evaporate}} PdCl_2 + 2HCl$$

whereas one by Shell Oil Co.[272] uses the route:

$$Pd + O_2 + HCl + NO \xrightarrow{\text{AcOH/30°}} PdCl_2\downarrow$$

$PdCl_2$ exists in at least two forms: of these, the α form prepared by chlorination above 500° has the linear chain structure (**I-42a**) as determined by Wells.[250]

† For $(R_3E)_2PtCl_2$ Chatt and Wilkins[266–268] showed that the cis isomer was more stable and its isomerization to the trans was endothermic ($\Delta H = 2.5, 1.2,$ and 2.4 kcal/mole for E = P, As, and Sb, and R = Et). The isomerization was also catalyzed by R_3E, possibly according to:

$$cis\text{-}(R_3E)_2PtCl_2 + ER_3 \rightleftharpoons [(R_3E)_3PtCl]^+Cl^- \rightleftharpoons trans\text{-}(R_3E)_2PtCl_2 + ER_3$$

Another very plausible intermediate is the 5-coordinate $(R_3E)_3PtCl_2$, however, see Ref. 268a.

This form appears to be metastable at ambient temperature. The stable β form is the one available commercially and is obtained when the temperature used in the preparation is below 500°. It is isomorphous and probably isostructural[273] with $(PtCl_2)_6$, (I-42b), the crystal structure of which has been determined by Brodersen, Thiele, and Schnering,[274] and shown to be a cluster analogous to $Ta_6Cl_{12}{}^{2+}$.† The metal atoms form an octahedron and are linked by chlorines arranged in a square plane around each metal atom. Each chlorine also bridges one edge of the octahedron. α-$PtCl_2$ has also been prepared, but is not isotypic with α-$PdCl_2$.[275] In agreement with this, β-$PdCl_2$ shows only one resonance in the chlorine NQR spectrum, indicating all chlorines to be equivalent‡; analysis of the data also indicates that the charges on Pd and Cl are virtually identical for α- and β-$PdCl_2$.[276]

(I-42a)

(I-42b)

In the mass spectrometer both α- and β-$PdCl_2$ gave rise to $Pd_6Cl_{12}{}^+$ ions. However, Pt_6Cl_{12} gave ions corresponding to $Pt_4Cl_8{}^+$, $Pt_5Cl_{10}{}^+$, as well as $Pt_6Cl_{12}{}^+$.[273] Schäfer et al. in 1967 also proposed a mechanism whereby α-$PdCl_2$ could be easily transformed into Pd_6Cl_{12} by breaking only one Pd–Cl bond and then performing the appropriate rotations of Cl–Pd–Cl groups. Bond angles remained unchanged during this operation.[273]

This result was already foreshadowed by the work of Bell, Merten, and Tagami[277] who in 1961 postulated Pd_5Cl_{10}, possibly mixed with Pd_6Cl_{12}, as the predominant species in the vapor below 980°. Above this temperature $PdCl_2$ monomer predominated.

Soulen and Chappell[278] in a DTA study of $PdCl_2$ found endotherms due to crystal transitions at 401° and 504°. On cooling, an exotherm at 440° (corresponding to the 504° endotherm) was observed, but not one corresponding to the 401° endotherm. The X-ray powder pattern of this material was now

† In contrast to $Ta_6Cl_{12}{}^{2+}$, however, metal–metal bond lengths in Pt_6Cl_{12} are long (3.32, 3.40 Å) and interactions therefore must be minimal.

‡ Pt_6Cl_{12} shows two resonances; this is ascribed to the small distortion of the structure with two chlorines at 2.34 and two at 2.39 Å.

identical to that of α-PdCl$_2$, but after several months at room temperature lines belonging to β-PdCl$_2$ began to appear.

Schäfer et al. have suggested that both Pd$_6$Cl$_{12}$ and Pt$_6$Cl$_{12}$ may be able to bind electron donors,[273,275] and Paiaro et al.[279] reported the formation of an olefin complex, (cyclododecatriene)$_4$Pt$_6$Cl$_{12}$, which may be structurally related to Pt$_6$Cl$_{12}$.

Although the energies of the α- and β-PdCl$_2$ are probably very similar (Schäfer et al.[273] report Pd–Cl bond energies of 48 and 46 kcal/mole, respectively, for α- and β-PdCl$_2$), and it would not seem likely that these species remain intact in solution, Moiseev and Grigor'ev[162] have reported that the two forms behave quite differently in their abilities to isomerize 1-hexene. In particular, β-PdCl$_2$ was more soluble and virtually inactive by comparison with the α form (see Chapter III,B1, Volume II).

Palladium chloride is rather insoluble in water and completely so in non-complexing organic solvents. It readily dissolves in hydrochloric acid and in aqueous solutions of chloride ions to give PdCl$_4$$^{2-}$. Stability constants for

$$Pd^{2+} + Cl^- \;\rightleftarrows\; PdCl^+ \qquad K_1$$
$$PdCl^+ + Cl^- \;\rightleftarrows\; PdCl_2 \qquad K_2$$
$$PdCl_2 + Cl^- \;\rightleftarrows\; PdCl_3^- \qquad K_3$$
$$PdCl_3^- + Cl^- \;\rightleftarrows\; PdCl_4^{2-} \qquad K_4$$

are reported by a number of authors, who find values for log β_4 of between 11 and 12.[169] Grinberg et al.[280] quote values of 4.3, 3.54, 2.68, 1.68 for log K_1, K_2, K_3, K_4, and 12.2 for log β_4. By comparison, log β_4 for PdBr$_4$$^{2-}$ is 16.1 and that for PdI$_4$$^{2-}$ is 24.9. This explains why I$^-$ will always replace Br$^-$ which will, in turn, replace Cl$^-$ and why mixed halide complexes are not known.

Numerous complexes of the type L$_2$PdCl$_2$ are easily formed from the appropriate ligand L. One of the best known and most useful is (PhCN)$_2$PdCl$_2$, originally prepared by Kharasch et al. in 1938 by heating palladium chloride with benzonitrile.[281] On cooling, yellow-brown crystals of the complex are formed. The complex is readily soluble in a wide variety of organic solvents (benzene, alcohol, chloroform, etc.), but it decomposes in solution to give, after a while, PdCl$_2$. Even a freshly prepared solution shows an infrared ν_{CN} band due to free PhCN of the same intensity as that of the ν_{CN} due to complexed PhCN.[282,283,283a]

Since it is readily soluble in organic solvents and the PhCN is so readily displaced, (PhCN)$_2$PdCl$_2$ reacts with other ligands, L, to L$_2$PdCl$_2$.[284] These are also conveniently prepared from tetrachloropalladate(II) salts in aqueous organic solvents in many cases.

A further type of complex which is well known is L$_2$Pd$_2$Cl$_4$; Chatt and Hart concluded from their dipole moments that they existed largely as (I-43).

Crystal structure determinations of $(Me_2SPdBr_2)_2$ and $(Me_3AsPdBr_2)_2$ show them to exist as (I-44)[240,243] (see also Baenziger et al.[247,248]).

$$\underset{Cl}{\overset{L}{>}}Pd\underset{Cl}{\overset{Cl}{<}}\underset{Cl}{\overset{Cl}{>}}Pd\underset{L}{\overset{Cl}{<}}$$

(I-43)

$$\underset{Br}{\overset{L}{>}}Pd\underset{Br}{\overset{Br}{<}}\underset{Br}{\overset{Br}{>}}Pd\underset{L}{\overset{Br}{<}}$$

(I-44)

These complexes are prepared by mixing the correct stoichiometric quantities of ligand and palladium halide, or by reacting $(NH_4)_2PdCl_4$ with L_2PdCl_2. In some cases the halo bridge in (I-43) can be broken by other ligands, L', to give the mixed complexes $LL'PdCl_2$ (L = R_3P, R_3As; L' = piperidine or p-toluidine).[285] Complexes of type (I-43), L = Cl (i.e., $Pd_2Cl_6^{2-}$), are also known and can be isolated in the presence of suitable large cations, e.g., Ph_4As^+, Et_4N^+, or Bu_4N^+.[286]

Recent work has shown that the metal–halogen stretching frequencies in the far-infrared are observed as strong bands and can give valuable structural information as their positions are relatively insensitive to cis substituents but very sensitive to changes in trans ligands.[257] Bands due to terminal Pd–Cl (ν_{PdCl_t}) occur mainly in the higher part of the range 287–360 cm^{-1}.[287] Trans-L_2PdCl_2 usually shows one band and cis-L_2PdCl_2 shows two. Bands due to bridging PdCl (ν_{PdCl_b}) [in complexes such as (I-43)], are usually found to lower frequencies. Again two bands are expected. Adams and Chandler[288] have suggested the ranges, ν_{PdCl_t} 339–366, ν_{PdCl_b} 294–308 and 255–301 cm^{-1} for $L_2Pd_2Cl_4$. These authors have also pointed out that considerable coupling exists between terminal and bridging Pd–Cl stretching modes, and have estimated the force constants for bridging and terminal Pd–Cl bonds in $Pd_2Cl_6^{-2}$ as 1.36 and 1.59 mdynes/Å, respectively.[286] This confirms that bridging Pd–Cl bonds are weaker than terminal ones.

Metal–halogen vibrations have also been discussed by Adams,[289] and useful compilations of references have been published.[289a]

iii. *PdBr₂*. The preparation and properties of $PdBr_2$ have been discussed by Thiele and Brodersen.[290] At least three modifications (γ, stable up to 550°; β, stable from 550°–670°; and α, stable from 670° to the decomposition temperature, 800°) are known. Both the α and β forms are metastable at room temperature. The structure of α-$PdBr_2$ has been determined and shown to be related to that of α-$PdCl_2$ [nearly square planar arrangement of four bromines about the Pd at 2.42 Å (average)], but the chain is angled, (I-45), the angle between two $PdBr_4$ units being 137°. Pd–Pd distances are 3.29 Å and no interaction is likely. $PtBr_2$ appears to be similar in structure to Pt_6Cl_{12}.

The properties of $PdBr_2$ are very similar to those of $PdCl_2$ except that it is more insoluble; it also forms complexes L_2PdBr_2 and $L_2Pd_2Br_4$, again of lower solubility. Adams and Chandler[288] give values of 268–283 for ν_{PdBr_t} and 189–194 and 166–185 for ν_{PdBr_b} in $L_2Pd_2Br_4$.

(I-45)   (I-46)

iv. *PdI₂*. The preparation and structures of PdI_2 have also been discussed by Thiele and Brodersen.[290] Three forms have again been found:

$$\gamma\text{-}PdI_2 \xrightarrow{340°} \beta\text{-}PdI_2 \underset{560°}{\rightleftharpoons} \alpha\text{-}PdI_2 \xrightarrow{630°} Pd + I_2$$

Thiele, Brodersen *et al.*[172, 253] have determined the crystal structures of α- and β-PdI_2. That of α-PdI_2 is similar to α-$PdCl_2$ but interactions between palladium atoms of one chain and the halides of adjoining chains are greater than in α-$PdCl_2$. β-PdI_2 again shows each Pd to be coordinated by four iodide ions in a square plane, but the structure is best regarded as a chain of Pd_2I_6 units arranged in a zigzag (I-46).

v. *PdO, PdO·xH₂O*. Palladium forms the black oxide PdO on heating in air, but this decomposes above 830°. It is completely insoluble in acids and even in aqua regia; hydrogen readily reduces it to the metal at ambient temperatures. On treatment of aqueous solutions of Pd(II) compounds with base, a precipitate varying in color from yellow to brown is obtained. Its composition, $PdO \cdot xH_2O$, is not constant and it contains varying amounts of water, which is difficult to remove completely. No $Pd(OH)_2$ is known. For comparison NiO and $Ni(OH)_2$ are both known, but Pt does not form a properly characterized oxide, PtO.

vi. *PdSO₄·2H₂O*. This compound forms very deliquescent and water-soluble brown crystals of unknown structure. On heating, it loses first water and then SO_2 and leaves a residue of Pd. It can be prepared by dissolving the nitrate or the hydrated oxide in sulfuric acid or by dissolving the metal in aqueous sulfuric acid under 50 atm of oxygen.[291]

vii. *Pd(NO₃)₂·xH₂O*. This is obtained as brown crystals from a solution of palladium in nitric acid. Field and Hardy[292] reported that on treatment of this hydrate with N_2O_5 they obtained a volatile $Pd(NO_3)_2$, but a compound obtained in a similar fashion by Addison and Ward[11] was shown to be $Pd(NO_3)_4$. These authors also regarded the hydrate as a Pd(IV) hydroxonitrato complex. (See also p. 52.)

viii. *Pd(OCOR)₂*. The palladium (and platinum) carboxylates [Pd(OCOR)₂]ₙ were prepared by Wilkinson *et al.*[293, 294] and Hausman *et al.*[295] The former workers prepared $Pd(OAc)_2$ by dissolving palladium sponge in a mixture of hot nitric and acetic acids or by heating commercial $Pd(NO_3)_2$ in acetic acid, while the latter workers made it by dissolving $PdO \cdot xH_2O$ in acetic acid.

Since $Pd(OAc)_2$ is now of considerable importance as a catalyst, for example, in the vinyl acetate synthesis (Chapter II, Volume II), it should be noted that great care must be taken to avoid the presence of nitrogen oxides when it is made by the former route. Only strict adherence to the prescribed method yields a nitrogen-free product. (See Chapter IV,B, Volume II). Materials containing nitrogen, even in small amounts, behave quite differently. An alternative route to $Pd(OAc)_2$, from the metal and acetic acid at 100° in the presence of oxygen, has been described by Brown *et al.*[291]

$Pd(OAc)_2$ forms purple crystals, and is monomeric in benzene at 80° but trimeric in benzene at 37°. $Pd(OCOEt)_2$ behaves similarly, but $Pd(OCOPh)_2$ is trimeric at 37° and 80° in benzene and at 132° in chlorobenzene. The perfluoroalkylcarboxylates $(R_fCOO)_2Pd$ are all monomeric in ethyl acetate at 37°. The carboxylates form relatively stable adducts $L_2Pd(OCOR)_2$ with amines (Et₃N, pyridine, etc.) for which the trans-square planar structure (with carboxylate acting as a monodentate ligand) has been suggested.[294] The

$$= \quad O \text{----} CMe \text{----} O$$

(I-47)

structure (**I-47**), based on a triangular arrangement of metal atoms with all six acetate groups bridging, has been found for $[Pd(OAc)_2]_3 \cdot H_2O$ by X-ray methods.[295a]

The carboxylates are air-stable and soluble in organic solvents; they decompose to give metal in alcohols on warming. The Pd–O bond is very labile and accounts for some of the unusual reactions which $Pd(OAc)_2$ undergoes. For example, the acetate reacts with acetylacetone or salicyaldehyde in the cold.

Palladium acetylacetonate [Pd(acac)$_2$] is probably isostructural with Cu(acac)$_2$.[296]

ix. *(Ph$_3$P)$_2$PdX and (Ph$_3$P)$_2$PdY.* Although many of the simple palladium complexes of oxy anions are either not known or poorly characterized, a number of well-defined complexes, (Ph$_3$P)PdX, (Ph$_3$P)$_2$PdY$_2$ (X = divalent, Y = monovalent anionic ligand) are known. These include (Ph$_3$P)$_2$PdCO$_3$, (Ph$_3$P)$_2$Pd(NO$_3$)$_2$, (Ph$_3$P)$_2$Pd(NO$_2$)$_2$, and (Ph$_3$P)$_2$PdSO$_4$, and have been made by treating the peroxy compound (Ph$_3$P)$_2$PdO$_2$ with the appropriate oxide, e.g.,[141, 143]

$$(Ph_3P)_2PdO_2 + SO_2 \rightarrow (Ph_3P)_2PdSO_4$$

or in different ways,

$$(Ph_3P)_2PdCl_2 + Ag_2CO_3 \rightarrow (Ph_3P)_2PdCO_3$$

Analogous reactions have been carried out for platinum.

x. *Miscellaneous Complexes.* Cationic complexes [L$_4$Pd]$^{2+}$ are also known for L = NH$_3$, PPh$_3$, and other strong ligands. Recent reports have also described a new way to prepare such complexes where L is a weak ligand such as MeCN.[297, 298]

$$Pd + 2NO^+BF_4^- \xrightarrow{\text{MeCN}} 2NO + [(MeCN)_4Pd](BF_4)_2$$

Other complexes of Pd(II) with weak ligands have been prepared from this complex.

Palladium also forms an insoluble white gelatinous cyanide, Pd(CN)$_2$, on treatment of a solution of a Pd(II) compound with mercuric cyanide. This is soluble in aqueous solutions of alkali metal cyanides to give Pd(CN)$_4{}^{2-}$ (see also Chapter II, Section C,1).

Pd(SCN)$_2$ is brick-red and also insoluble. A number of adducts, L$_2$Pd(SCN)$_2$, have been studied to determine the mode of attachment of the thiocyanate to the metal. Isomerism has been observed for L = Ph$_3$As and L$_2$ = bipy by Basolo *et al.*[299, 300] The nature of the ligand, steric factors and solvent all play a role in determining whether Pd–NCS or Pd–SCN is preferred.[301, 302] The structure of K$_2$Pd(SCN)$_4$ (S-bonded) has been determined.[303]

g. *Comparisons with Other Metals of* d^8 *Configuration*

The d^8 configuration includes M(0) for Fe, Ru, and Os; M(I) for Co, Rh, and Ir; and M(III) for Cu, Ag, and Au. The greatest similarities to Pd(II) are found for Rh(I) and Ir(I). Both usually exhibit the square planar geometry characteristic of Pd(II), whereas the iron group in the zero-oxidation state and Co(I) have a high tendency to 5-coordination. Paramagnetic complexes for these oxidation states are virtually unknown, and in fact these (0) and (I) oxidation states are, as usual, only stable when good π-acid ligands are present. There is no chemistry corresponding to the simple binary complexes of Pd(II). One important characteristic of Pd(II) and Pt(II) shared by Rh(I) and Ir(I) is their relatively facile reversible two-electron oxidation to octahedral d^6 complexes. Rh(I) and particularly Ir(I) undergo oxidative addition more easily than Pd(II).

Only for Au is the M(III) (d^8) configuration common, and gold shares with Pt(II) and Pd(II) the square planar geometry. A further point in common is the ease of oxidative addition to Au(I) (d^{10}) to give the Au(III) complexes. Further oxidative addition to Au(III) does not occur.

5. The M(III) Oxidation State

The (III) oxidation state for Ni, Pd, and Pt is easily the most dubious. The d^7 configuration is not very widespread; Co(II) is the best example and Rh(II), for instance, is rather rare and is only known in the dimeric carboxylates and in some phosphine complexes.[304, 305]

The best characterized complexes are those of Ni(III), e.g., K_3NiF_6 and $(Et_3P)_2NiBr_3$. The pitfall of placing too much emphasis on formal oxidation states is clearly illustrated by the recent crystal and ESR study of *trans*-[(diars)$_2$NiCl$_2$]$^+$Cl$^-$ by Gray *et al.*[15] This complex is paramagnetic and, as expected for a low-spin d^7 configuration, shows a magnetic moment indicating the presence of one unpaired electron and a tetragonally distorted octahedral structure. The ESR spectrum however indicates that far from the unpaired electron being located on the metal atom it is largely concentrated on the diarsine ligand as shown by a comparison of this spectrum with one of the diarsine radical cation. Gray *et al.* term this ligand a "σ-radical" and say that the complex has not an authentic Ni(III) ground state.

A true complex of Ni(III) has recently been reported by Meek *et al.*[16] The crystal structure of $(Me_2PhP)_2NiBr_3$ shows it to contain trigonal bipyramidally coordinated Ni, with the phosphines axial and the three bromine ligands equatorial. The magnetic moment of 2.17 B.M. is that expected for a low-spin d^7 state and the complex does not show an ESR spectrum for a radical cation.

The complex $PtBr_3$ has recently been shown to be a Pt(II)Pt(IV) complex by Thiele and Woditsch.[306] PdF_3 has been shown to be $Pd(PdF_6)$.[270] Garif'yanov *et al.*[307] claim that bis(2-benzyldioximato)palladium(II) is oxidized by bromine to bromobis(2-benzyldioximato)palladium(III). The ESR spectrum of this is reported to be in agreement with a low-spin d^7 state with a square pyramidal geometry, but this has been disputed for some analogous diphenylglyoximato complexes.[308]

Warren and Hawthorne[309] have prepared the bis[(3)-1,2-dicarbollide]-palladium complex, formally of Pd(III)

$$[(\pi\text{-}(3)\text{-}1,2\text{-}B_9C_2H_{11})_2Pd^{II}]^{2-} + [(\pi\text{-}(3)\text{-}1,2\text{-}B_9C_2H_{11})_2Pd^{IV}]^0 \rightarrow 2[(\pi\text{-}(3)\text{-}1,2\text{-}B_9C_2H_{11})_2Pd^{III}]^-$$

The complex is unstable and paramagnetic ($\mu_{eff} = 1.68$ B.M.). Its nickel analog is also known.[13] In both cases, however, it is not possible, on present evidence, to say whether these are really true M(III) complexes.

6. The M(IV) Oxidation State

a. *General*

The (IV) oxidation state (d^6) plays a much smaller role in the chemistry of palladium than it does for platinum, but is more important and prevalent than Ni(IV). The (IV) state for these metals invariably has octahedral geometry and is diamagnetic (low-spin d^6). In these respects it is very similar to the (III) oxidation states for Co, Rh, and Ir which are again octahedral and diamagnetic (except for CoF_3, which has a high-spin d^6 configuration). A further feature common to Pt(IV) and the cobalt subgroup in the (III) oxidation state is that they are all kinetically inert and undergo substitution reactions slowly. For this reason the reactions of Pt(IV) and especially Co(III) have been intensively studied.[43] The kinetic inertness increases from Co(III) to Ir(III), and a similar trend is probable for Pd(IV) and Pt(IV).

For this reason and because higher oxidation states are more favored for the 5d elements, the known chemistry of Pt(IV) is extensive. That of Pd(IV), however, is still comparatively little explored.

b. *Compounds of Pd(IV)*

The most common complexes of Pd(IV) are the hexahalopalladates PdX_6^{2-} (X = F, Cl, Br, I). Of the binary halides only PdF_4 is known; by comparison, all four platinum(IV) halides appear to exist.[290]

i. *PdF₄*. A compound, originally formulated as PdF_3, was prepared by fluorination of $PdCl_2$ at 200°–250° by Ruff and Ascher[9] and it is probable that the fluoride isolated by Moissan[7] from high temperature fluorination of palladium was an impure form of this material. The crystal structure of PdF_3 was determined by Hepworth *et al.*[245] who found it isostructural with RhF_3 and IrF_3 with each metal atom octahedrally coordinated by six fluorides at

2.04 Å in an ionic lattice. Bartlett and Rao,[270] however, showed that the para-magnetism ($\mu_{eff} = 2.88$ B.M.) was better explained by the formulation Pd^{II} ($Pd^{IV}F_6$) where the Pd(II) was in a high-spin d^8 state and the Pd(IV) in the low-spin d^6 state. The two palladium atoms appear however to be nearly identical in the crystal lattice. On treatment with fluorine at 7 atm and 300°, PdF_4 was formed.[270]

$$Pd^{II}Pd^{IV}F_6 + F_2 \rightarrow 2PdF_4$$

PdF_4 has the low-spin d^6 configuration and was found to be only very weakly paramagnetic.† It had the UCl_4 structure,[310] with four F^- closely co-ordinated, and was very easily reduced back to $PdPdF_6$.[270] Other hexafluoro-palladates, $M^{II}PdF_6$, have been prepared by Henkel and Hoppe[311]:

$$(NH_4)_2PdCl_6 + MO \text{ or } MCO_3 \xrightarrow{F_2} MPdF_6 \quad (M = Mg, Cu, Zu, Cd)$$

and Na_2PdF_6 has been shown to be isotypic with Na_2SiF_6.[311] Whereas the hexafluoroplatinates could be recrystallized from water, the hexafluoropalladates were immediately decomposed, with the formation of a brown $PdO_2 \cdot xH_2O$.[10]

ii. M_2PdCl_6. The hexachloropalladates are formed either by chlorination of the tetrachloropalladates(II) or, more commonly, by the addition of alkali halides to a solution of palladium in aqua regia (which contains $PdCl_6^{2-}$). K_2PdCl_6 is isostructural with K_2PtCl_6, and both have a regular octahedral arrangement of Cl^- about the metal.[6] It has also been shown that the radius of Pd(IV) [2.300(7) Å] in $(NH_4)_2PdCl_6$ is virtually identical to that of Pd(II) [2.299(4) Å] in $(NH_4)_2PdCl_4$.[198]

The solid K_2PdCl_6 begins to lose chlorine on heating above 175°; this is complete at 280°. It is not very soluble in water, and in solution it also slowly dissociates to chlorine and $PdCl_4^{2-}$ even in the cold.[6]

Oxidation potentials for $PdX_4^{2-} \rightarrow PdX_6^{2-}$ are summarized in Table I-1.

The nuclear quadrupole resonance spectra of a series of hexahalometal-lates(IV) have been measured; the results point to palladium and platinum having almost identical electronegativities.[312]

iii. $Pd(NO_3)_4$. On treatment of commercial palladium nitrate with N_2O_5 at −78°, the above tetranitratopalladium(IV) was obtained. It was diamagnetic and not volatile and was a surprisingly poor oxidizer. For example, Fe(II) was not oxidized, although it did oxidize iodide to iodine. It was also reported to be inert to hydrocarbons. In water it formed a material identical to that obtained on dissolving palladium in nitric acid and which was postulated to be a hydroxonitratopalladium(IV) complex.[11]

† As the crystal field splitting parameters, Δ_0, increase sharply with formal charge on the metal, it is not surprising that even PdF_4 is diamagnetic.

Chapter II

Compounds with Palladium–Carbon Sigma Bonds and Palladium Hydrides

A. INTRODUCTION

Organotransition metal compounds can broadly be divided into two categories, those where the metal–carbon bond is one of σ symmetry (i.e., the electron density in the M–C bond is concentrated along the M–C axis) and those where a π (or μ) bond is present. The latter topic will be dealt with in the succeeding chapters.

Ligands of the type to be considered here are usually monodentate and can be either neutral or anionic. This classification is again a formality; although we regard methyl, phenyl, ethynyl, hydride, etc., as anionic ligands, it is *not* to be inferred that the complexes are ionic in nature. The reverse is, in fact, true. The so-called neutral ligands are carbonyls, isonitriles and carbenes.

B. NEUTRAL LIGANDS

1. Carbonyls

Simple binary carbonyls $[M_x(CO)_y]$ of the type so well known for most of the other transition metals are only known for nickel. The analogs to $Ni(CO)_4$,

$Pd(CO)_4$ and $Pt(CO)_4$, have not, to date, been made despite many attempts to prepare them. A polymeric carbonyl $[Pt(CO)_2]_n$, of unknown structure has been briefly reported for platinum.[20,21]

Possible reasons why neither platinum nor palladium form $M(CO)_4$ have been discussed by Nyholm[313] in terms of the first ionization potential of metallic Ni (5.81 eV) compared to Pd and Pt (8.33 and 8.20 eV). This is taken to imply that it is easier for an electron from nickel to be promoted into a metal–carbon π orbital (equivalent to back-donation from a filled metal d orbital to a π^* orbital of the CO) than for palladium or platinum. The corollary to this is that when the ligand is a better donor, the ionization energy is decreased and complexes such as $(Ph_3P)_nM$ become stabilized. However, it is at least possible that the failure to isolate $Pd(CO)_4$ and $Pt(CO)_4$ may arise from their high kinetic lability, rather than from a lower thermodynamic stability than $Ni(CO)_4$.

An argument in favor of Nyholm's suggestion is the existence of mixed carbonyl phosphine complexes for platinum(0). A large number of such complexes are known including $(Ph_3P)_3PtCO$,[25, 26, 314] $(Ph_3P)_2Pt(CO)_2$,[22, 314] and a number of cluster complexes, for example, $[(PhMe_2)P]_4Pt_4(CO)_5$, $(Ph_3P)_4Pt_4(CO)_5$, and $(Ph_3P)_4Pt_3(CO)_3$,[315] as well as others of as yet undetermined structure.[20, 87]

Misono et al. have recently reported the synthesis of the first zero-valent palladium carbonyls.[92]

$$Pd(acac)_2 + 3PPh_3 + CO \xrightarrow{Et_3Al/PhMe/-50°} (Ph_3P)_3PdCO$$

$$Pd(acac)_2 + PPh_3 + CO \xrightarrow{Et_3Al/PhMe/-50°} [Ph_3PPdCO]_n$$

$$3(Ph_3P)_3PdCO \underset{20°}{\rightleftarrows} Ph_3P + (Ph_3P)_4Pd_3(CO)_3$$

(II-1)

$(Ph_3P)_3PdCO$ and $(Ph_3P)_4Pd_3(CO)_3$ appear to have the same stoichiometries as the corresponding platinum complexes,[20, 315] and may be presumed to be isostructural. $(Ph_3P)_3PtCO$ shows a single ν_{CO} at 1903 cm^{-1}, $(Ph_3P)_3PdCO$ at 1955 cm^{-1}; $(Ph_3P)_4Pt_3(CO)_3$, (II-1), shows ν_{CO} at 1788(s), 1803(s), and 1854(w) cm^{-1}, and $(Ph_3P)_4Pd_3(CO)_3$ at 1820(vs) and 1845(s) cm^{-1}. In each case there is an increase in ν_{CO} of 40–50 cm^{-1} in going from Pt to Pd; this

implies stronger metal to carbonyl back-bonding for the platinum complexes.†
The mononuclear complex, $(Ph_3P)_3PdCO$, is more stable to air than the
cluster complexes. There does not at the moment appear to be a platinum
complex comparable to $[Ph_3PPdCO]_n$, but it is probably not impossible
that the stoichiometry reported is incorrect and that this is an analog of
$(Ph_3P)_4Pt_4(CO)_5$.[315]

It is clear, from the isolation of zero-valent cluster complexes containing CO
for both Pd and Pt, that simple arguments based on ionization potentials of
the metals have only a limited value.

Another sharp point of contrast between Ni, Pd, and Pt is that while platinum forms several types of carbonyl halides, $[Pt(CO)X_2]_2$, $[Pt(CO)_2X]_2$,
$[Pt(CO)X_3]^-$, and $LPt(CO)X_2$ (where $X = Cl$, Br, and I), nickel apparently
does not.[19] The position for palladium is complex; a number of carbonyl
chlorides have been reported, but their nature is far from clear. Manchot and
König prepared a complex, either by heating a methanolic suspension of palladium chloride in an atmosphere of CO or by passing CO saturated with
methanol vapor over $PdCl_2$ at $20°$,[316] which was analyzed as $PdCOCl_2$. They
noticed that when the first method was used, uptake of one mole of CO per
$PdCl_2$ was very rapid. In the presence of water this reaction failed and only
palladium metal was obtained; however, traces of water were necessary in order
to obtain $PdCOCl_2$. $PdCOCl_2$ was relatively stable up to $60°$, but was decomposed by water according to

$$PdCOCl_2 + H_2O \rightarrow Pd + 2HCl + CO_2$$

Irving and Magnusson[317] measured the infrared spectrum of $PdCOCl_2$.
This showed ν_{CO} at 1976 cm^{-1}, whereas a large number of platinum carbonyl
chlorides of the type $[PtCOCl_2]_2$, $PtCOX_3^-$, and $LPtCOX_2$ all had ν_{CO} between 2100 and 2150 cm^{-1}. The very high ν_{CO} in the latter complexes implies,
by the usual argument, a high C–O bond order and hence a weak metal–
carbon π bond. This is reasonable as the metal is in a relatively high formal
oxidation state. By the same argument, if $PdCOCl_2$ has the same structure as
$[PtCOCl_2]_2$,

as is usually assumed,[317] this implies a *much* stronger metal–carbon π bond for
the palladium complex. However, cases are known (see below) where carbonyl

---

† The bonding in metal carbonyls is usually viewed as involving a weak σ-donor bond from
the carbon to the metal and a strong back-bond (d_π–p_π) from filled metal orbitals of appropriate symmetry and energy to the π^* orbitals of the CO; this decreases the CO bond order.
Since the σ-donor ability of CO is very low, the greater part of the stabilization of the bond
arises by back-donation; therefore only metals in low oxidation states form carbonyls as a
rule.

stretching frequencies of around 2100 cm^{-1} are observed for palladium(II) complexes. Furthermore, other evidence strongly suggests (see above and Chapter III) that palladium is poorer at back-bonding than platinum. Present evidence therefore allows no conclusion to be drawn about the nature of PdCOCl$_2$.

Fischer and Vogler obtained a yellow diamagnetic Pd$_2$(CO)$_2$Cl by passing CO into a chloroform solution of bis(benzonitrile)palladium chloride or (ethylene)palladium chloride dimer.[318]

$$(PhCN)_2PdCl_2 + CO \xrightarrow{\text{CHCl}_3} Pd_2(CO)_2Cl \longleftarrow (C_2H_4PdCl_2)_2 + CO$$

This material was again very insoluble; its infrared spectrum showed a very strong carbonyl band at 1975 cm^{-1} together with a shoulder at 1936 cm^{-1}. The probable absence of terminal Pd–Cl bonds was also concluded from the spectrum.

Treiber[319] found that PdCOCl$_2$ in methanol in the presence of moist air underwent reaction to give a red-violet PdCOCl.

$$PdCOCl_2 + H_2O \rightarrow CO_2 + HCl + PdCOCl$$

The constitution of this material was inferred from its decomposition at 250° to give CO, and palladium metal and palladium chloride in a 1:1 ratio.

$$2PdCOCl \xrightarrow{\Delta} PdCl_2 + Pd + 2CO$$

PdCOCl was again a very insoluble polymer and showed bands ascribed to metal carbonyl bonds at 1975, 1925 (shoulder), and 1892 (shoulder) cm^{-1}.

Schnabel and Kober[320] have repeated the preparation of the complex reported as Pd$_2$(CO)$_2$Cl and obtained a material with identical properties; however, the formulation of this *gray-green* solid was PdCOCl. An X-ray powder photograph of one sample showed some crystallinity and the absence of both metal and PdCl$_2$.

The presence of very strong bands at 1975 cm^{-1} in the infrared spectra of all these complexes suggests that they all contain a common entity which might well be a polynuclear carbonyl in a lower oxidation state [+I or (0)]. Gel'man and Meilakh[321] reported the formation of a material, which they formulated as Pd$_2$(CO)$_2$Cl$_4^{2-}$, from the action of carbon monoxide on a solution of ammonium tetrachloropalladate(II) in hydrochloric acid. Irving and Magnusson[317] reinvestigated this complex and isolated it using [ethylenediamine H$_2$]$^{2+}$ as the cation. This material had ν_{CO} at 2056 cm^{-1} and they suggested [enPdCOCl]$^+$[PdCOCl$_3$]$^-$ as the structure, following a suggestion by Anderson. It is not clear, however, why such a formulation should only give rise to the one observed carbonyl stretching frequency. Moiseev and Grigor'ev[162] have suggested the compound prepared by Gel'man and Meilakh is a Pd(I) complex,

$$\left[\begin{array}{ccc} & \overset{\displaystyle Cl}{|} & \overset{\displaystyle Cl}{|} \\ OC{-} & Pd{-} & Pd{-}CO \\ & \underset{\displaystyle Cl}{|} & \underset{\displaystyle Cl}{|} \end{array} \right]^{2-}$$

Kingston and Scollary[322] observed that carbonylation of palladium chloride in methoxyethanol gave a yellow solution from which large cations, especially Ph_4As^+, precipitated brown diamagnetic solids which were 1:1 electrolytes in DMF. The anion was formulated as $PdH(CO)Cl_2^-$. Two bands in the infrared spectrum of $Ph_4As[PdH(CO)Cl_2]$ were assigned to ν_{CO} (1900 cm^{-1}) and ν_{PdH} (weak, 1960 cm^{-1}), respectively. If this formulation is correct, it is possible that some of the other complexes hitherto considered to be carbonyls may also contain metal-bonded hydrogen.

A last curiosity is the stable red complex isolated by Burianec and Burianova[323] from the action of carbon monoxide on [(o-phen)$_2$Pd]Cl$_2$ in aqueous solution. This material was analyzed as [(o-phen)PdCO]·2H$_2$O, and showed a strong band at ca. 1800 cm^{-1}, which the authors ascribed to a bridging carbonyl. They proposed the dimeric structure

A number of complexes are known where ν_{CO} bands at around 2100 cm^{-1} are observed. Perhaps the best characterized is trans-[(Et$_3$P)$_2$PdCOCl]BF$_4$, prepared by carbonylation of the bridged dimeric complex [(Et$_3$P)$_2$PdCl$_2$Pd-(PEt$_3$)$_2$](BF$_4$)$_2$ by Clark et al.[324] They report ν_{CO} as 2135 cm^{-1}. The carbonyl complex is very labile and easily reverts to the dimer; the platinum analogs are more stable, whereas the nickel ones could not be isolated. Again here, ν_{CO} for the Pd complex is significantly higher (25 cm^{-1}) than for the isostructural platinum complex.†

Medema et al.[326] have reported observing a band at 2100 cm^{-1} when the complex (II-2) was carbonylated, which, they suggested, was due to a 5-coordinate complex (II-3). The 2100 cm^{-1} band decayed and two new bands were observed, one at 1910 cm^{-1}, due to another intermediate which also decayed with time, and the other at 1750 cm^{-1}, corresponding to ν_{CO} (ester) in the final product, (II-4).

† However, the recently reported anions, [MCO(SnCl$_3$)$_2$Cl]$^-$, show almost identical ν_{CO} for M = Pd (2054) cm^{-1}) and Pt (2058 cm^{-1}).[325]

$$
\left[
\begin{array}{c}
\text{MeOCH} \overset{\displaystyle CH_2}{\underset{\displaystyle CH_2}{}} Pd \overset{Cl}{\underset{}{}} \\
NMe_2
\end{array}
\right]_2 \;+\; CO \;\rightleftharpoons\;
\left[
\begin{array}{c}
\text{MeOCH} \overset{\displaystyle CH_2}{\underset{\displaystyle CH_2}{}} \overset{CO}{Pd} \overset{Cl}{\underset{}{}} \\
NMe_2
\end{array}
\right]_2
$$

(II-2) $\qquad\qquad\qquad$ *insertion* (II-3)

$\downarrow$ EtOH $\quad\longleftarrow$

$$Me_2NCH_2CH(OMe)\cdot CH_2COOEt + Pd$$

(II-4)

A high frequency ν_{CO} (2121 cm^{-1}) has also been reported for the product of carbonylation of π-allylpalladium chloride dimer, $(\pi\text{-}C_3H_5PdCl)_2$, under mild conditions. This product was unstable and has not been fully characterized.[327]

π-Allylpalladium chloride dimer reacted with palladium chloride and CO in methanolic HCl to give an amorphous material with infrared spectrum similar to that of $(\pi\text{-}C_3H_5PdCl)_2$, but also showing the presence of a metal carbonyl at 1940 cm^{-1} (and 1910 cm^{-1}, said to be due to ν_{13CO}). This material was analyzed as $[C_3H_5Pd_2Cl_2(CO)_2]_2$.[328]

Numerous Russian workers have investigated the oxidation of carbon monoxide to carbon dioxide, either by Pd(II) salts or by quinones using Pd(II) salts as catalysts.[329–336] Lower oxidation state carbonyls have been proposed as intermediates, but the evidence is incomplete. Fasman et al.[331] noted that in the reaction of CO with $PdBr_4{}^{2-}$ in water three moles of CO were absorbed and one mole of CO_2 was produced before precipitation of metal began. In a later paper these workers[334] studied the effects of small amounts of water on the reaction of K_2PtBr_4 with CO in dioxane. With 1 % water a yellow precipitate was obtained which showed ν_{CO} at 2122 cm^{-1}; with increasing amounts of water precipitates with new bands at 1910 and then 1950 and 2010 cm^{-1} were obtained. A complex from $PdBr_2$ and CO in methanol corresponding to that of Manchot and Konig, and formulated as $PdBr_2CO$, showed ν_{CO} at 1950 cm^{-1}.

Many of the reports on the preparation of the palladium carbonyl halides specify that small traces of water are necessary, and it seems likely that a reduction of part of the palladium to a lower oxidation state is occurring.

This is confirmed by the unexpectedly low ν_{CO} observed for many of the carbonyl halides. Other explanations for this are that bridging carbonyls, or anionic carbonyls, both of which are expected to exhibit low ν_{CO}, are present. It is also possible that some of the observed bands are due to ν_{PdH}. More work in this field and some X-ray structures are highly desirable.

2. Reactions of Palladium Carbonyl Chlorides

Only a few such reactions have been reported. Fischer and Werner reacted $PdCOCl_2$ with chelating dienes (1,5-cyclooctadiene, norbornadiene) to give the diene–$PdCl_2$ complexes, e.g.,[337]

However, with 1,3-cyclohexadiene and 1,3-cycloheptadiene, π-allylic complexes, e.g., (II-5), were obtained.[337–339]

(II-5)

Svatos and Flagg have reported the synthesis of $Pd(PF_3)_4$ from $PdCOCl_2$.[340]

3. Isonitriles and Carbenes

Isonitrile complexes are known for Pd(II) and Pd(0). They have been investigated by Malatesta and Fischer and their co-workers.[34, 46, 47, 60, 61, 341] Reduction of $(RNC)_2PdI_2$, to the formally zero-valent $(RNC)_2Pd$ (R = Ph, p-MeC_6H_4, p-$MeOC_6H_4$) is caused by an excess of the isonitrile in strongly alkaline solution. These and other reactions have been discussed in Chapter I, Sections B,2,b,ii and B,2,d.

The structures of the formally two-coordinated Pd(0) complexes $(RNC)_2Pd$ are of interest, but have not yet been determined. The complexes where R is aryl, prepared by Malatesta, are very insoluble, dark brown, and metallic looking. Malatesta[61] has suggested that they are polymers with metal–metal bonds, but has rejected the idea of bridging RNC groups. The $(RNC)_2Pd$ complexes prepared by Fischer and Werner had R = cyclohexyl and isopropyl[46]; these were more soluble, but the molecular weights increased with time to a value close to that for $[(C_6H_{11}NC)_2Pd]_4$. After two weeks at 25° a completely insoluble brown solid precipitated. Malatesta has also mentioned the existence of unstable yellow complexes which may be $(RNC)_3Pd$ or $(RNC)_4Pd$[76] (see also Section C,2).

Considerable interest has recently been focused on isonitrile complexes of Pd(II) and Pt(II) by the discovery that they are convenient starting materials for the synthesis of complexes containing carbene-type ligands. These are prepared by the addition of alcohols or amines to the isonitrile complexes,[341a] for example,

$$L(RNC)MCl_2 + XH \rightarrow \begin{array}{c} L \\ | \\ Cl-M-Cl \\ | \\ C \\ \diagup \diagdown \\ HNR \quad X \end{array}$$

(M = Pd, Pt; L = Ph₃P, RNC; X = R'O or R"R"'N)

Isonitriles also insert into Pd–methyl and Pd–allyl bonds to give new types of σ-bonded complexes.[341b]

Another route to complexes containing carbene-like ligands is by the solvolysis of cationic monoacetylene complexes of Pt(II),[341c] a reaction which can probably also be applied to Pd(II).

C. COMPLEXES WITH ANIONIC LIGANDS

Broadly the anionic ligands can be divided into those where an sp or sp² carbon is attached to the metal and back-bonding from filled metal to empty π^* ligand orbitals is possible and probably occurs, and those where the metal is attached to sp³ carbon (alkyl) and this back-donation cannot take place. Since H (hydride) is analogous in many respects to an alkyl group, it will also be considered; formally a hydride is merely the lowest member of the $M(CH_2)_nH$ where $n = 0$.

1. Cyanides

Although not usually regarded as organic ligands, cyanides should be mentioned here for comparative purposes. Palladium cyanide is a white insoluble material, presumably a two-dimensional polymer with Pd—C≡N—Pd bridges. It is soluble in excess cyanide to give $Pd(CN)_4^{2-}$; $K_2Pd(CN)_4$ could be reduced to $K_4Pd(CN)_4$ by potassium in liquid ammonia (Chapter I, Section B,2,b,ii, this volume). This was a highly reactive material, significantly more so than $K_4Ni(CN)_4$. Structures are not known, but the $M^0(CN)_4^{4-}$ is probably a tetrahedral anion. The nickel complex exhibits ν_{CN} at 1985 cm⁻¹, a very low value, which suggests that considerable back-donation to π^* orbitals of the CN is occurring, as expected for a low oxidation state and four negative charges.[149]

Beck and Schuierer[342] have reported fulminato complexes, $Pd(CNO)_4^{2-}$ and $(Et_3P)_2Pd(CNO)_2$, which are similar to the cyanides.

2. Alkynyls (Acetylides)

Alkynyl complexes of Pd(0) and Pd(II) are known. The first, of the type $K_2[Pd(C\equiv CR)_2]$ (R = H, Me, Ph), were prepared by Nast and Horl in liquid ammonia.

$$K_2Pd(CN)_4 + RC\equiv CK \longrightarrow K_2Pd(CN)_2(C\equiv CR)_2 \xrightarrow{\text{K}} K_2Pd(C\equiv CR)_2$$

They were diamagnetic and pyrophoric, but not shock-sensitive. With water they gave largely the acetylene, $RC\equiv CH$, and some $RCH=CH_2$.[73] Nast and Horl suggested they were very low polymers because of their solubility in liquid ammonia, but this is a very poor criterion since ammonia is a base which would be able to depolymerize the polymer to form species such as $(NH_3)_2Pd(C_2R)_2^{2-}$.

Nast and Heinz[74] have also prepared alkynylplatinum complexes.

$$K_2Pt(SCN)_4 + 4KC_2R \longrightarrow K_2Pt^{II}(C_2R)_4 \downarrow$$

$$K_2Pt(CN)_4 + 2KC_2R \longrightarrow [K_2Pt(CN)_2(C_2R)_2] \xrightarrow{\text{K}} K_2Pt^0(C_2R)_2$$

A little more information about possible structures is available here as the infrared spectra of $K_2Pt(C_2R)_4$ and $K_2Pt(C_2R)_2$ were quoted. For the former $\nu_{C\equiv C}$ spanned a large range (1931 cm^{-1} for R = H, to 2116 cm^{-1} for R = Me) but only one band was observed. In the zero-valent complexes $K_2Pt(C_2R)_2$, on the other hand, the range was even larger and now at least three bands were observed for every complex. Therefore, although the $\nu_{C\equiv C}$ of $K_2Pt(C_2R)_4$ are compatible with a simple square planar structure, those of $K_2Pt(C_2R)_2$ appear to indicate very much greater complexity.

Since their stoichiometries and properties are so similar, this argument presumably also applies to the palladium complexes. While a discussion of the structures must await a complete crystal structure determination, it is worth pointing out that Pd(0) and Pt(0) are isoelectronic with M(I) for the copper subgroup. These latter metals also form alkynyl compounds of the type $(LMC\equiv CR)_m$ and $(MC\equiv CR)_n$ of considerable complexity. The structures of

$(Me_3PCuC_2Ph)_4$, (II-6), and $(Me_3PAgC_2Ph)_n$, (II-7), determined by Corfield and Shearer[343, 344] are illustrated:

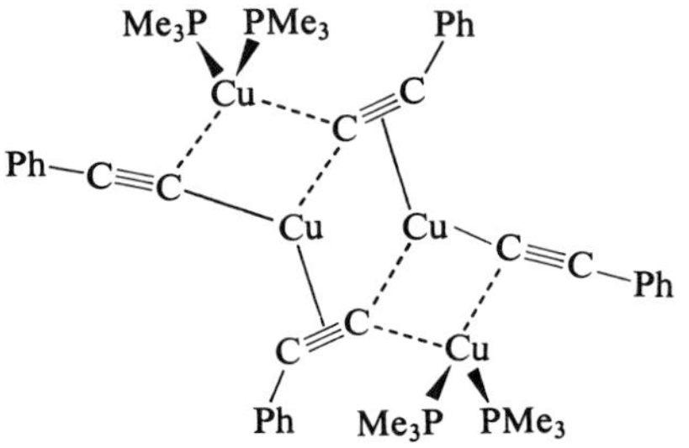

(II-6) 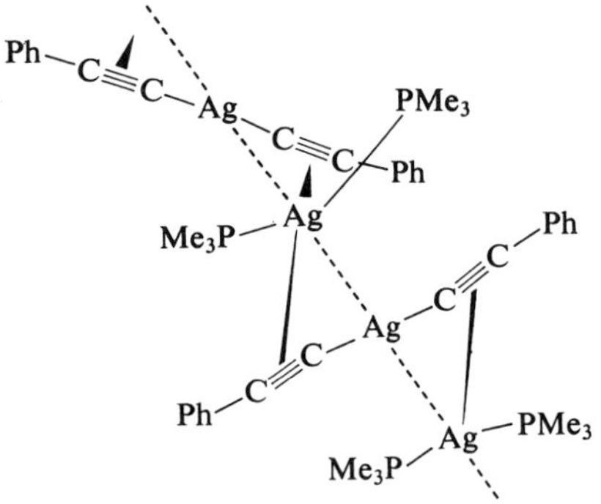   **(II-7)**

Considerable delocalization of electrons over the molecules must be supposed. Although $K_2Pd(C_2R)_2$ and $K_2Pt(C_2R)_2$ need not be the same, structures of comparable complexity may well occur here too.

Since CN^-, CNR, and C_2R^- are formally isoelectronic to CO, the existence of $Ni(CO)_4$, $Ni(CN)_4^{4-}$, $Ni(CNR)_4$, and $Ni(C_2R)_4^{4-}$ and the absence (except for $K_4Pd(CN)_4$) of the tetracoordinated Pd and Pt analogs is obviously of some significance.

Alkynyl complexes of nickel(II), palladium(II), and platinum(II) containing other ligands are also known, for example, $Pd(CN)_2(C_2R)_2^{2-}$, $(R_3P)_2M(C_2R')_2$, (II-9), and $(R_3P)_2M(C_2R')X$, (II-8). All these are significantly less reactive than the zero-valent anionic alkynyls, for example, in their stability to air. The preparation of the first has been mentioned above; the others were prepared by the action of alkali metal alkynyls (or, giving somewhat poorer yields, alkynyl Grignard reagents) on the dihalobis(t-phosphine)metal complexes.[49, 345–347]

$$(R_3P)_2MX_2 + R'C{\equiv}CM' \longrightarrow (R_3P)_2M(C_2R')X$$

(II-8)

$$\Big\downarrow R'C_2M'$$

$$(R_3P)_2M(C_2R')_2$$

(II-9)

(M = Ni, Pd, Pt; M' = Li, Na, Mg;
R = Et, Ph; R' = H, Me, CF_3, Ph, etc)

The nickel complex $(Et_2PhP)_2Ni(C_2Ph)_2$ adds another mole of the phosphine to give the 5-coordinate red $(Et_2PhP)_3Ni(C_2Ph)_2$[346]; similarly $(Et_2PhP)_2Ni(CN)_2$ adds Et_2PhP to give $(Et_2PhP)_3Ni(CN)_2$.[348]

Other modes of synthesis which have been employed include oxidative addition to $(Ph_3P)_4Pt$,[119,349]

$$(Ph_3P)_4Pt + PhC{\equiv}CBr \longrightarrow (Ph_3P)_2Pt(C_2Ph)Br$$

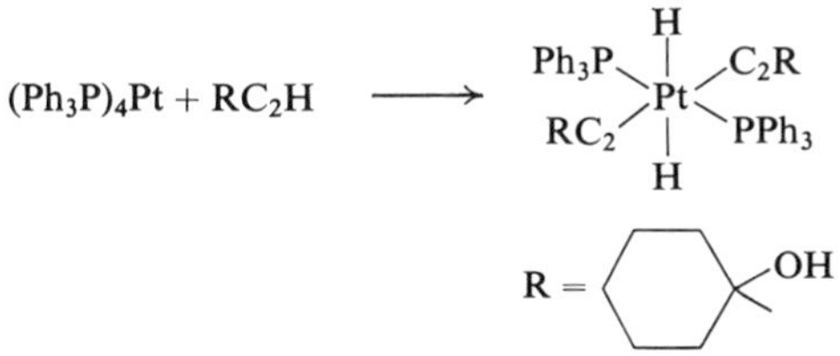

Reactions of this type are also known for $(R_3P)_2IrCOCl$.[350]

A reaction which may be related is[49]

$$(Et_3P)_2PdMe_2 + p\text{-}NO_2C_6H_4C{\equiv}CH \xrightarrow{C_6H_6} (Et_3P)_2Pd(C_2C_6H_4\text{-}p\text{-}NO_2)_2$$

and Glockling and Hooton[351] suggested that a six-coordinate Pt(IV) complex, **(II-10)**, may be intermediate in the following reaction,

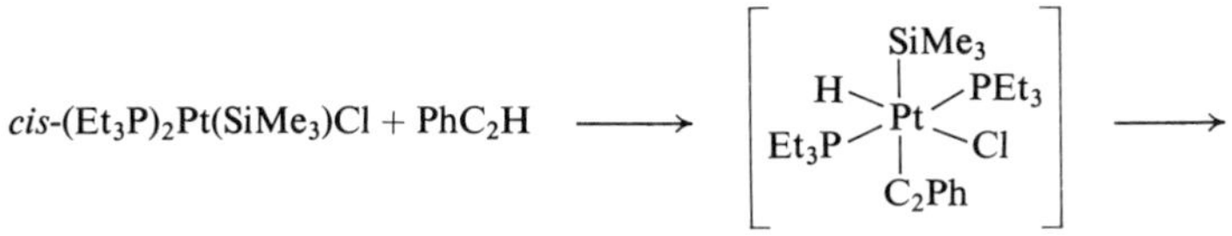

(II-10)

$trans\text{-}(Et_3P)_2PtH(C_2Ph) + (Et_3P)_2PtCl(C_2Ph) + Me_3SiCl + Me_3SiH + (Et_3P)_2Pt(SiMe_3)Cl$

All the bisalkynyl complexes **(II-9)** appear to have the trans configuration.[345,346] Few properties have been mentioned; Chatt and Shaw[347] reported the $\nu_{C{\equiv}C}$ in $(Et_3P)_2Pt(C{\equiv}CR)_2$ to be only about 100 cm^{-1} less than for $RC{\equiv}CR'$. Crystal structure determinations of the square planar trans-bis(triethylphosphine)di(phenylethynyl)nickel, **(II-9)** (R = Et, M = Ni, R' = Ph) have been reported by Owston et al.[352] and Amma et al.[353] The results of these two groups of workers agree in the main except for the Ni–P distances, though their conclusions differ. They also show that the Ni–C distance (1.87, 1.88 Å) is close to that expected for a Ni–C_{sp} bond (1.85 Å) and that the $C{\equiv}C$ bond length (1.18, 1.22 Å) is not significantly different to that in acetylene (1.20 Å). In this respect it is of interest that Ni–C σ-bond lengths seem to be invariant for a range of compounds, at least within the limits of the accuracy of measurements,[255,354,355] [acacNiC$_8$H$_{13}$, 1.95(2)† Å; π-C$_5$H$_5$Ni(C$_6$H$_5$)PPh$_3$,

† See Glossary.

1.919(13) Å; π-$C_5H_5Ni(C_6F_5)PPh_3$, 1.915(14) Å]. Only in $Ni(CO)_4$ is the Ni–C distance [1.82(1) Å] significantly shorter. The implication is that d_π–p_π bonding is not important in the Ni–C bonds of any of these complexes except $Ni(CO)_4$. Both Owston and Amma and their co-workers agree that the Ni–P bond length has been significantly reduced below that expected on the basis of the sum of covalent radii, although the exact extent of this reduction is not clear. This appears to point to a considerable degree of d_π–d_π bonding between the metal and the phosphorus.

Stone et al.[345] have reported cis addition of HCl to both triple bonds in trans-bis(triethylphosphine)bis(2,2,2-trifluoropropynyl)platinum, giving (II-11). However, $(PhEt_2P)_2Ni(C_2Ph)_2$ reacted with dry hydrogen chloride in

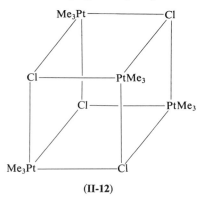

$$(Et_3P)_2Pt(C{\equiv}CCF_3)_2 + HCl \longrightarrow$$

(II-11)

ether to give trans-$(PhEt_2P)_2Ni(C_2Ph)Cl$.[346]

3. Alkyls, Acyls, Vinyls, and Aryls

a. General

The tetrameric iodotrimethylplatinum(IV), one of the first organic derivatives of the transition metals, was prepared by Pope and Peachey in 1909 from

(II-12)

$PtCl_4$ and MeMgI in benzene.[356] The structure is similar to that of the chloride (II-12), determined by Rundle and Sturdivant.[357] Many other tri-

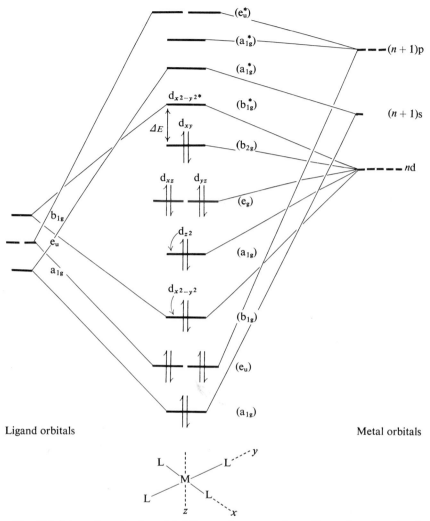

FIG. II-1. Schematic molecular orbital diagram for square planar d^8 complexes. Reprinted, with permission, from Green.[359]

methylplatinum(IV) complexes [and also some of Au(III)] were prepared, but attempts to prepare other alkyl (or aryl) transition metal compounds largely failed until the work of Chatt and Shaw in 1959. The usual results from these attempts using Grignard reagents (RMgX) and metal halide were coupled products, R_2, or olefins or saturated hydrocarbons, RH.

In contrast to organic derivatives of the more electropositive elements, complexes containing Pd–C (and also Ni–C and Pt–C) σ bonds show only low reactivity to air, water, and other protic reagents. This simplifies experimental

procedures to some extent, but precautions such as using protective atmos-
pheres (usually of nitrogen) are still normally taken. These complexes are
characterized by low thermal stability and it is good practice when preparing
them to work at low temperatures. In this way coupling and disproportionation
reactions are minimized.

Chatt and Shaw[358] argued that the distinguishing feature of the transition
metals was the presence of partially filled valence-shell d orbitals. The lowest
unoccupied orbital, at an energy ΔE above that of the highest energy occupied
orbital, is a metal–ligand σ^* (antibonding) orbital. This is illustrated for a
square planar d^8 complex in Fig. II-1. The ligand σ orbitals are combined into
four molecular orbitals (a_{1g}, e_u, b_{1g}), two of which are degenerate, and these
then combine with the metal nd, $(n + 1)$s, and $(n + 1)$p orbitals to give a series
of molecular orbitals for the complex. No π interactions are included and the
relative energies of the molecular orbitals are reasonable estimates.[359] Alto-
gether 16 electrons (8 from the four ligands and 8 from one metal atom) have
to be accommodated in the eight lowest energy orbitals. Only one d orbital,
$d_{x^2-y^2}$, takes part in σ bonding, and the antibonding combination of this with
the ligand orbital, $(b_{1g})^*$, is the lowest unoccupied molecular orbital of the
complex. ΔE is the energy difference between this orbital and the highest
occupied orbital (b_{2g}).

If this energy gap is small, say of the order of thermal energies, then promo-
tion of an electron from $b_{2g} \rightarrow b_{1g}^*$ will be easy. This would then effectively
put an electron into a metal–ligand σ^* orbital. If the ligands are normal anionic
ligands (e.g., halide), where the anion is stable, then the process is reversible
and merely corresponds to a momentary dissociation.

$$M–X \rightleftharpoons M^+ + X^-$$

Similarly if the ligand is neutral (e.g., R_3P) and stable the dissociation is also
reversible.

$$M–PR_3 \rightleftharpoons M + PR_3$$

However, if the "anionic" ligand atom is carbon, rupture of the M–C bond by
the above mechanism leads to the formation of a free radical or to a carbanion.
With few exceptions these latter species are very reactive and therefore once
formed, even momentarily, they will react further (with each other or the
medium) rather than revert to the organometallic compound, namely,

$$M–R \rightleftharpoons M\cdot + R\cdot$$

$$R\cdot \rightarrow R–R$$

$$\downarrow \text{solvent}$$

$$R–H$$

The result is decomposition.

According to this hypothesis, therefore, decomposition can be avoided by working at low temperatures and/or by effectively increasing the activation free energy of homolytic (or heterolytic) decomposition by increasing ΔE. In general, ΔE is increased by any of the following:

(a) Increase in formal oxidation state of the metal (hence stable complexes of Pt(IV), $[Me_3PtX]_4$).

(b) The presence of strong-field ligands, i.e., ligands which normally cause large d-orbital splittings as measured spectroscopically, such as CN^-, bipy, R_3P, CO, etc.

(c) A change from a 3d to the corresponding 4d or from a 4d to the 5d element. For octahedral complexes values of Δ_0 increase by 30–50% on going from a 3d element to the 4d, and by as much again on going to the 5d.

A further factor leading to an increase in ΔE would be the presence of empty orbitals on the ligand(s) of correct energy and symmetry to interact with the d_{xy} orbital. In this case a decrease in the energy of the d_{xy} orbital would occur. This situation was envisaged to occur for the alkynyl complexes $L_2M(C_2R)_2$, where, if the energies were compatible, overlap between acetylenic orbitals and d_{xy} would be possible. A similar situation would also hold for σ-aryl, and presumably (to a lesser extent) for σ-vinyl compounds.

The criteria which were used to judge the stability or otherwise of these complexes included melting and/or decomposition points, ease of synthesis, and occasionally reactivity toward air, water, and other reagents. More recently other criteria such as bond lengths, NMR coupling constants, and metal–ligand stretching frequencies and their associated force constants have been used. The difficulties entailed in separating the first type of criteria, which involve largely kinetic factors, from the second, which are more closely associated with thermodynamic properties, are formidable and at the time of writing no clear picture can be said to have emerged. One particular problem is that it is not always obvious how to interpret the physical data.

The broad predictions of the Chatt–Shaw hypothesis hold true. In other words, for a series of complexes of Ni, Pd, and Pt containing σ-bonded organic ligands, those of Pt(II) are the most readily obtained and the most stable and vice versa for Ni(II). Strong-field ligands such as bipy, R_3P, etc., do certainly give complexes whereas weak-field ligands such as NH_3, halide, etc., do not give isolable complexes. Alkynyls, where the possibility of d_π–p_π bonding in principle exists, do form more "stable" complexes than most other organic ligands. However, when the complexes are regarded in more detail, serious discrepancies are revealed. For example, $L_2M(aryl)X$ and $L_2M(aryl)_2$ are very much stabilized for M = Ni when the aryl group bears a bulky ortho substituent[346] and is therefore perpendicular to the coordination plane. This geometry allows maximum overlap of the aryl π^* orbitals with the metal d_{xy}

orbital.† However, for palladium this effect appears less important than the presence of an electronegative substituent on the aryl (which can be p-). Then again, the crystal structures of trans-$(Et_3P)_2Ni(C_2Ph)_2$ discussed above (p. 63) and of other Ni(II) complexes appear to indicate that strengthening of the Ni–C bonds by d_π–p_π bonding is very small or even negligible.‡

Matsuzaki and Yasukawa[361,362] have reported ESR measurements on a series of cobalt(II) complexes $(Et_2PhP)_2Co(aryl)_2$, the first examples of which were prepared by Chatt and Shaw.[363] The Japanese workers found that para substitution on the aryl group had no effect on the ESR spectrum, whereas substitution on the phenyl attached to phosphorus did. The conclusion here is that the unpaired electron is not significantly delocalized over the aryl group and that hence there cannot be much d_π–p_π bonding. On the other hand, there does appear to be evidence for P–Co d_π–d_π bonding. This is also borne out by the crystal structure of $(Et_2PhP)_2Co(Mes)_2$ (Mes = 2,4,6-trimethylphenyl).[364]

A number of complexes both for platinum and palladium are also now known, largely due to the work of Cope and his co-workers, in which the organic ligand is part of a chelate ring (and in which no strong-field ligands are present) which are very "stable"; certainly more so than $(R_3P)_2Pd(alkyl)X$. Similarly, although Calvin and Coates[49] note that dimethyl(1,5-cycloocta-diene)palladium, (II-13), is thermally very unstable and difficult to isolate, the cyclooct-4-enyl complexes, (II-14), prepared many years ago by Chatt and co-workers,[365] are exceedingly easy to prepare and are thermally relatively stable.

(II-13)   (II-14)   (II-15)

Numerous complexes of the type (II-15) (M = Ni, Pd, Pt; R = alkyl, aryl, etc.) have been isolated and characterized,[366–370,370a,370b] apparently with greater ease than $(R_3P)_2MR_2'$ or $(R_3P)_2MR'X$.

It is also becoming increasingly apparent that a major mode of thermal decomposition of σ-alkyl complexes is by a β-elimination to give a metal hydride and an olefin and that homolytic, and particularly heterolytic, cleavage

† Shaw[360] has shown by an NMR method that in $(PhMe_2P)_2Ni(Aryl)X$ and $(PhMe_2P)_2Ni$ (o-tolyl)$_2$ the aryl groups are perpendicular to the coordination plane of the metal.

‡ It can, not unreasonably, be argued that the "normal" Ni–C bond length observed is caused by cancellation of the opposing effects of the trans influence of one Ni–C bond on the other (causing lengthening) and the d_π–p_π overlap. All the same, neither of these effects can be large, since a large degree of d_π–p_π bonding would imply a significant increase in the C≡C bond length over that normally observed for acetylenes. This is not observed.

of the M–C bond is frequently of little importance. Quite stable complexes can be obtained when this type of elimination is not possible, as for MCH_2CR_3 or MCH_2SiR_3.[370c]

The presence of a vacant coordination site on the metal, or of a ligand which is a good leaving group and which easily gives rise to a vacant site, also increases the reactivity of a complex. In this connection it is of interest that of the three metals, Ni(II) is the most, and Pt(II) the least, labile. This would lead to an expected order of reactivity, Ni $\gg$ Pd > Pt. Furthermore, particularly in bimolecular reactions, steric effects are likely to be important, and bulky ligands will obviously protect the metal from attack more than smaller ones.

Another factor for greater thermal stability (and probably for lower reactivity) is the presence of electron-withdrawing groups on the organic ligand. Complexes containing σ-perfluoroalkyl, -aryl, or -vinyl groups are particularly noted for their ease of handling. This effect may be explained in valence bond terms as due to the contribution of canonial forms:

$$M-\underset{\underset{F}{|}}{\overset{\overset{F}{|}}{C}}- \longleftrightarrow \overset{+}{M}=\underset{\underset{F^-}{|}}{\overset{\overset{F}{|}}{C}}- \longleftrightarrow \overset{+}{M}=\underset{\underset{F}{|}}{\overset{\overset{F^-}{}}{C}}-$$

with consequent shortening of the M–C and lengthening of C–F bonds.[371, 372]

In summary, therefore, it appears that, if lack of reactivity is used as a criterion of stability for a complex, a number of effects other than those considered by Chatt and Shaw must be invoked. Only more detailed investigations of a large number of complexes and their reactions can hope to unravel these problems.

Information on metal–carbon bond strengths is obviously crucial. Thermochemical data in nontransition metals has been reviewed by Skinner.[373] This shows the trend that, for the alkyls of elements in one group, the *mean* dissociation energy decreases from the lighter to the heavier elements. Little data is available for the transition metal compounds, except for some of the carbonyls. Ashcroft and Mortimer[374] have, however, determined the heat of reaction for

$$\text{trans-}(Et_3P)_2PtPh_2 + HCl \rightarrow \text{trans-}(Et_3P)_2PtClPh + C_6H_6$$

as -28.7 ± 1.6 kcal/mole and from this have estimated the energy of the Pt–Ph bond to be about 60 kcal/mole. This compares with a mean value of 32.4 ± 2.1 kcal/mole for Hg–Ph (in $HgPh_2$).[373] No data are available for palladium complexes.

b. *Preparation of Palladium Alkyls, Aryls, Vinyls, and Acyls*

i. *From Grignards, Organolithium, Organoaluminum, and Organomercury Compounds.* This represents the most obvious and direct, though not always the most convenient, route to R–Pd. The reaction is defined as

$$L_2PdX_2 \xrightarrow{\text{RM}} L_2PdRX \xrightarrow{\text{RM}} L_2PdR_2$$

where RM represents an organolithium or organomagnesium reagent. The former, being the more reactive, will usually give L_2PdR_2 as the product; the latter, less reactive reagent, only L_2PdRX unless present in great excess. Both types of complexes usually have the trans configuration except when L_2 is a chelating ligand. The complexes L_2PdR_2 are also apparently cis when R = Me, L = PPh_3, PEt_3, $AsEt_3$[49]; and R = C_6F_5, L = Et_3P.[375]

Palladium complexes of both types for R = Me, aryl, vinyl, alkynyl, perfluorophenyl, and perfluorovinyl have been prepared by these routes.[49, 345, 376-378] L is usually a strong ligand such as a t-phosphine, t-arsine, bipyridyl, etc., Et_3P being especially favored. Attempts to prepare complexes without such ligands result in decomposition if carried out at normal temperatures. For example, Moiseev and Vargaftik[379] obtained approximately equal amounts of ethane and ethylene from reaction of palladium chloride with ethylmagnesium bromide in ether.

$$PdCl_2 + 2EtMgBr \xrightarrow{\text{Et}_2\text{O}} C_2H_4 + C_2H_6 + Pd^0 + 2MgClBr$$

They explained this in terms of the following mechanism:

$$EtMgBr + PdCl_2 \rightarrow EtPdBr + MgCl_2$$

$$EtPdBr \rightarrow C_2H_4 + Pd^0 + HBr$$

$$HBr + EtMgBr \rightarrow C_2H_6 + MgBr_2$$

However, in a chelate ring even an amine will stabilize a Pd–C bond.[380]

Diphenyl- (but not dimethyl-) mercury will also rapidly react with a Pd– or Pt–Cl bond to give a monosubstituted complex,[380a]

$$Ph_2Hg + L_2MCl_2 \rightarrow L_2M(Ph)Cl + PhHgCl$$

Heck, in an important series of papers,[381-387] has utilized the high reactivity of RPdX (formed *in situ* from R_2Hg or RHgCl and Li_2PdCl_4 or $LiPdCl_3$, where R is usually aryl), for the arylation of olefins (see Chapter I, Volume II, Section A). Analogous reactions have been reported by Henry,[388] e.g.,

$$PhHgCl + PdCl_2 + CO \xrightarrow{MeCN} PhCOCl + Pd + HgCl_2$$

In the absence of an organic substrate with which the PhPdCl can react, decomposition, largely to biphenyl (Ph_2) and palladium, occurs. Moiseev and Vargaftik[379] suggested that the reactions

$$HOCH_2CH_2HgCl + PdCl_2 \xrightarrow{Et_2O} CH_3CHO + Pd + Hg + HgCl_2 + HCl$$

$$EtOCH_2CH_2HgCl + PdCl_2 \xrightarrow{Et_2O} EtO \cdot CH{=}CH_2 + Pd + Hg + HgCl_2 + HCl$$

proceeded via an intermediate $ROCH_2CH_2PdCl$ (see Chapter II, Volume II).

Organotin and organolead compounds have also been used in place of organomercury compounds in some of these reactions.[381–388]

Very similar methods have also been applied to the syntheses of L_2MR_2 and L_2MRX, where M = Ni or Pt. For platinum, the starting complex of choice is cis-L_2PtX_2, since the ligands L usually have a strong trans effect and activate the halogens, X. The greater lability of L_2MX_2, where M = Ni or Pd makes this unimportant here.

An interesting variant has been applied to the synthesis of some nickel complexes using an alkylaluminum as alkylating agent. This could undoubtedly be applied to the palladium (and platinum) complexes too.

Wilke and Herrmann[389] noted that whereas $(R_3P)_2NiMe_2$ complexes were very unstable for R = phenyl and alkyl, quite stable complexes could be prepared when R_3P was a sterically hindered phosphite, e.g., $(o\text{-}PhC_6H_4O)_3P$.

$$(o\text{-}PhC_6H_4O)_3P + Ni(acac)_2 + (AlMe_3)_2 \rightarrow [(o\text{-}PhC_6H_4O)_3P]_2NiMe_2$$

Bipyridyl reacts with trialkylaluminums; therefore, for the synthesis of bipy-NiR_2 these authors and Saito et al.[390–392] used the dialkylaluminum alkoxide,

bipy + Ni(acac)₂ + Et₂AlOEt ⟶

The intermediacy of diethylnickel was postulated by Tsutsui and Zeiss[393] in the reaction of ethyl Grignard with nickel halide at low temperature in the presence of diphenylacetylene; 1,2,3,4-tetraphenyl-1,3-cyclohexadiene and hexaphenylbenzene were the products.

PtCl$_4$ reacts with methylmagnesium iodide in benzene to give iodotrimethyl-platinum tetramer; further reaction to (Me$_4$Pt)$_4$, despite earlier reports,[357] is not possible.[394] However, Ruddick and Shaw have reported the synthesis of derivatives of tetramethylplatinum, (II-16), by a series of reactions, one step of which involves the oxidative addition of MeBr to a square planar Pt(II) complex[395, 396] (see also Section C,3,d,i)

$$cis\text{-}(R_3P)_2PtCl_2 + MeLi \rightarrow cis\text{-}(R_3P)_2PtMe_2 \xrightarrow{\text{MeBr}}$$

$$(R_3P)_2PtMe_3Br \xrightarrow{\text{MeLi}} cis\text{-}(R_3P)_2PtMe_4$$

$$\textbf{(II-16)}$$

ii. *By Oxidative Addition to Pd(0) Complexes.* As already discussed (Chapter I, Section A,2,d), complexes of metals with certain electron configurations, notably d^8 and d^{10}, very readily undergo two-electron oxidations by organic halides and similar species. The d$^{10} \rightarrow$ d^8 oxidations have been most studied for Pt(0), but some reactions are now also known for palladium,

$$(Ph_3P)_4Pd + RX \rightarrow \quad \begin{array}{c} Ph_3P \diagdown \diagup R \\ Pd \\ X \diagup \diagdown PPh_3 \end{array}$$

$$\textbf{(II-17)}$$

Complexes of type **(II-17)** have been prepared for R = CF$_3$, C$_2$F$_5$, C$_3$F$_7$ and X = I;[118] R = Ph or Me and X = I; R = –CH$_2$CMe=CH$_2$, –COMe, –COOEt, –Ph, –CH$_2$Ph and X = Cl.[117, 397, 398] Chloroolefins also react with (Ph$_3$P)$_4$Pd to give the σ-vinyl complexes (X = Cl, R = –CCl=CCl$_2$, –CH=CCl$_2$, *cis-* and *trans*-CH=CHCl). No isomerization was observed when RX was either *cis-* or *trans*-1,2-dichloroethylene.[88] In the analogous reactions of (Ph$_3$P)$_4$Pt chloroolefin complexes of the type (Ph$_3$P)$_2$Pt(CCl$_2$=CCl$_2$) have been isolated as intermediates and the rearrangement to the σ-vinyl studied[121, 133, 135, 399] (Chapter I, Section B,2,d, this volume).

Complexes of the chelating phosphine, diphos, are not quite so straight-forward in their reactions[118]:

$$(diphos)_2Pd + CF_3I \rightarrow diphosPdI_2 + diphosPd(CF_3)I$$

$$\qquad\qquad\qquad\qquad 28\% \qquad\qquad 47\%$$

(Bu^{t}NC)$_2$Pd(0) will also add methyl iodide at low temperatures to give *trans*-(Bu^{t}NC)$_2$PdMeI[78] (see Section C,3,b,iv).

A closely related reaction is that of (Ph$_3$P)$_4$Pd with polar double-bonded species R$_2$C=Y, for example, SC=S,[23, 120] (CF$_3$)$_2$C=O,[127] or (CF$_3$)$_2$C=S, to give complexes which can be represented by either **(II-18a)** or **(II-18b)**.

As is discussed in Chapter III, when Y = CR$_2$', the structure of the adduct is perhaps better represented as **(II-18a)**; here, where Y is a more electronegative

$$
\underset{\textbf{(II-18a)}}{\overset{\displaystyle R \quad R}{\underset{\displaystyle Y}{\overset{\displaystyle C}{\underset{Ph_3P}{\overset{Ph_3P}{>}}Pd{-}\|}}}} \qquad\qquad \underset{\textbf{(II-18b)}}{\overset{\displaystyle R \quad R}{\underset{\displaystyle Y}{\overset{\displaystyle C}{\underset{Ph_3P}{\overset{Ph_3P}{>}}Pd{<}}}}}
$$

element, **(II-18b)** may perhaps be more appropriate. Hence complexes such as $(Ph_3P)_2PdCS_2$,

$$
\underset{Ph_3P}{\overset{Ph_3P}{>}}Pd\underset{C=S}{\overset{S}{<}}
$$

can be regarded as having a Pd–C σ bond.†

These reactions are well known for the Ni(0) and Pt(0) complexes too. Halpern *et al.*[95] have studied the rates of a very similar reaction:

$$
(Ph_3P)_2Pt{-}\|\underset{CH_2}{\overset{CH_2}{}} + RX \;\rightarrow\; (Ph_3P)_2PtRX + C_2H_4
$$

$$(R = Me, X = I; R = PhCH_2, X = Br)$$

in which the first step is the fast establishment of the equilibrium,

$$
(Ph_3P)_2Pt{-}\|\underset{CH_2}{\overset{CH_2}{}} \;\rightleftharpoons\; (Ph_3P)_2Pt + C_2H_4
$$

followed by the second, fast step.

$$(Ph_3P)_2Pt + RX \;\rightarrow\; (Ph_3P)_2PtRX$$

Presumably a similar mechanism, involving a fast dissociation, applies to the reactions of $(R_3P)_4M$ with organic halides.

Related reactions are those of bis(1,5-cyclooctadiene)nickel(0) and hexaphenylethane to give **(II-19)**, and of 1,5,9-cyclododecatrienenickel(0) and

$$
\left[\text{COD}\right]Ni\left[\text{COD}\right] + Ph_3C{-}CPh_3 \;\rightarrow\; Ph_3CNiCPh_3
$$

(II-19)

† Support for this formulation in $(Ph_3P)_2PdCS_2$, and in its platinum analog, comes from the long bond to the phosphorus which is "trans" to the carbon [Pd–P 2.42(1) Å,[23] Pt–P 2.35 Å[400]], compared to the other bond, "trans" to sulfur, which is more normal [Pd–P 2.32(1) Å, Pt–P 2.24 Å]. This suggests that the carbon in CS_2 has a high trans influence owing to the σ-bond character of the M–C bond.

triphenylmethylchloride to give **(II-20)**.[401] The exact formulation of the Ni–CPh$_3$ bond in these complexes, whether σ- or π-benzyl, is not certain.

$$+ Ph_3CCl \rightarrow Ph_3CNiCl$$

(II-20)

iii. *Addition of Nucleophiles to Olefin–Pd(II) Complexes.* This is a reaction of considerable importance in organopalladium and organoplatinum chemistry, but as yet is rather rare in its simpler forms for other organotransition metal complexes. The reactions were first discovered by Hofmann and von Narbutt in 1908,[402] but not recognized as such until the work of Chatt, Vallarino, and Venanzi in 1957.[365, 403]

Chatt *et al.* were able to isolate diene–MX$_2$ complexes, **(II-21)**, using the chelating 1,5-dienes, dipentene, dicyclopentadiene, and, most important, 1,5-cyclooctadiene. These complexes reacted with varying facility with alkoxide ion (usually the alcohol in the presence of sodium carbonate or acetate) to give the complexes **(II-22)** in which M and OR$^-$ had been added to one double bond and one Cl$^-$ removed from the metal. The complexes **(II-22)** were dimers with asymmetric Cl bridges.[282, 404, 405]

(II-21)   **(II-22)**   **(II-23)**

(M = Pd, Pt)

On treatment with other anions, X$^-$ (bromide, iodide, thiocyanate), exchange occurred and the complexes **(II-23)** were formed, while HCl gave back the diene complexes **(II-21)**. The reactivities of **(II-22)** and **(II-23)** in these reactions were much higher for M = Pd than Pt, as expected.

The crystal structure of a complex, derived from *endo*-dicyclopentadiene and PtCl$_2$, has been shown to be **(II-24)** by Whitla, Powell, and Venanzi[200] (Chapter I, Section B,4,e, this volume). The palladium analog has a very similar structure.[406]

The crystal structure of **(II-24)** shows the methoxy group to be exo to the metal. This conclusion also applies to the reaction of 1,5-cyclooctadiene-platinum chloride with methoxide ion, where the pyridine adduct **(II-26)** was shown to have the structure illustrated[407] (see also Ref. 201a).

Similar conclusions resulted from the degradation of the complexes (II-24) and (II-27) by hydrogen or borohydride which gave octahydro-*exo*-5-methoxy-4,7-*endo*-methanoindene, (II-25), and *exo*-2-methoxynorbornane, (II-28), respectively.[408] Rearrangement is presumed not to occur during these reactions. A careful NMR study of (II-27) (M = Pd) and (II-24) (Pd in place of Pt) also indicated the alkoxy substituent to be exo.

(II-24)   (II-25)   (II-26)

(II-27)   (II-28)

This implies that, in these cases at least, there has been direct attack of alkoxide ion exo on the coordinated double bond of the ligand,

$+ Cl^- \rightarrow$ (II-22)

Reaction paths involving primary attack of OR^- on the metal, followed by an insertion of the C=C into the metal–OR bond,

appear to be ruled out for rigid organic ligands since this would cause the alkoxy group to be endo to the metal in the product.

If the diene has an acyclic double bond, this is attacked in preference,[408, 409] for example,

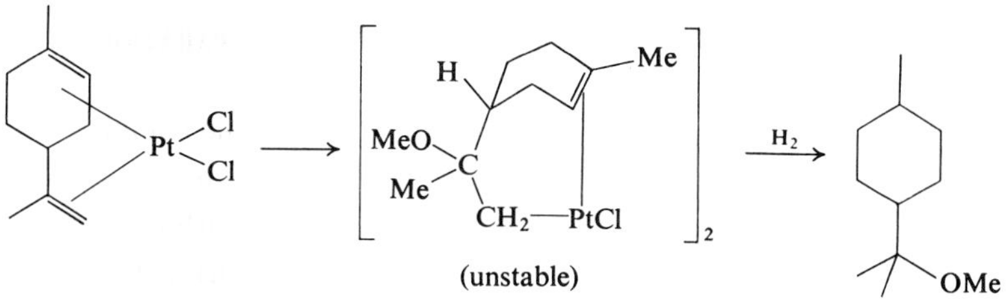

(unstable)

Additions to other dienes are also usually assumed to be exo; however, Anderson and Burreson[410] have disputed this for (II-29).

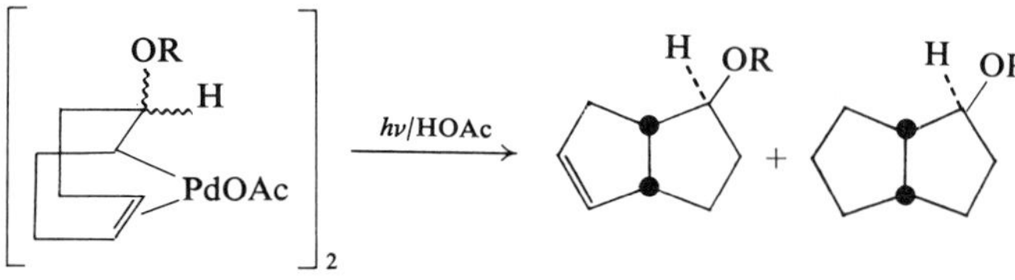

(II-29)   (II-30)   (II-31)

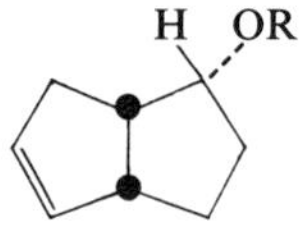

(II-32)

(R = Me, Ac)

Their argument is based on the isolation of a 55% yield of the *exo*-bicyclo-octene (II-30) [plus some of the bicyclooctane (II-31)] on irradiation. The formation of the exo isomer (II-30) [rather than the endo isomer (II-32)] can be rationalized if (II-29) has the OR group endo to the metal and the reaction proceeds via an intermediate such as (II-33).

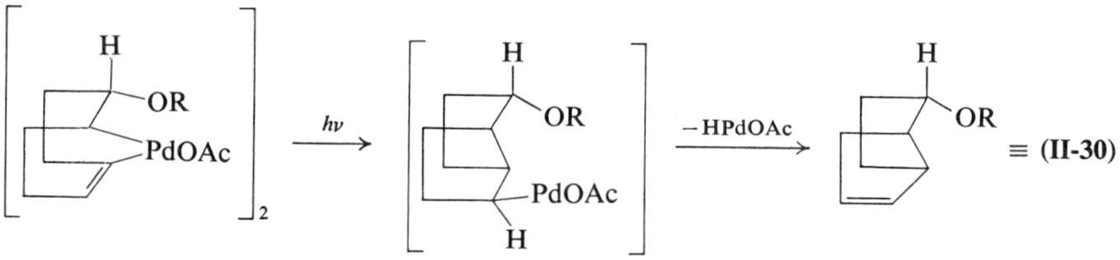

(II-33)

This argument is, however, intriguing rather than compelling, as we know very little about the processes which occur on irradiation of organometallic complexes. It is by no means inconceivable that the overall reaction is much more complex than depicted and that an isomerization occurs. This has also been suggested by White[411] who regards the probable reaction path as involving a photochemical isomerization.

Many other nucleophiles react similarly with these diene–MCl_2 complexes, for example, malonates, $CH(COOR)_2^-$ [412–414]; β-diketonates, RCOCH COR'$^-$ [415]; acetate (as AgOAc)[410]; acetylacetonate, $CH_3COCHCOOR^-$ [413]; primary amines,[409,416] ammonia, and azide ion.†[417] Since it forms the most stable complexes, 1,5-cyclooctadiene is frequently the diene, but many others including 4-vinylcyclohexene,[409] 1,5-hexadiene,[409] and the Diels–Alder adduct of cyclooctatetraene and diethyl maleate,[419] have been used. It is also possible to obtain the alkoxy complexes (II-35) or (II-36) directly from dicyclopenta-diene and sodium tetrachloropalladate in alcohol.[402,420] The 1,5-cycloocta-diene complexes react with thallium(I) β-diketonates to give (II-34).[415]

(II-34)

Attempts to add more than one nucleophile usually result in decomposition of the metal complex[414] (see also Schultz[421]), but a recent report has stated that the mono-adduct (II-24) (M = Pt) in the presence of methoxide and excess PPh_3 gave a di-σ-bonded complex.[421a]

The facility with which the alkoxy complexes (II-22) revert to the diene complexes (II-21) varies. Palladium complexes react much more readily with

† Not all nucleophiles react thus. Diene metal halides react with methyl Grignard reagents to give disubstitution at the metal,[418]

(M = Pd, R = Me; M = Pt, R = Me, Et, Ph, o-tolyl)

The reasons for this important difference are by no means clear, but it does emphasize the fact that several possible reaction routes do exist.

HCl than platinum complexes.[365] The one derived from dicyclopentadiene reverts particularly easily, commercial chloroform contains enough HCl for this reaction to proceed.[408] Furthermore, treatment of (II-35) with ethanol in chloroform gives the ethoxy complex (II-36).

(II-35)   (II-36)

In all these reactions the first step is presumably protonation of the ether which then rearranges to the diene complex via the cation (II-37). This implies

(II-37)

a considerable stability for a cation such as (II-37). The stabilization of cations at carbon atoms alpha to π-complexed ligands is well established in ferrocene chemistry and also in π-allylic palladium complexes (see p. 229).

These "oxypalladation" reactions are similar to the oxymercuration reaction,[422]

$$C_2H_4 + Hg(OAc)_2 \underset{H^+}{\overset{OH^-}{\rightleftarrows}} AcOCH_2CH_2HgOAc$$

which is very easily reversed by addition of acid. The mercury complexes appear to be much more labile than either the palladium or platinum complexes (II-22).

Since the diene complexes (II-21) are not known for nickel, such reactions do not occur here. A somewhat reminiscent reaction, however, is[255]

The tetraphenylcyclobutadiene complexes, $[Ph_4C_4MCl_2]_2$ and $[Ph_4C_4 MC_5H_5]^+$ (M = Ni, Pd), also undergo very facile reactions with alkoxide and other nucleophiles to give the cyclobutenyl complexes. These are very similar to those described above and are considered in Chapter IV, this volume.

A further reaction which can be regarded as a variant of this was described by Cope, Kliegman, and Friedrich.[423] The relation to the above reactions is

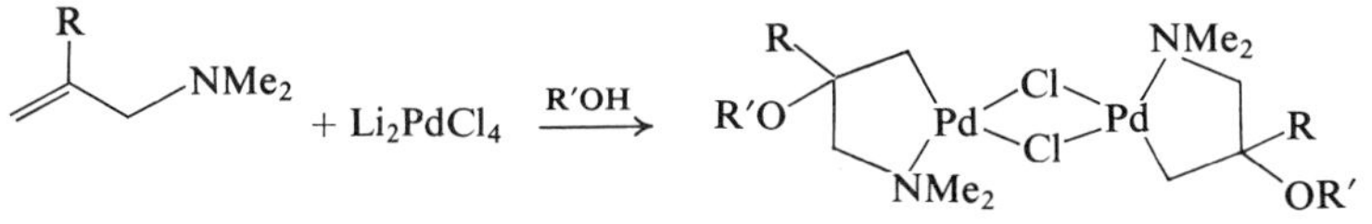

(II-38)

(R = H, Me; R′ = Me, Et, HOCH$_2$CH$_2$–)

obvious and an intermediate π complex, **(II-39)**, which is then readily attacked by alkoxide to give **(II-38)**, can be postulated, but has not yet been isolated.

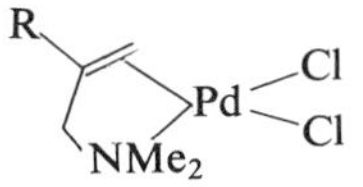

(II-39)

Kasahara et al.[424] have carried out a similar reaction with 2-vinylpyridine in place of the allylamine. It is interesting that coordination to an amine does not diminish the facility with which these reactions occur; in fact, it may even increase it.

These reactions are very similar to the oxypalladation reactions to be described in Chapter II, Volume II.

iv. *Insertion of Organic Groups into Pd–Cl and Pd–X Bonds.* A group of recently reported reactions testify to the fact that Pd–Cl (and also, presumably Pd–OR, etc.) will "add across" triple bonds and carbenes. For example, Yukawa and Tsutsumi[425] reacted the acetylenes, **(II-40)**, with palladium chloride and an excess of lithium chloride in methanol to obtain **(II-41)**.

RC≡CCR$_2$′NMe$_2$ + PdCl$_2$ + LiCl + MeOH

(II-40)

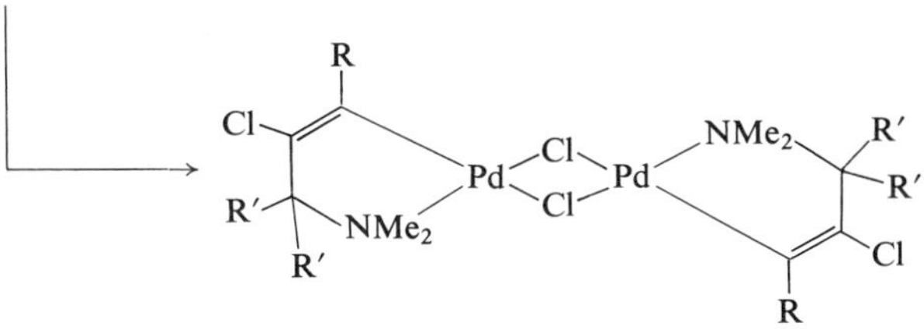

(II-41)

(R = Ph, R′ = H; R = H, R′ = Me)

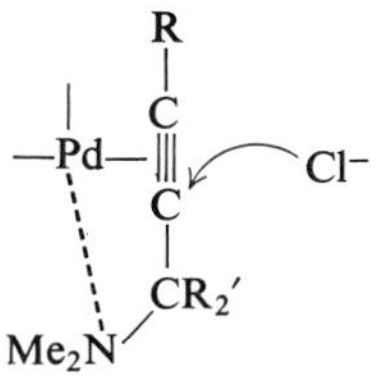

(II-42)

The mechanism for the formation of (II-41) is not known, but it may well involve attack by Cl^- (synchronous with Pd–C bond formation) on an intermediate such as (II-42) rather than a cis addition of Pd–Cl to the triple bond. This would explain the observed trans stereochemistry of (II-41).

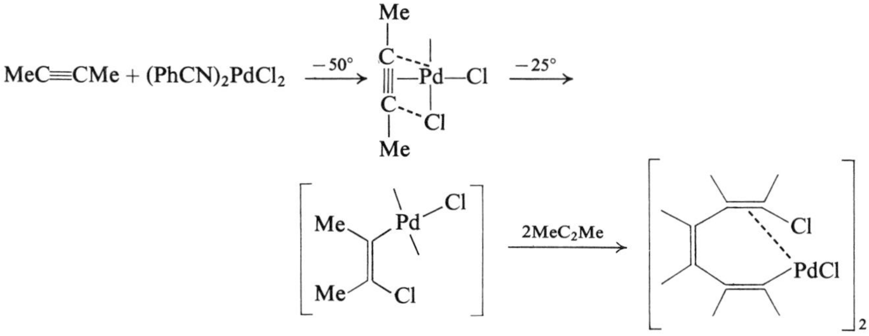

(II-43)

An example of a reaction in which cis addition of Pd–Cl to a coordinated acetylene has been postulated as the rate-determining step is in the formation of the complex (II-43) from 2-butyne and palladium chloride[282, 404] (see Chapter I, Section C,3,d, Volume II).

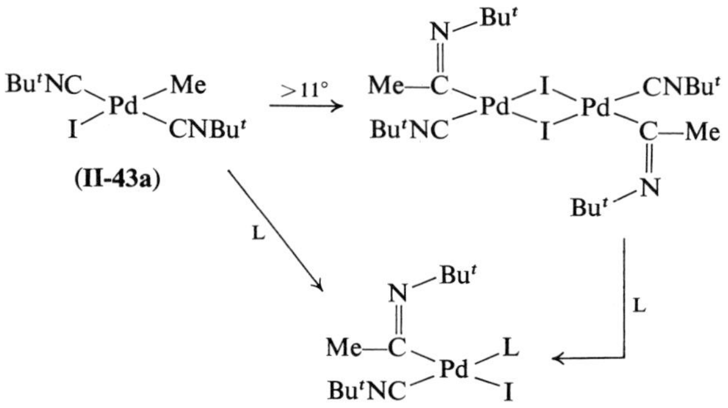

(II-43a)

(L = PPh₃, BuᵗNC)

Olefins and acetylenes readily insert into Pd–C bonds, but examples where stable σ-bonded complexes are isolated are rare apart from the ones mentioned here. These reactions are more fully discussed in Chapter I, Volume II.

An unusual example of insertion into a Pd–Me bond, in (II-43a), has been described by Otsuka et al.[78] (see also p. 25).

Ni(CNBu^t)$_4$ reacts analogously with MeI and PhCOCl to give complexes containing oligomers, e.g.,

$[R = Me, X = I; R = PhCOC(=NBu^t)C(=NBu^t), X = Cl]$

A very similar palladium complex has been obtained from (Ph$_2$MeP)$_2$Pd-(Me)I and cyclohexylisonitrile. Intermediates in which one and two C$_6$H$_{11}$NC molecules had been incorporated were also isolated.[341b]

Matsumoto, Odaira, and Tsutsumi[426] have reported that diazoacetonitrile decomposes in the presence of a palladium chloride complex such as [Ph$_3$PPdCl$_2$]$_2$ to give a product formulated as (II-44).

(II-44)

Ashley-Smith et al.[427] have described a very similar reaction.

$(PhCN)_2PdCl_2 + (CF_3)_2CN_2 \longrightarrow$

No such reactions have yet been reported for Pt; but a reaction recently described by Miller et al.[428] appears to fit into this picture if one accepts the view that benzyne is an intermediate which inserts in the nickel–chlorine bond.

v. *Aromatic Substitution by Pd(II)*. A most interesting and important property of palladium(II) is its ability, under suitable conditions, to effect electrophilic aromatic substitution. This was first noted by Cope and Siekman,[429] who reacted azobenzene with potassium tetrachloropaliadate (or potassium tetrachloroplatinate) in dioxane and obtained the complex (II-45) (M = Pd or Pt).† Similar reactions were later reported by Cope and Friedrich,[430] and others.[431,432,432a]

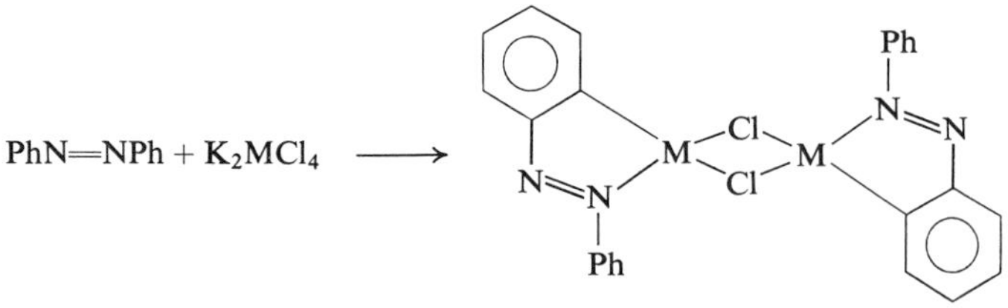

$$PhN{=\!=}NPh + K_2MCl_4 \longrightarrow$$

(II-45)

In all cases the position of attack was checked by decomposition of the complex with LiAlD₄ to give the appropriate *o*-deuterophenyl compound. Cope *et al.* in their investigation of the reaction giving (II-46) showed that $PdCl_4{}^{2-}$ was much more reactive and gave higher yields than $PtCl_4{}^{2-}$. Furthermore, the reaction with $PdCl_4{}^{2-}$ was quite sensitive to substituents on the benzene ring and on the amine. Both $4\text{-}MeOC_6H_4CH_2NMe_2$ and $3,5\text{-}Me_2C_6H_4CH_2NMe_2$ gave the analogs of (II-46), but $4\text{-}NO_2C_6H_4CH_2NMe_2$ only gave the normal donor complex $(4\text{-}NO_2C_6H_4CH_2NMe_2)_2PdCl_2$; this suggested that Pd(II) acted as a

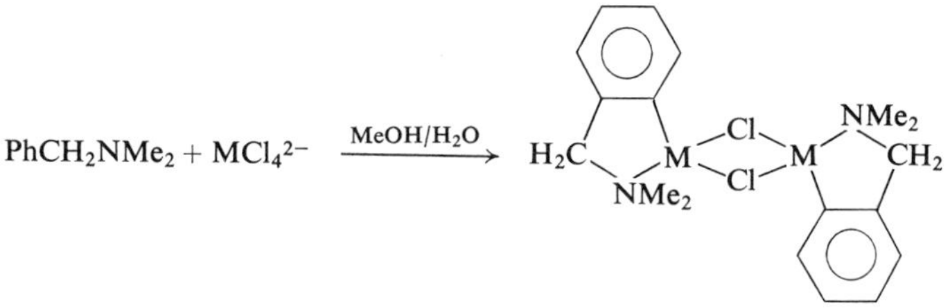

$$PhCH_2NMe_2 + MCl_4{}^{2-} \xrightarrow{\text{MeOH/H}_2\text{O}}$$

(II-46)

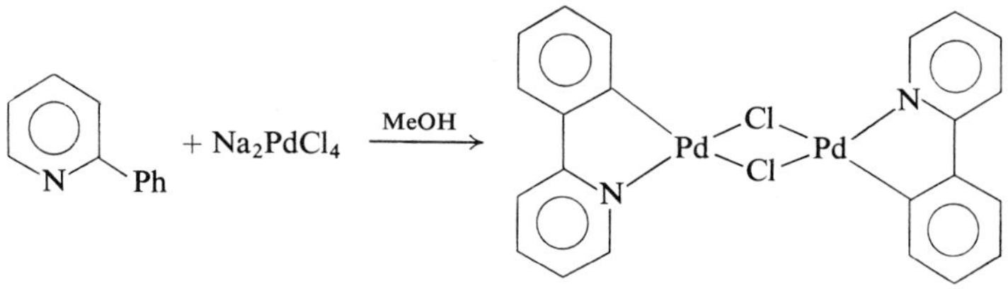

† The formation of five-membered rings is strongly favored in these reactions.

weak electrophile. PhCH$_2$NMeH also only gave the normal donor complex (PhCH$_2$NMeH)$_2$PdCl$_2$.[430]

Takahashi and Tsuji[433] confirmed this suggestion by showing that the major products from asymmetrically substituted azobenzenes, p-XC$_6$H$_4$N=NPh, were those to be expected on the basis of electrophilic attack by Pd(II) on the benzene ring. For example, (II-47) was the major and (II-48) the minor product (ratio 3:1) from 4-methylazobenzene.

(II-47)   (II-48)

Davidson and Triggs[159] have investigated the Pd(II)-catalyzed coupling of benzene to biphenyl, and have proposed that the first step in the reaction is electrophilic substitution to give a phenylpalladium(II) species which then reacts further to give biphenyl and a Pd(I) species.

$$C_6H_6 + Pd(II) \rightarrow C_6H_5Pd(II) + HCl$$

$$2C_6H_5Pd(II) \rightarrow C_6H_5C_6H_5 + 2Pd(I)$$

The reaction was normally carried out using palladium acetate in acetic acid and it was found that HClO$_4$ had a large catalytic effect. The rate-determining first step was first-order in both Pd(II) and C$_6$H$_6$. It showed a large primary isotope effect when C$_6$D$_6$ was substituted ($k_H/k_D \sim 5.0$) implying that transfer of hydrogen from an intermediate containing Pd(II) and benzene (presumably a σ complex) was rate-determining (see Chapter I, Section D, Volume II).

vi. *Miscellaneous.* Maitlis and Stone[434] argued, on the general similarity in chemical behavior between perfluoroalkyl iodides and iodine, that R$_f$I ought to cleave Pd–alkyl bonds with the formation of Pd–R$_f$ bonds. In fact, the reaction of bipyridyldimethylpalladium (II-49) with an excess of perfluoropropyl iodide did give some bipyridylbis(heptafluoropropyl)palladium (II-50). Ethane was the only volatile product. However, a closer analysis showed this reaction to be more complex than that expected for a simple cleavage, as, for example, by iodine. Reaction of one mole of perfluoropropyliodide with (II-49) gave largely bipyPdMe(C$_3$F$_7$) and another material which decomposed on purification but which may have been bipyPdMeI.

$$bipyPdMe_2 + C_3F_7I \rightarrow bipyPdMe(C_3F_7) + bipyPdMeI(?)$$

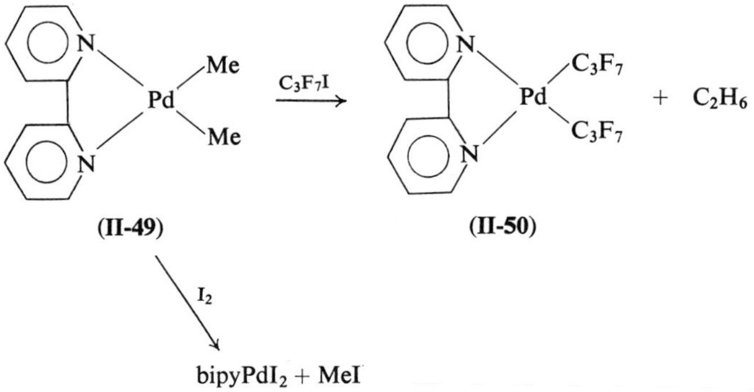

(II-49)   **(II-50)**

bipyPdI$_2$ + MeI

A possible mechanism could involve oxidative addition of C$_3$F$_7$I to **(II-49)** giving **(II-51)** [Pd(IV), stereochemistry unknown], which can then eliminate either MeI or, presumably, MeC$_3$F$_7$. It is not clear whether the ethane and

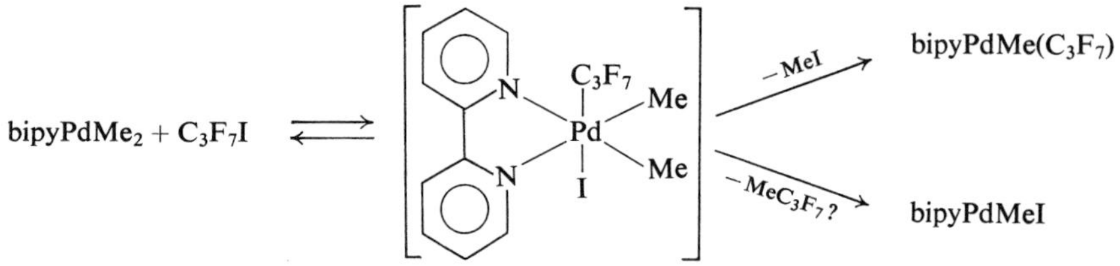

(II-51)

(II-50) in the experiment with excess perfluoropropyliodide arise from the same reaction. Reactions described by Kistner et al.[435] are of interest in this connection.

cis-(py)$_2$Pt(tolyl)$_2$ + RI → (py)$_2$Pt^{IV}(tolyl)R(I)$_2$

(tolyl = o-, p-CH$_3$C$_6$H$_4$–; R = Et, Pr)

(py)$_2$Pt(o-tolyl)$_2$ + PhI → (py)$_2$Pt(o-tolyl)I

Coulson[397] has reported the following most unusual reaction in which P–phenyl is cleaved and converted into a Pd–phenyl, giving (Ph$_3$P)$_2$PdPhCl and **(II-52)**. The unusual structure of **(II-52)** was assigned partly by ^{31}P NMR.

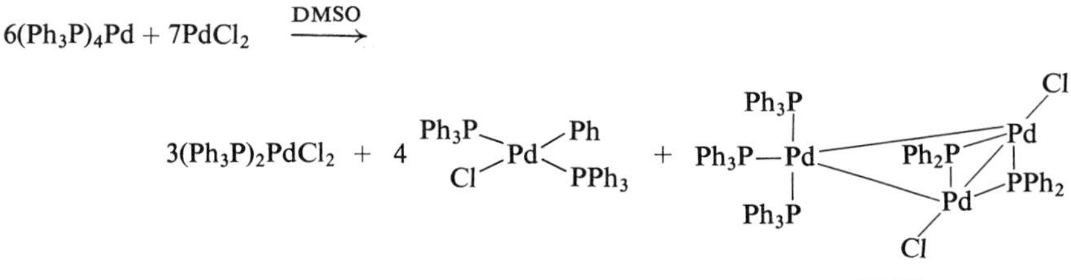

6(Ph$_3$P)$_4$Pd + 7PdCl$_2$ $\xrightarrow{\text{DMSO}}$

(II-52)

The reaction proceeded best in DMSO at 130°, but was also said to go at 20°. A mechanism has been proposed, but more details are necessary before its validity can be established.

$$(Ph_3P)_4Pd + PdCl_2 \longrightarrow \begin{array}{c} Ph_2P\!-\!Ph \\ \downarrow \;\;\;\nearrow \;\;\; PPh_3 \\ Cl\!-\!Pd\!-\!Pd\!-\!Cl \\ \downarrow \;\;\;\searrow \\ PPh_3 \;\;\; PPh_3 \end{array} \longrightarrow (Ph_3P)_2PdPhCl + Ph_3PPd(PPh_2)Cl$$

$$2Ph_3PPd(PPh_2)Cl + (Ph_3P)_4Pd \rightarrow \textbf{(II-52)} + 3Ph_3P$$

vii. *Formation of Palladium Acyl Bonds.* Although palladium complexes play a very important role in catalytic carbonylation, very little is known about intermediates. The preparation and properties of acylpalladium (and acyl-platinum) complexes have only been reported by Booth and Chatt[436, 437] who investigated the reversible reaction,

$$\begin{array}{c} Et_3P\!\diagdown \;\;\;\diagup Me \\ M \\ X\!\diagup \;\;\;\diagdown PEt_3 \end{array} \underset{}{\overset{CO}{\rightleftarrows}} \begin{array}{c} Et_3P\!\diagdown \;\;\;\diagup COMe \\ M \\ X\!\diagup \;\;\;\diagdown PEt_3 \end{array}$$

$$\textbf{(II-53)} \qquad\qquad\qquad \textbf{(II-54)}$$

$$(M = Pd, Pt;\; X = halide,\; etc.)$$

The palladium complexes **(II-53)** reacted readily at 20° and 1 atm; *trans*-$(Et_3P)_2PtMeX$ needed higher temperature and pressure, and *cis*-$(Et_3P)_2PtMeX$ also gave the *trans*-acetyls **(II-54)**. Triphenylphosphine complexes gave less stable products. The reactions were reversed on heating the solid acetyls above their melting points. $(Et_3P)_2PdMe_2$ decomposed on carbonylation, but *cis*-$(Et_3P)_2PtMe_2$ gave biacetyl and a triethylphosphineplatinum carbonyl complex.

$$cis\text{-}(Et_3P)_2PtMe_2 + CO \xrightarrow{90°} MeCOCOMe + (Et_3P)_2Pt_3(CO)_{3 \text{ or } 4}$$

The chelated complex **(II-55)** gave an insoluble monoinsertion product.

$$\begin{array}{c} \diagup PEt_2\!\diagdown \;\;\;\diagup Me \\ \Big[\;\;\;\;\;\;\;\;\; Pt \\ \diagdown PEt_2\!\diagup \;\;\;\diagdown Me \end{array} \xrightarrow{CO} \begin{array}{c} \diagup PEt_2\!\diagdown \;\;\;\diagup COMe \\ \Big[\;\;\;\;\;\;\;\;\; Pt \\ \diagdown PEt_2\!\diagup \;\;\;\diagdown Me \end{array}$$

$$\textbf{(II-55)}$$

The ν_{CO} for **(II-54)** ranged from 1661 to 1675 cm^{-1} for M = Pd and from 1628 to 1636 cm^{-1} for M = Pt (X = halide, NO_2, NO_3, etc.). This was interpreted to mean that there was less contribution of type **(II-56b)** structures

$$\begin{array}{c} O \\ \parallel \\ M\!-\!C\!\diagdown \end{array} \qquad\qquad \begin{array}{c} O^- \\ \mid \\ \overset{+}{M}\!=\!C\!\diagdown \end{array}$$

$$\textbf{(II-56a)} \qquad\qquad \textbf{(II-56b)}$$

(i.e., less metal to ligand back-bonding) than (**II-56a**) for M = Pd, than for M = Pt.[438]

An alternative route to acyl and aroyl complexes is by oxidative addition of RCOCl to the zero-valent complexes. This has been carried out for both palladium[117] and platinum.[120]

$$(Ph_3P)_4Pd + MeCOCl \rightarrow trans\text{-}(Ph_3P)_2Pd(COMe)Cl$$

$$(Ph_3P)_3Pt + PhCOCl \rightarrow trans\text{-}(Ph_3P)_2Pt(COPh)Cl$$

Evidence to support the formation of acylpalladium intermediates from palladium metal and acyl halides or aldehydes is discussed in Volume II, Chapter I, Section B and Chapter V, Section B.

viii. *Potential Reactions for the Formation of Pd–C Bonds.* A number of reactions have been used for the formation of Ni–C and Pt–C and not, as yet, for Pd–C σ bonds. In many cases there is no obvious reason why these should not also be used for the formation of Pd–C bonds and these methods are briefly reviewed.

The most important general method which has not yet been used to form Pd–C bonds (to give stable compounds, as distinct from the possibility that these reactions occur in catalytic processes) is the reaction of Pd–H with an olefin or acetylene.

For example, *trans*-chlorohydridobis(triethylphosphine)platinum(II) reacts with a number of olefins and acetylenes by addition of Pt–H across the unsaturated organic compound. Sometimes, however, as with the fluoroolefins (e.g., hexafluorocyclobutene), a σ-vinyl complex is formed by elimination of HF. The mechanisms of these reactions are not known, but probably involve the intermediacy of a Pt(II)–olefin or Pt(II)–acetylene complex.† The

$$trans\text{-}(Et_3P)_2PtHCl + C_2H_4 \underset{180°}{\overset{95°/40\ atm}{\rightleftharpoons}} trans\text{-}(Et_3P)_2Pt(C_2H_5)Cl \quad [35]$$

(R = CF₃, H)

† Clark and Puddephat[438a] have shown that (Me₂PhAs)₂PtMeCl gives an adduct with CF₃C≡CCF₃ which rearranges slowly to give the product in which cis insertion of the acetylene into the Pd–Me bond has occurred.

$$trans\text{-}(Et_3P)_2PtHCl + (CF_3)_2CN_2 \rightarrow (Et_3P)_2Pt[C(CF_3)_2H]Cl \quad [442]$$

mode of addition is *cis* in at least one example. Propylene reacts to give a low yield of the *n*-propyl (not the isopropyl) derivative; this may be a thermodynamic rather than a kinetic effect. The higher alkenes do not form alkyl derivatives with $(Et_3P)_2PtHCl$.[439]

Deeming *et al.*[442a] have shown that $trans\text{-}(Et_3P)_2PtH(NO_3)$ reacts readily with diolefins to give enyl complexes, the reaction being facilitated by the excellent leaving group property of nitrate. With ethylene only a replacement occurred to give $trans\text{-}[(Et_3P)_2PtH(C_2H_4)]^+$, further reaction being inhibited by the trans disposition of the hydride and π-ethylene ligands.

$(Et_3P)_2PdHCl$ is known and some of its reactions with olefins have been briefly investigated, but no well-characterized products have as yet been reported.

Wright[443] has made use of the reducing power of formic acid (which also acts as a source of CO) to form, presumably, a hydridoplatinum intermediate from $PtCl_4{}^{2-}$. This reacts with terminal olefins to give the alkyl complexes (**II-57**).

$$RCH{=}CH_2 + Li_2PtCl_4 + HCOOH \xrightarrow{\text{DMF}/100°} [RCH_2CH_2PtCOCl]_2$$

$$\text{(II-57)}$$

Elimination of N_2 from $R\text{—}N{=}N\text{—}Pt$ provides another route to R–Pt, namely,

c. *Physical Properties of Palladium Alkyls, Aryls, and Acyls*

As already discussed in Section C,3,b, despite earlier failures to prepare them, a whole range of complexes, containing Pd–alkyl, Pd–aryl, Pd–acyl, and Pd–vinyl bonds are now known. The factors that determine which can be made successfully at ambient temperatures are not all known in detail, but it is evident that the presence of ligands which are not readily replaced (for example, in a chelate ring) will considerably stabilize the complexes. The apparent greater "stability" of complexes containing Pt–C compared to these in which Ni–C or Pd–C bonds are present may well be due largely to kinetic factors. One way in which this could be explained is the greater ease with which palladium, and *especially* nickel, increase their coordination number. This, in turn, has been correlated by Nyholm with the $(n - 1)d{\rightarrow}np$ energy difference for the metal ions.[313] A smaller separation should facilitate rehybridization, and the separation increases from 3d to 4d to 5d for the transition metals.

Although only a few X-ray crystal structure determinations have been carried out on σ-bonded organopalladium complexes, quite a number have been done for platinum. If the covalent radius of Pd(II) [or Pt(II)]† is taken to be 1.30 Å and that of carbon taken to be the appropriate value for the hybridization considered (0.60 Å for sp, 0.67 Å for sp², and 0.77 Å for sp³), then the sum of the covalent radii observed is usually very close to that calculated. For example, that in (**II-58**) is 1.998(13) Å[201] (calculated, 1.97 Å), while that in the platinum complex from *endo*-dicyclopentadiene (**II-24**) is 2.07 Å (calculated, 2.07 Å)[200] (see also p. 40 and Ref. 201a).

(**II-58**)   (**II-59**)   (**II-60**)

Discrepancies do, of course, occur; an interesting one is found in complexes of the type (**II-59**) (R = R' = *n*-propyl; R = Me, R' = OEt) where the Pt–Me distances are normal [2.02(3) and 2.08(9) Å, respectively], but where the Pt–CH bonds are significantly longer [2.39(3) and 2.56(9) Å respectively],[446,447] This lengthening probably arises in part from the high trans influence of the methyls. Another example of a long Pt–C bond is in (**II-60**) where Pt–CH₂ is 2.15(4) Å[448]; presumably the long bonds relieve some of the strain in the four-membered ring.

The wide variety of complexes known with Pd–C σ bonds and the paucity of physical data reported makes it inappropriate to discuss the physical properties further at this stage. Calvin and Coates[49] have measured the infrared spectra

† For the purposes of this discussion it is assumed that the covalent radii of Pd(II), Pd(IV), Pt(II), and Pt(IV) are all equal within the limits of accuracy currently attainable.

of a number of complexes L_2MMe_2 (M = Pd, Pt) and ascribe bands in the range 457–534 cm^{-1} to ν(Pd–C) and 508–555 cm^{-1} to ν(Pt–C).

d. *Reactions of Complexes Containing σ-Bonded Alkyls, Aryls, and Acyls*

Since palladium–carbon σ bonds are strongly implicated in very many catalytic reactions, it is of great importance to know what sort of reactions are typical of these linkages in order to understand the catalyzed reactions better. Unfortunately very few studies have been carried out on the behavior of palladium–carbon bonds in isolated complexes and only a few more on platinum–carbon bonds. The simple chemistry, in this sense, of transition metal–carbon bonds has been sadly neglected and needs considerably more attention.

Perhaps the most important single reaction characteristic of complexes containing platinum(II)–carbon σ bonds are those which involve oxidative addition either to give a stable adduct or, via an intermediate, to give a substitution product. The formation of stable 6-coordinate adducts [Pt(IV)] is well documented for Pt(II), but is unknown for Pd(II). It remains, however, a possible reaction path for Pd(II), particularly with reagents such as HCl. These reactions are considered first.

i. *Oxidative Addition to Platinum(II)*. These were first reported by Chatt and Shaw in 1959.[358] Examples are:

$$(Et_3P)_2PtMeI + MeI \underset{80°/vac.}{\overset{100°}{\rightleftharpoons}} (Et_3P)_2PtMe_2I_2 \xrightarrow{100°} (Et_3P)_2PtI_2 + C_2H_6$$

$$\textbf{(II-61)}$$

$$cis\text{-}(Et_3P)_2PtMe_2 + Cl_2 \rightarrow (Et_3P)_2PtMe_2Cl_2$$

$$\textbf{(II-62)}$$

The configurations of **(II-61)** and **(II-62)** are not known; from the dipole moment of **(II-62)**, however, a trans addition of chlorine to the the cis complex seems likely. The complex **(II-61)** decomposed on heating at 100° to ethane and the diiodide. In contrast, the *cis* dimethyl complex gave the diiodide in a two-step reaction.

$$cis\text{-}(Et_3P)_2PtMe_2 + I_2 \rightarrow trans\text{-}(Et_3P)_2PtMeI + MeI$$

$$trans\text{-}(Et_3P)_2PtMeI + I_2 \rightarrow trans\text{-}(Et_3P)_2PtI_2 + MeI$$

No Pt(IV) intermediates were isolated here, but Stone *et al.*[449] obtained an adduct in the analogous reaction with the diphenylplatinum complex.

$$cis\text{-}(Et_3P)_2PtPh_2 + I_2 \rightarrow (Et_3P)_2PtPh_2I_2$$

The complex (**II-16**) (Section C,3,b,i, this Chapter) and its phosphine analog lost ethane on heating and regenerated the dimethylplatinum starting complex.[396]

$$cis\text{-}(Me_2PhAs)_2PtMe_4 \xrightarrow{160°} cis\text{-}(Me_2PhAs)_2PtMe_2 + C_2H_6$$
(**II-16**)

The reactions of the complexes L_2PtR_2 with HCl also probably go via intermediate Pt(IV) complexes, but these have not been isolated; only cleavage of the organic group is observed.

$$cis\text{-}(Et_3P)_2PtMe_2 + HCl \rightarrow cis\text{-}(Et_3P)_2PtMeCl + CH_4 \quad \text{[358]}$$

$$cis\text{-}(Et_3P)_2PtMeCl + HCl \rightarrow cis\text{-}(Et_3P)_2PtCl_2 + CH_4$$

$$cis\text{-}(Et_3P)_2PtPh_2 + HCl \rightarrow cis\text{-}(Et_3P)_2PtPhCl + C_6H_6$$

In contrast, the pentafluorophenyl groups in $cis\text{-}(Et_3P)_2Pt(C_6F_5)_2$ are not cleaved by HCl or I_2. This may well reflect the greater difficulty with which pentafluorophenyl complexes undergo oxidative addition.

Belluco et al. have determined the kinetics for the reactions[450, 451]

$$trans\text{-}(Et_3P)_2PtMeX + HCl/MeOH \rightarrow (Et_3P)_2PtXCl + CH_4 \quad (X = Cl, I)$$

$$trans\text{-}(Et_3P)_2PtPh_2 + HCl/MeOH \rightarrow trans\text{-}(Et_3P)_2PtPhCl + C_6H_6$$

In each case a two-step mechanism is invoked to explain the results most satisfactorily. The first step is a fast solvent-assisted addition of HCl to form a 6-coordinate Pt(IV) hydride, and the rate-determining slow second step is loss of the hydrocarbon to give the product. An alternative formulation of the intermediate is $[(Et_3P)_2Pt^{IV}HMeCl \cdot S]^+Cl^-$ (or $[(Et_3P)_2Pt^{IV}HPh_2 \cdot S]^+Cl^-$) where S is a neutral solvent molecule; the exact formulation of the intermediate depends on the relative nucleophilicities of S and Cl^- toward Pt(IV).

Chatt and Shaw have noted that chlorohydridobis(triethylphosphine)-platinum added HCl to give an unstable adduct, probably of Pt(IV), which readily lost HCl to regenerate starting material.[35]

$$trans\text{-}(Et_3P)_2PtHCl + HCl \rightleftarrows (Et_3P)_2PtH_2Cl_2$$

Polar molecules such as MeI or HCl are not the only ones which will add oxidatively to metals in d^{10} or d^8 states. H_2 and O_2 will also add and the mechanism of cleavage of M–R by these reagents may well involve the primary formation of adducts of the type $(R'_3P)_2MR_2H_2$ or $(R'_3P)_2MR_2O_2$. Olefins, particularly those bearing electron-withdrawing substituents, also undergo reactions analogous to the oxidative additions to give adducts (Chapter III). These adducts appear to destabilize any M–C bonds present in the original complex. For example, Saito et al.[390] showed that the green diethyl-

(bipyridyl)nickel. (**II-63**), was stable to $100°$ when it decomposed to *n*-butane, ethylene, and ethane. Yamamoto and Ikeda[391] later showed that in the presence of acrylonitrile, this complex decomposed at $25°$ to give the same mixture of gases. An intermediate orange olefin π complex, probably (**II-64**), was isolated.

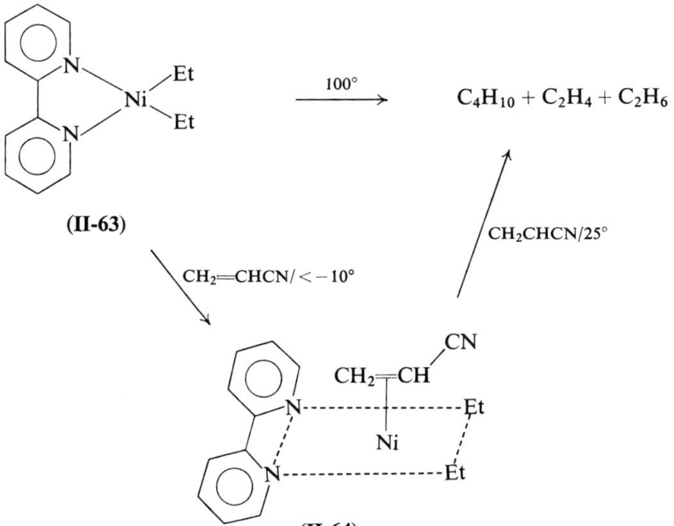

(**II-63**)

(**II-64**)

A similar sort of process probably occurs during cis-insertion reactions; the stereochemistry of the complex intermediate and the nature of the ligands and metal are critical.

A different type of oxidative addition takes place during the bromination and dehydrobromination of dibromobis(*o*-allylphenyldimethylarsine)platinum, (**II-65**).[452] Very complex rearrangements occur and the final product has been shown by an X-ray structure determination to be (**II-66**), p. 92.[196] The detailed mechanisms of the reactions are not clear.

ii. *Decomposition of L_2PdR_2 to R–R.* This type of decomposition can be effected by reactions such as those described above, or thermally or photo-chemically. When R = alkyl other than methyl, a competing reaction to give the olefin also occurs (see below).

The first reaction of this type was the thermal decomposition of $(Et_3P)_2PdMe_2$ reported by Calvin and Coates.[49] This complex probably has the cis configuration and it would seem reasonable that for dimerization to occur the two alkyls need to be adjacent.

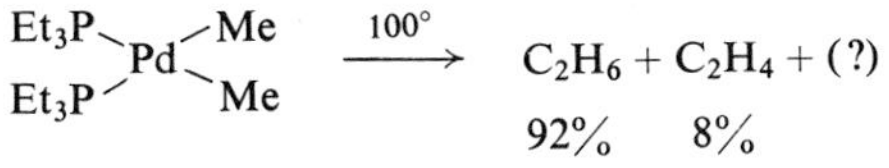

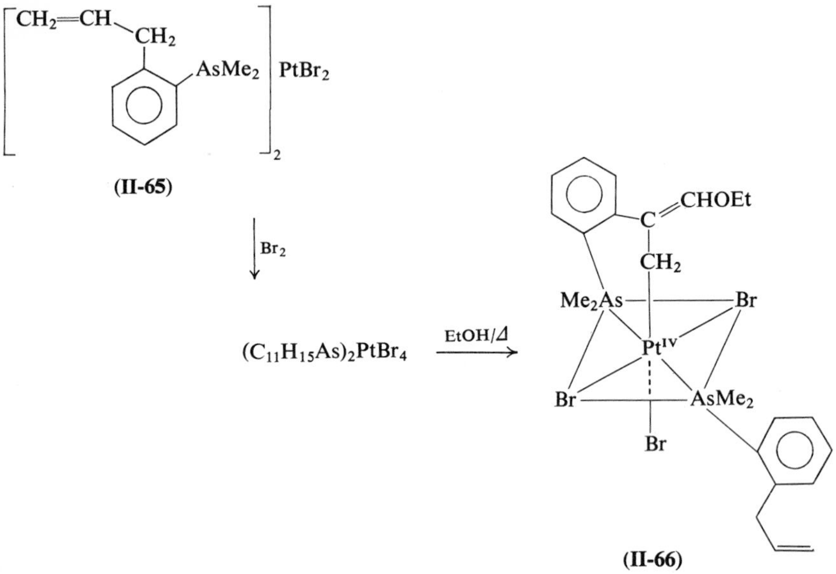

(II-65)

(II-66)

These reactions are probably more complex than they appear to be; for example, Calvin and Coates in a preliminary report on $(Et_3P)_2PdPh_2$[376] described it as being fairly stable to heat *in vacuo* but decomposing rapidly to biphenyl and metal on heating in air.

Keim[453] has reported that methyltris(triphenylphosphine)rhodium decomposes on heating as follows:

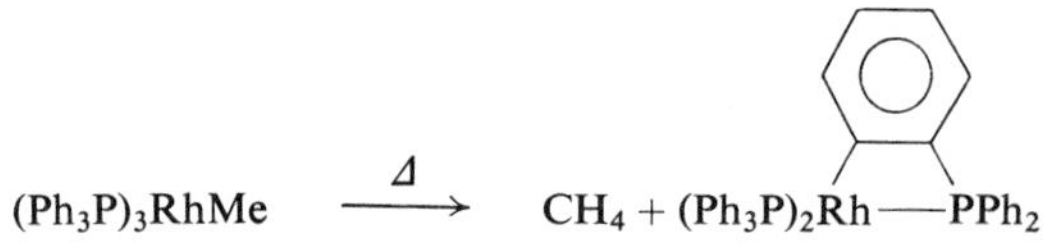

Similar reactions may also occur with palladium and platinum. (See also Ref. 453a).

The thermal decomposition reactions of some Pt(IV) complexes, $L_2PtMe_{4-n}(hal)_n$, have been described by Ruddick and Shaw.[396]

Heck[381] and Henry[388] have noted that biphenyl is the main product of the reaction of diphenylmercury (or other phenylating agents such as PhHgX, Ph_4Sn) and palladium chloride. Biphenyls also arise from the reaction of benzenes or substituted benzenes with palladium acetate or palladium chloride and sodium acetate. These reactions all probably proceed via a phenylpalladium intermediate,[159] though other reaction paths have been suggested.[454]

As already mentioned, the only volatile product of the reaction of bipyridyl-(dimethyl)palladium (II-49) and an excess of heptafluoropropyliodide was found to be ethane,[434] see pp. 84–85.

One example of a photochemical decomposition has been reported by Müller and Göser.[455]

$$\text{(cyclooctadiene)Pt(Pr}^i\text{)}_2 \xrightarrow{h\nu} \text{(cyclooctadiene)Pt}$$

This may proceed by dimerization of two isopropyl radicals or by a more complex path involving elimination of propylene.

Bis(triphenylmethyl)nickel, $(Ph_3C)_2Ni$, reacted with triphenylphosphine to give hexaphenylethane and $(Ph_3P)_4Ni$.[401]

iii. *β-Elimination from Alkylpalladium Complexes.* The general form of this very important reaction in organometallic chemistry is:

$$M\text{-}CH_2CHXR \rightarrow M\text{-}X + RCH{=}CH_2$$

and in general appearance, if not in detailed mechanism, it is an almost universal reaction for main-group as well as transition metals.

For the elements considered here, it has been documented by Chatt *et al.*[439] who had earlier found that the reaction of *trans*-chlorohydridobis(triethylphosphine)platinum with ethylene was reversed at higher temperatures.[35]

$$\textit{trans-}(Et_3P)_2PtHCl + C_2H_4 \underset{180°}{\overset{90°/200 \text{ atm}}{\rightleftharpoons}} \textit{trans-}(Et_3P)_2Pt(C_2H_5)Cl$$

In their later studies they showed that solvent exerted no influence on the reaction course and that on heating the dideuteroethyl complex a 1:1 mixture of the hydrido- and the deuteridoplatinum complexes was obtained.

$$\begin{array}{ccc}
\overset{Et_3P}{\underset{Br}{>}}Pt\overset{CD_2CH_3}{\underset{PEt_3}{<}} & \xrightarrow{\Delta} & \overset{Et_3P}{\underset{Br}{>}}Pt\overset{H}{\underset{PEt_3}{<}} & + & \overset{Et_3P}{\underset{Br}{>}}Pt\overset{D}{\underset{PEt_3}{<}}
\end{array}$$

This result was explained in terms of the following equilibria.

$$-Pt\text{-}CD_2CH_3 \rightleftharpoons -Pt\text{-}H \ (CD_2{=}CH_2) \rightleftharpoons -Pt\text{-}CH_2CD_2H \rightleftharpoons -Pt\text{-}D \ (HCD{=}CH_2)$$

$$-Pt\text{-}H + C_2H_2D_2 \qquad\qquad -Pt\text{-}D + C_2H_3D$$

A considerable stability for the 5-coordinate π-complex intermediate is also implied; alternatively a phosphine ligand may dissociate reversibly to give a 4-coordinate intermediate.

The only product of the reaction of $(Et_3P)_2PtCl_2$ with cyclohexyl- or iso-propyl-Grignard reagent is the hydride. For these and the higher olefins the equilibrium,

$$RCH_2\!\!=\!\!CH_2 + (Et_3P)_2PtHCl \rightleftharpoons (Et_3P)_2PtCl(CH_2CH_2R)$$

lies far over to the left-hand side. Experiments with 1-octene and the deuterido complex show that H–D exchange does occur.[439] However, diisopropyl(π-1,5-cyclooctadiene)platinum has been prepared from dichloro(π-1,5-cyclooctadiene)platinum and isopropylmagnesium bromide.[455]

Keim[456] has reported the formation of $(Ph_3P)_3RhH$ by the action of tri-isopropylaluminum on $(Ph_3P)_3RhCl$.

Simple reactions of this type have not yet been described for palladium, but they are very probably involved in a number of important catalytic processes. The reactions which are known involve complexes derived from 1,5-cyclo-octadiene, and have been studied largely by Tsuji and his co-workers.[412–414] A number of products can be isolated from the reaction of adducts of malonates (or acetylacetonates or other anions) and dichloro(1,5-cyclooctadiene)pal-ladium, for example, (II-67), with various bases. This complex reacted with weak bases to give the 1-substituted 3,5-cyclooctadiene (II-68) and with a stronger base ($MeSOCH_2^-$) to give the 2-disubstituted bicyclo[6.1.0]non-6-ene (II-69). Both these reactions can be visualized as base-catalyzed eliminations of HPdCl from (II-67). (See also p. 100.)

Similar reactions occur when the methoxycyclooctenyl complex, (II-67) [OMe in place of $CH(COOR)_2$], is heated in methanol,[421] and when the amino- or azidocyclooctenyl complexes (II-70) are treated with base.[417] Here solvolysis of the amine also occurs to give cycloocten-5-one. Partial or complete reduction can also occur (see following section).

An interesting variant has been described by Johnson et al.[50] for the complex (II-71), which reacts with triphenylphosphine, triphenylarsine, or triphenyl-

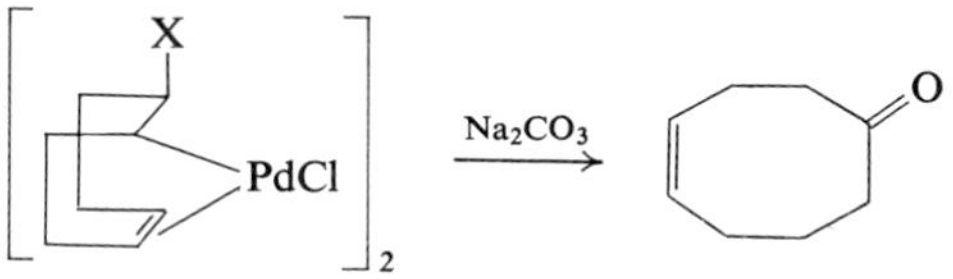

(II-70)

(X = NH₃ or N₃)

stibine to give a zero-valent palladium complex and an acetylacetonylcyclo-octadiene.

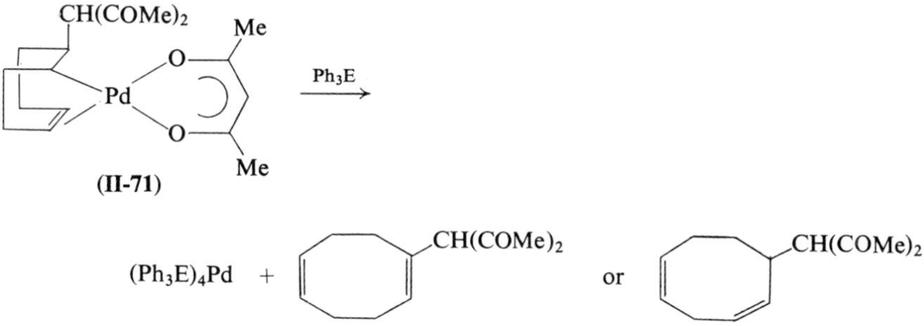

(II-71)

(Ph₃E)₄Pd   +   or

By contrast the platinum analog, (II-72), undergoes a remarkable reaction with triphenylphosphine to give the novel complex (II-73) in which an O-bonded acetylacetonate has been transformed into a C-bonded acetylacetonyl group and a new metal–carbon σ bond has been formed (see also Ref. 421a).

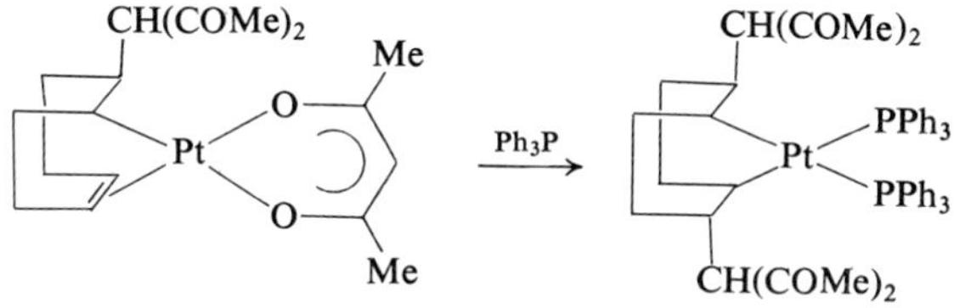

(II-72)   (II-73)

The formation of the bicyclo[3.3.0]oct-2-ene (II-30) by irradiation of the cyclooctenylpalladium acetate complex (II-29) has already been discussed in Section C,3,b,iii. This involves the elimination of HPdOAc.

Unfortunately insufficient work has as yet been done to establish the method by which HPdX is lost, in particular, whether the H eliminated is cis or trans to the palladium. Green, Haszeldine, and Lindley have suggested that it must be cis in a cyclohexane ring.[457] Although β elimination is most frequently observed, γ elimination can also occur, e.g., in the formation of (II-69). Other examples of elimination reactions are discussed in Chapter V, Volume I, and Chapters I and III, Volume II.

iv. *Replacement of –PdX by –H.* Acids, bases, and hydridic reagents have all been used to cleave M–C bonds. The reactions of $(Et_3P)_2PtRX$ with HCl have been discussed (Section C,3,d,i); the kinetics indicate first oxidative addition to Pt(II), followed by elimination of HR. Similarly, Calvin and Coates[49] reported that dimethylbis(triethylphosphine)palladium reacted quantitatively with aqueous ethanolic hydrobromic acid to give the dibromide and methane.

$$(Et_3P)_2PdMe_2 + 2HBr \rightarrow (Et_3P)_2PdBr_2 + 2CH_4$$

Other acids, even weak ones, acted similarly; thiophenol gave $(Et_3P)_2Pd(SPh)_2$ and *p*-nitrophenylacetylene gave $(Et_3P)_2Pd(C_2C_6H_4NO_2\text{-}p)_2$. Ethanol also reacted with $(Et_3P)_2PdMe_2$ below room temperature. The reaction proceeded in two stages—a fast one, leading to the liberation of one mole of methane, in which no metal was formed and no free radicals were present; the second, much slower step, was accompanied by the formation of some methane, ethane, ethylene, and acetaldehyde and the deposition of palladium. This was interpreted in terms of a free radical decomposition of an intermediate formulated as $(Et_3P)_2Pd(OEt)Me$, but other mechanisms are possible. In the presence of an excess of triphenylphosphine, $(Et_3P)_2PdMe_2$ decomposed in ethanol to give $(Ph_3P)_4Pd$. Some aryl–Pd bonds are also cleaved to give the arene by methanol in the presence of PPh_3.[457a]

Keim has reported the reaction of $(Ph_3P)_3RhR$ with phenol to give $(Ph_3P)_3RhOPh$ and RH (R = Me, Ph, H).[453,456] Wilke *et al.* have reported cleavage of Ni–R bonds by various types of acid, e.g.,[389,458]

$$bipyNiEt_2 + PhOH \rightarrow bipyNi(OPh)Et + C_2H_6$$

$$bipyNiEt_2 + HCl \rightarrow bipyNiCl_2 + 2C_2H_6$$

However, cleavage of M–C by acid is not a universal reaction, and is also dependent on the nature of the organic group. Metal–phenyl bonds are particularly easily split by acid. For example, Ph_3PAuEt was stable to dilute HCl, but attempts to prepare Ph_3PAuPh by a reaction, the last step of which involved a dilute acid hydrolysis, gave only Ph_3PAuCl.[459] Miller *et al.* have shown that a trichlorovinyl group is less easily detached from nickel than a phenyl.[428]

One methyl–gold bond in Me_3Au is easily split by HCl to give $[Me_2AuCl]_2$; phenol and organic acids do not cleave the Me–Au bond, but thiols do.[460]

$$Me_3Au \cdot OEt_2 + PhSH \rightarrow \tfrac{1}{2}[Me_2AuSPh]_2 + CH_4$$

A few reactions are also known where attack of the proton is not at the M–C bond; for example,[365, 403, 461–463]

$$[M = Pd, Pt; R = alkoxy, CH(COMe)_2]$$

Bases, especially aqueous cyanide, will also cleave M–C bonds; this reaction is particularly efficient since the tetracyanometallates, $M(CN)_4{}^{2-}$, formed are so stable. An example is[415]

Cyanide also releases cyclopropane from the Pt(IV) complex **(II-73a)** (see p. 191),[448, 464, 465]

(II-73a)

Sodium ethoxide in ethanol also cleaves Pd–C bonds,[414]

as will hydrogen, even in the absence of a catalyst.[408, 409] Other double bonds are also reduced, e.g.,

$$\left[\text{(norbornene-PdCl, MeO)} \right]_2 \xrightarrow{\text{H}_2} \text{MeO-norbornane}$$

Analogously, $(Ph_3P)_3RhR$ is reduced by hydrogen at 40 atm and 25° to $(Ph_3P)_3RhH$ and RH (R = Ph, Me).[456]

Perhaps the most common reducing agent has been sodium borohydride or occasionally $LiAlH_4$.[408, 410, 414, 462] Cope *et al.*[423, 429, 430] and Kasahara[431] have used $LiAlD_4$ to replace the Pd by deuterium and have thus been able to determine the position of attachment of the palladium. Some examples include:

$$\left[\begin{array}{c} \text{CH}_2\text{—PdCl} \\ \text{MeOCMe} \\ \text{CH}_2\text{—NMe}_2 \end{array} \right]_2 \xrightarrow{\text{LiAlD}_4} \text{Me}_2\text{NCH}_2\text{—}\overset{\overset{\text{OMe}}{|}}{\underset{\underset{\text{Me}}{|}}{\text{C}}}\text{—CH}_2\text{D}$$

$$\left[\text{(benzo ring, N≡NPh, PdCl)} \right]_2 \xrightarrow{\text{LiAlH}_4} \text{PhNHNHPh}$$

Borohydride usually reduces any double bond present in complexes as well as breaking O–C, N–C, and Pd–C bonds, e.g.,[417]

$$\left[\text{(NH}_2\text{, PdCl)} \right]_2 \xrightarrow{\text{BH}_4} \quad \underset{60\%}{\text{(cyclooctene-NH}_2\text{)}} \;+\; \underset{27\%}{\text{(cyclooctene)}} \;+\; \underset{8\%}{\text{(cyclooctadiene)}}$$

By using smaller quantities of borohydride Stille and Morgan were able to obtain partially reduced species[408] (p. 99).

Reduction of double bonds under these conditions is due to the formation of B_2H_6; at −40° in the presence of norbornene Vedejs and Salomon[465a] were able to trap the B_2H_6 and obtain stereospecific reduction of enyl complexes to the olefins.

The use of hydridic reducing agents has not been reported for the simpler phosphine complexes $(R_3P)_2MR'_2$.

v. *Insertion into the Pd–C Bond.* The reaction usually involves "insertion" of coordinated CO or C=C into a Pd–C bond, and is very important in catalytic reactions in which C–C bonds are formed. It is discussed fully in Chapter I, Volume II.

Coulson[466] has recently reported the ligand-assisted insertion reaction in which a norbornenyl complex is transformed into a nortricyclene complex. (See also p. 140 and Refs. 201a, 421a.)

Another example of this reaction is the formation of vinylpentamethyl-cyclopentadiene by the action of PPh_3 (or $AsPh_3$) at low temperatures on the complex (**II-43**) (Section C,3,b,iv) obtained from trimerizing 2-butyne with $PdCl_2$.[282, 467]

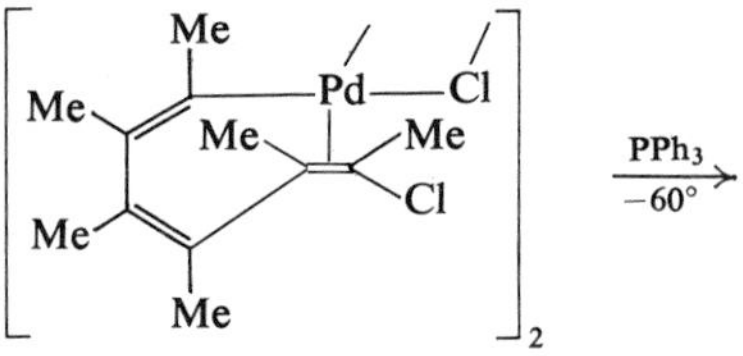

(II-43)

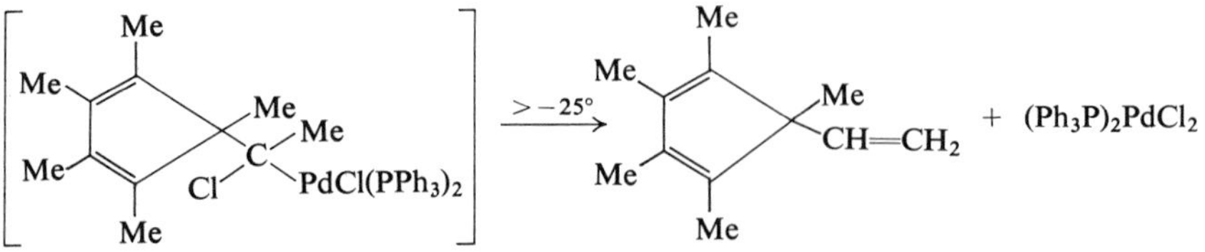

Insertion of a coordinated C=C bond into Pd–C is also involved in the formation of (II-30) from (II-29) (Section C,3,b,iii), and probably also in the formation of (II-74) from reaction of (II-67) with excess malonate ion in DMSO.[414] This reaction may be envisaged to proceed as shown:

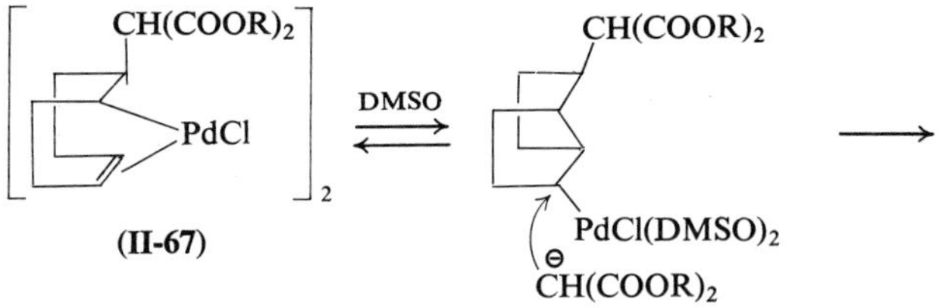

(II-67)

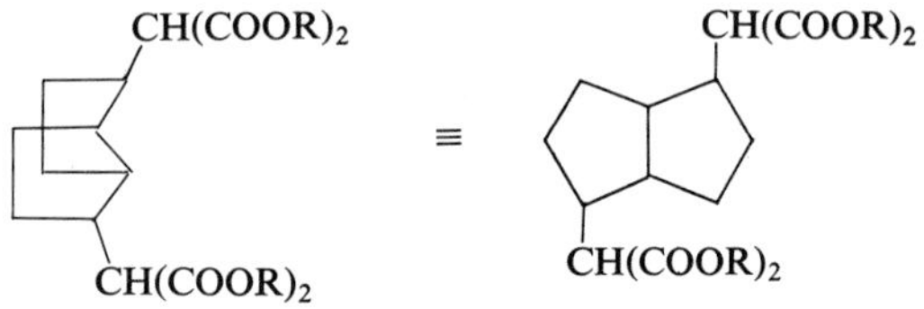

(II-74)

vi. *Replacement of Pd in Pd–C by X.* Palladium in a Pd–C bond can formally be displaced by H (see above) or by a number of other groups. The general form of the reactions is:

$$Pd–C + X–Y \rightarrow Pd–Y + X–C$$

Perhaps the simplest (at least overall, if not in detail) such reaction is[434]

$$bipyPdMe_2 + I_2 \rightarrow bipyPdI_2 + 2MeI$$

Similar reactions are known for Pt and Ni (see also Chapter IV, Section A,2, Volume II).

Fitton et al.[398] have reported that benzylpalladium complexes react with silver (and, to a lesser degree, with potassium) acetate to give benzylacetate and benzylidene acetate. Perhaps the most remarkable feature of this reaction is the inertness of the Pd–benzyl bond here; even under optimum conditions (AgOAc/HOAc/100°) only 57% of the acetates are obtained.

$$\text{Ph}_3\text{P} \diagdown \text{Pd} \diagup \text{CH}_2\text{Ph}$$
$$\text{Cl} \diagup \quad \diagdown \text{PPh}_3$$

or

$$\text{Ph}_3\text{P} \diagdown \text{Pd} \diagup \text{Cl} \diagdown \text{Pd} \diagup \text{CH}_2\text{Ph}$$
$$\text{PhCH}_2 \diagup \quad \diagdown \text{Cl} \diagup \quad \diagdown \text{PPh}_3$$

$$\longrightarrow \quad \text{PhCH}_2\text{OAc} + \text{PhCH(OAc)}_2$$

Heck[468] has replaced the palladium in the complex (II-45) by other metals on reacting it with the appropriate metal carbonylate anion.

(II-45)   (M = Co, x = 4; M = Mn, Re; x = 5)

vii. *Reactions in Which Pd–C Bonds Are not Broken.* Some complexes, especially those where the Pd–C forms part of chelate ring such as (II-45), are surprisingly unreactive toward reagents which will otherwise cleave Pd–C bonds. For example, Heck found that on reaction of (II-45) with aqueous cyanide only replacement of Cl by CN occurred.

These reactions are characteristic to some extent of many compounds of the type *trans*-LL′RPdCl since R will exert its well-known trans-labilizing effect, and provided that the Pd–C bond will withstand the conditions needed, many such metathetical reactions are possible.

Crociani et al. have investigated the reactions of (II-75) with a number of amines, *t*-phosphines, (EtO)$_3$P, Ph$_3$As, and Ph$_3$Sb.[405] Only when L was a

(II-75)   (II-76)   (II-77)

t-phosphine, *p*-toluidine, pyridine, or 4-methylpyridine was an adduct isolated. They claimed, on the basis of the far-infrared spectra of the adducts (notably from the position of bands ascribed to $\nu_{\text{Pd–Cl}}$ and $\nu_{\text{Pt–Cl}}$) that trans adducts, (II-77), were obtained when L was an amine and cis adducts, (II-76),

were the products when L was a *t*-phosphine. A combination of thermodynamic and steric factors was invoked to explain this. The trans structure was
shown to be correct for (II-77) (M = Pt, L = py) by an X-ray structure determination. Pyridine in (II-77) (L = py) could only be displaced by other neutral
ligands, and Cl in (II-77) (L = py) could only be displaced by anionic ligands
such as Br⁻. (See also ref. 421a).

Green and Hancock,[462] in contrast, have reported that (II-78) reacted with
pyridine to give a cationic complex. A comparative study by Johnson *et al.*[415]

(II-78)

has shown that (II-79) gave the adduct (II-80). When the palladium analogs
were subjected to the same reactions only the complexes $(Ph_3E)_2Pd(hal)_2$ were
isolated.

(II-79)   (II-80)   (E = P, As)

Bromine in $(Et_3P)_2PdMeBr$ has been replaced by CN and SCN[49]; and bridging chlorines can frequently be replaced by acetylacetonate, acetate, or cyclopentadienyl. Thallium acetylacetonate and cyclopentadienide are very useful
reagents for these reactions, requiring only very mild conditions, e.g.,[408, 469]

Halogens can usually be substituted for each other in the sense that I replaces Br, which replaces Cl. Organometallic fluorides are virtually unknown.

The most revealing study of these reactions has been that of Basolo, Chatt, Gray, Pearson, and Shaw[193] who studied the rates of replacement of chloride by pyridine in

$$trans\text{-}(Et_3P)_2MClR + py \rightarrow [trans\text{-}(Et_3P)_2MRpy]^+ + Cl^-$$

for different M and R. When R = o-tolyl the relative rates were 5×10^6(Ni), 10^5(Pd), and 1(Pt). The importance of steric effects was shown for the platinum complexes, where the relative rates for R = Ph, o-tolyl, and mesityl were 30:6:1. Even greater effects were noted for the cis complexes and this has been explained in terms of greater steric hindrance there in the transition state for substitution.[195] The order of trans effect for R was H $\gg$ Me > Ph $\gg$ Cl.

Other studies on L_2PtRX have been reported.[470–472] An interesting manifestation of the trans effect not involving halogen and which illustrates the high stability of Pt–carbon bonds is due to Doyle and co-workers,[418]

$$dienePtR_2 + 2L \rightarrow L_2PtR_2$$

where diene = 1,5-cyclooctadiene, cyclooctatetraene, norbornadiene; R = Me, Et, Ph; and L = py and PPh_3.

D. PALLADIUM HYDRIDES

Although the interaction of palladium metal with hydrogen has been the subject of innumerable studies, very little is known about the molecular hydrides. This is in sharp contrast to platinum where these complexes are well characterized and easy to prepare and study. It is convenient to review here what is known about palladium hydrides as they are frequently postulated as intermediates in reaction schemes.

$trans$-Chlorohydridobis(triethylphosphine)platinum was first prepared by Chatt, Duncanson, and Shaw in 1957.[473] It and some closely related hydrides are remarkable for their unusual stability; the chemistry of these compounds has been reviewed by Cross.[39]

$$cis\text{-}(Et_3P)_2PtCl_2 \rightarrow trans\text{-}(Et_3P)_2PtHCl$$

The usual methods of preparation for the platinum hydride, which involve reduction of the cis-dihalide by alcoholic potash, hydrazine, or metal hydrides, were not successful for the preparation of the pure palladium analog.

Chatt, Duncanson, and Shaw[474] in a later note mentioned that they had obtained solutions of what was believed to be $trans$-$(Et_3P)_2PdHCl$, as indicated by the band at 2035 cm^{-1} in the infrared spectrum ascribed to ν_{PdH}. Brooks and Glockling in a series of papers[475–477] described the preparation and the properties of the pure palladium hydride by the rather unusual reaction:

$$trans\text{-}(Et_3P)_2PdCl_2 + Me_3GeH \xrightarrow{40°} trans\text{-}(Et_3P)_2PdHCl + HCl + Me_3GeCl + Me_6Ge_2$$

Me_3GeH was preferred to Ph_3GeH, used earlier, since the products were easier to separate; neither Me_3SiH nor Me_3SnH gave the complex nor could pure products be isolated from other t-phosphine–palladium complexes. $trans$-$(Et_3P)_2PdBr_2$ reacted more slowly than the chloride, but $(Et_3P)_2PdHBr$ was less sensitive to oxygen. $(Et_3P)_2NiBr_2$ also underwent the reaction, but the nickel–hydride product was very unstable indeed. Brooks and Glockling[476] have shown that the formation of the hydridochloride was catalyzed by palladium metal and have proposed a free radical chain mechanism.

X-Ray studies of two platinum hydrides, $(Et_3P)_2PtHBr$[197] and $(Et_2PhP)_2$ $PtHCl$,[478] have been carried out; these complexes have trans structures and the H occupies a normal coordination site in square planar platinum.

A recent report of an X-ray structure determination of $(Et_3P)_2PdHCl$ by Schneider and Shearer (quoted by Glockling and Brooks[228]) shows the complex to have the expected trans configuration. The hydrogen was not located, but the geometry of the molecule clearly showed that it occupied one site in the coordination sphere of the metal. The Pd–Cl (trans to the H^-) was very long (2.43 Å), and the Pd–P bond lengths were 2.31 Å.

The ^1H NMR spectrum indicated the complex to be trans in solution too, and also showed the hydride at $\tau 23.6$. No coupling to ^{31}P was observed and this suggests that some exchange process is occurring. The trans assignment was made on the basis of the observed virtual coupling between the two $trans$-Et_3P groups. The phenomenon of virtual coupling in such complexes was discovered by Jenkins and Shaw,[269,479] who noted that the ^1H NMR spectra of the methyls in $(Me_2PhP)_2PdX_2$ appeared as a triplet in the trans isomer (equivalent coupling to both phosphorus nuclei) but only as a doublet in the cis isomer (coupling observed to only one phosphorus.)

$trans$-$(Et_3P)_2PdHCl$ decomposed at 55°; contrary to earlier speculation though, it was quite stable at 20° in the absence of oxygen, and could be sublimed in $vacuo$. It was also relatively stable to methanol. On reaction with KI in acetone, the impure iodide was obtained, but cyanide in methanol did not give the hydridocyano complex—only $(Et_3P)_2Pd(CN)_2$ and hydrogen. The authors believed this was due to the high trans effect of CN^-.

Reaction of the hydrido complex with HCl or CCl_4 led to H–halogen exchange.

$$(Et_3P)_2PdHCl + X\text{–}Cl \rightarrow trans\text{-}(Et_3P)_2PdCl_2 + XH \quad (X = H, CCl_3)$$

The former reaction should be contrasted with that of the platinum analog which merely added HCl reversibly (Section C,3,d,i).

The hydrido complex would be expected to react with olefins in the same manner as the platinum analog but rather more readily. In fact, reactions have been reported with phenylacetylene, butadiene, and acrylonitrile, but no products have been characterized.[476]

The chemistry of $(Et_3P)_2Pd(GePh_3)_2$, including its hydrogenation at 100 atm to trans-$(Et_3P)_2PdH(GePh_3)$ has been reported.[228,477,480] The latter complex showed ν_{PdH} at 1890 cm^{-1}, a very low value ascribed to the high trans influence of the Ph_3Ge group.

Kingston and Scollary[322] have reported the formation of hydridocarbonyl anions by carbonylation of $PdCl_2$ in methoxyethanol (Section B,1).

Hydrido complexes, L_2MHX, where L is a bulky trialkylphosphine, have very recently been obtained for both Pd and Ni by other routes.[481,482]

A very interesting series of stable 5-coordinate hydrido-nickel cations, e.g., $[(EtO)_3P]_4NiH^+$, have been described,[115,116] and it is possible that the palladium analogs may also be formed in a similar fashion and undergo analogous reactions.[482a]

Kudo et al.[114] have very recently reported that both $(Ph_3P)_3PdCO$ and $(Ph_3P)_4Pd$ reacted with HCl at $-50°$ in ether to give the hydride, e.g.,

$$(Ph_3P)_3PdCO + HCl \rightarrow trans\text{-}(Ph_3P)_2Pd(H)Cl$$

Chapter III
Monoolefin and Acetylene Complexes

A. INTRODUCTION

Historically the first olefin–transition metal complex was discovered by Zeise in 1827, who prepared the complex known as Zeise's salt, $K[C_2H_4 PtCl_3] \cdot H_2O$, by heating potassium tetrachloroplatinate(II) in ethanol.[483] It could also, with some difficulty, be synthesized directly from K_2PtCl_4 and ethylene and was later recognized as an ethylene complex.

The history of the various ideas of the metal–olefin bonding in this and similar complexes was summarized by Chatt and Duncanson in 1953.[484] These authors rejected most of the previous theories and instead developed an approach based on the model proposed by Dewar in 1951 to explain the bonding in olefin–Ag^+ complexes.[485] The Dewar–Chatt–Duncanson approach was essentially to regard the bond between platinum and the olefin as made up of two parts: (a) a σ-type bond, by overlap of a (vacant) $5d6s6p^2$ hybrid orbital of Pt(II) with the π orbital of the olefin, and (b) a π-type bond, by overlap of the filled $5d$ orbitals of the metal with the olefin π^* orbitals. The latter interaction could be strengthened if a $5d$–$6p$ hybrid, rather than a true $5d$ orbital was used. The situation is depicted in Fig. (III-1).

In principle, the olefin could bond to the metal either perpendicular or coplanar to the coordination plane of the platinum. In complexes such as Zeise's salt, nonbonding interactions with the *cis*-chlorines would make the coplanar arrangement higher in energy than the perpendicular. The latter arrangement is usually observed. However, the energy barrier to rotation about the Pt–olefin bond is not great, hence even the coplanar arrangement is energetically not too unfavorable. This picture has, without major modification, been used ever since and, to a first approximation at least, it explains the properties of such complexes very well. X-Ray crystal structures have also confirmed the essential correctness of this hypothesis. The bonding situation in olefin and acetylene complexes is discussed in more detail in Section D.

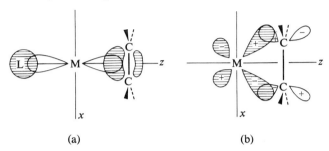

(a)   (b)

FIG. III-1. Bonding in olefin complexes, for example, Zeise's salt, $K[C_2H_4PtCl_3]$ (M = Pt, L = Cl^-). (a) "σ-type" bond; (b) "π-type" bond. Only one trans ligand, L, is shown.

Olefin complexes of palladium are much more reactive than those of platinum. They were first prepared by Kharasch, Seyler, and Mayo in 1938,[281] from reaction of the olefin with the bis(benzonitrile) complex.

$$(PhCN)_2PdCl_2 + \text{olefin} \xrightarrow{C_6H_6} (\text{olefinPdCl}_2)_n\downarrow + 2PhCN$$

The most stable, in a desiccator at 20°, were the cyclohexene and styrene complexes. Even the former slowly lost cyclohexene on standing; in the presence of olefin the complexes appeared relatively stable. The molecular weight of the styrene complex indicated that $n = 2$. X-Ray crystal structures of the ethylene and styrene complexes showed that they had the trans dimeric structures, **(III-3)** and **(III-4)** (p. 112).

The considerable attention which these complexes are now attracting is because they are intermediates in a wide variety of palladium-catalyzed reactions of olefins.

Recently some new olefin complexes of Pd(0) have been described which have quite different properties to those of Pd(II), for example,

$$\begin{matrix} Ph_3P \\ Ph_3P \end{matrix} Pd \begin{matrix} HC & \diagup CO \\ \| & \\ HC & \diagdown CO \end{matrix} O$$

A Pd(0) complex containing only olefinic ligands, bis(dibenzylideneacetone)-palladium, has also recently been prepared,[485a] see Chapter IV, Section D,4, this volume.

Acetylenes differ from olefins in having two π orbitals orthogonal to each other, and hence also in having two sets of π^* orbitals. This would permit an acetylene to bond equivalently to two metal atoms and indeed such complexes are known, for example,[486]

However, numerous examples are now known where the situation is much more complex and the acetylene is bonded to three or more metal atoms. A structure of this type has been proposed by King et al. for the product, $(CF_3C_2CF_3)_3Ni_4(CO)_3$, of the reaction of hexafluoro-2-butyne and $Ni(CO)_4$.[487]

The best known acetylene complexes, however, are those where only one metal atom is bonded to the acetylene. These were prepared by Chatt and his co-workers for both Pt(II)[488–490] and Pt(0),[491] and crystal structures have been published for both types, (III-1)[492] and (III-2).[493]

(III-1)   (III-2)

Nickel(II) complexes of the former type are unknown. They are likely to remain so in view of the high reactivity of nickel(II) complexes and even the Pt(II) complexes are only isolable when at least one substituent of the acetylene is a bulky electron-releasing group such as t-butyl.[489] The nickel(0) analogs of (III-2) are well known.[45, 494]

Until very recently Pd(II)–acetylene complexes were unknown; this was un-doubtedly due to the ease with which acetylenes are oligomerized with Pd(II) complexes. A complex of 2-butyne, formulated as $[MeC_2MePd_2Cl_4(solvent)_x]$ has been detected by NMR at $-50°$ by Maitlis and co-workers.[282, 404]

Hosokawa et al.[495] have isolated and characterized the stable $[(t\text{-}BuC_2t\text{-}Bu)PdCl_2]_2$ from di-t-butylacetylene and ethylene–palladium chloride dimer.

A number of acetylene complexes of Pd(0), $L_2Pd(RC_2R)$ ($L = PPh_3$, $R = CF_3$, COOMe; $L = Bu_3P$, Me_2PhP, $R = CF_3$) analogous to (III-2) have been reported by Greaves, Lock, and Maitlis.[45]

B. PREPARATION OF MONOOLEFIN AND MONOACETYLENE COMPLEXES

1. Monoolefin Complexes of Pd(II)

Since the olefin–palladium(II) complexes are very reactive, the choice of routes to them is limited, and only two methods have been used. These are displacement of a very loosely bound ligand, L, from $(LPdCl_2)_2$ or L_2PdCl_2 or direct reaction of the olefin with solid $PdCl_2$. Kharasch et al.[281] used $(PhCN)_2PdCl_2$.

$$(PhCN)_2PdCl_2 + olefin \rightarrow (olefinPdCl_2)_2 + 2PhCN$$

for ethylene, isobutene, styrene, cyclohexene, pinene, and camphene. Other olefin complexes have also been made by this route.[279, 496] Hüttel and his collaborators found that with short reaction times some olefins (cyclohexene, cycloheptene, cyclooctene, 1-methyl-cyclohexene, isobutene, and α-methyl-styrene) reacted with palladium chloride in 50% aqueous acetic acid at 20° to give the olefin complexes.[497–499] Yields were generally low owing to other competing reactions such as oxidation and the formation of π-allylic complexes.

Pestrikov et al.[500–502] and Henry[503] have determined equilibrium constants for

$$olefin + PdCl_4^{2-} \rightleftharpoons olefinPdCl_3^- + Cl^-$$

under conditions where further reaction was minimized. No complexes were isolated (see Section C).

Pregaglia et al.[504, 505] have reported the preparation of complexes from the liquid olefin and $PdCl_2$; however, isomerization frequently occurred and the complexes were difficult to free from excess $PdCl_2$ because of their low solubility. Complexes of 1-olefins were more soluble than those of internal olefins, and cis-olefins dissolved the $PdCl_2$ faster than trans-olefins. Moiseev and co-workers have reported that the solubility of $PdCl_2$ in olefins also depended on the thermal history of the $PdCl_2$, the α form apparently being the more soluble.[162]

Olefin complexes have also been made by exchange reactions[504]; an interesting application of this was the formation of complexes of 1,3-butadiene and

1,3-cyclooctadiene in which only one double bond was coordinated to the metal, e.g.,[506]

$$[1\text{-pentenePdCl}_2]_2 + 2 \quad \bigcirc \hspace{-1.2em}= \quad \longrightarrow \quad \left[\bigcirc \hspace{-1.2em}= \hspace{-0.2em}-\!PdCl_2 \right]_2 + 2C_5H_{10}$$

Ketley et al.[507] have studied the effects of various solvents on the reaction of $PdCl_2$ with ethylene to give $[C_2H_4PdCl_2]_2$ under pressure at temperatures from $20°$–$50°$. Some solvents, notably ethyl chloride and t-butyl chloride exerted a very strong catalytic effect, reaction being complete in less than 1 minute at $20°$. After 20 hours the red solution of $[C_2H_4PdCl_2]_2$ in chloroform containing a trace of ethanol gave a yellow precipitate of $(C_2H_4)_2PdCl_2$, which, however, was only stable under ethylene pressure. Extensive dimerization to butenes and their complexes also occurred.

Some complexes which, formally at least, have a monoolefin ligand are those derived from dienePdCl$_2$ and a nucleophile. They are discussed in Chapters II and IV, this volume.

2. Monoolefin Complexes of Pd(0)

These have all been prepared from tetrakis(triphenylphosphine)palladium(0) and the appropriate olefin.[62, 508]

$$(Ph_3P)_4Pd + \text{olefin} \quad \xrightarrow[\text{or THF}]{C_6H_6} \quad (Ph_3P)_2Pd(\text{olefin})$$

Complexes where the olefin was dimethyl maleate, dimethyl fumarate, maleic anhydride, p-benzoquinone, 1,4-naphthoquinone, and octafluoro-2-butene were reported.

3. Acetylene Complexes of Pd

The zero-valent acetylene complexes have been prepared by Greaves and Maitlis,[45, 509] who found that acetylenes with electron-withdrawing substituents (CF_3, COOMe) displaced two triphenylphosphines from $(Ph_3P)_4Pd$,

$$RC\!\equiv\!CR + (Ph_3P)_4Pd \rightarrow (Ph_3P)_2Pd(RC_2R) + 2Ph_3P$$

Complexes of hexafluoro-2-butyne and other phosphines ($R'_3P = Bu^nP$, Me_2PhP) were obtained by hydrazine reduction of the appropriate $(R'_3P)_2PdCl_2$ in the presence of excess phosphine, followed by reaction with $CF_3C_2CF_3$. The intermediates, presumably $(R'_3P)_{3,4}Pd$, were not isolated.

$$(R'_3P)_2PdCl_2 + CF_3C_2CF_3 \quad \xrightarrow{N_2H_4} \quad (R'_3P)_2Pd(CF_3C_2CF_3)$$

Hosokawa *et al.*[495] obtained a stable acetylene–Pd(II) complex from ethylenepalladium chloride dimer and di-*t*-butylacetylene.

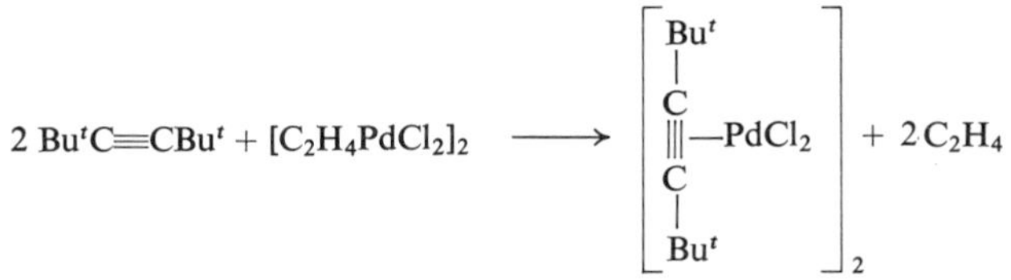

$$2 \; Bu^tC{\equiv}CBu^t + [C_2H_4PdCl_2]_2 \; \longrightarrow \; \begin{bmatrix} Bu^t \\ | \\ C \\ \||{-}PdCl_2 \\ C \\ | \\ Bu^t \end{bmatrix}_2 + 2 \, C_2H_4$$

The same complex, but in lower yield, was also obtained from $(PhCN)_2$ $PdCl_2$. Other acetylenes, even those with quite bulky substituents, did not give stable products of this type. For example, in the above reaction with $PhC{\equiv}CBu^t$, a π-allylic complex rather than an acetylene complex was isolated (see Chapter V, this volume).

As already mentioned, a complex of 2-butyne, formulated as $[Me_2C_2PdCl_4$ $(solvent)_x]$, has been detected by NMR at $-50°$ in a solution of 2-butyne and $(PhCN)_2PdCl_2$ in $CDCl_3$.[282, 404]

4. Preparation of Olefin and Acetylene Complexes of the Neighboring Metals

Olefin and acetylene complexes of Ni(II) have not yet been isolated. Those of Ni(0) and Pt(0) were prepared by methods very similar to those described above, involving reduction of $(R_3P)_2MX_2$ in the presence of the unsaturated ligand.[491, 494, 510, 511]

Olefin and acetylene complexes of Pt(II) are fairly inert and stable, and have been prepared by reaction of the appropriate ligand with K_2PtCl_4 under aqueous or aqueous–organic conditions.[484] Since Pt(II) complexes are much less labile than Pd(II) complexes, the reaction of K_2PtCl_4 with ethylene to give Zeise's salt is very slow. Cramer[512] has catalyzed this reaction with small amounts of $SnCl_2$. Presumably an intermediate complex, $K_2[PtCl_3SnCl_3]$, is formed in which the $SnCl_3$ exerts a very large trans effect which labilizes a Pt–Cl bond toward substitution by ethylene. Replacement reactions have also commonly been used, e.g.,

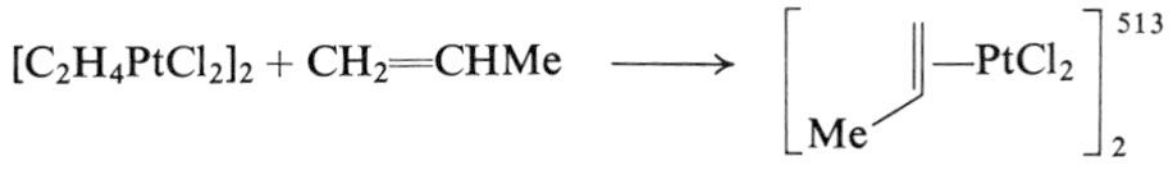

$$[C_2H_4PtCl_2]_2 + CH_2{=}CHMe \; \longrightarrow \; \begin{bmatrix} \||{-}PtCl_2 \\ Me \end{bmatrix}_2^{513}$$

$$(Ph_3P)_2PtC_2H_4 + PhC_2H \rightarrow (Ph_3P)_2Pt(PhC_2H) \;^{95, 139, 514}$$

The intermediate in the latter is $(Ph_3P)_2Pt$; the ethylene complex has also been reacted with olefins bearing electron-withdrawing substituents to give $(Ph_3P)_2Pt$(olefin) complexes.[515]

Nickel(0) complexes such as (1,5-cyclooctadiene)$_2$Ni and $Ni(CO)_4$ have been used for the preparation of some olefin complexes.[516, 517]

$$Ni(CO)_4 \text{ or } Ni(1,5\text{-}COD)_2 + CH_2{=}CHCN \longrightarrow \left[Ni\left(\left\| \,^{CN} \right. \right)_2 \right]_n$$

Dubini and Montino[153] have reported some complexes, [olefinNiBr] and [olefinNiBrPPh$_3$] (olefin = dimethyl maleate or fumarate) which appear to be complexes of Ni(I).

Rhodium(I) and iridium(I) (d^8) form a large number of olefin complexes of various types. $[(C_2H_4)_2RhCl]_2$ has been described by Cramer,[518] and both $(Ph_3P)_3RhCl$ and $(R_3P)_2IrCOCl$ react with olefins (and acetylenes), particularly those with electron-withdrawing substituents, to give complexes, $(Ph_3P)_2Rh$ (olefin)Cl and $(R_3P)_2IrCOCl$(olefin). Olefin complexes of Fe(0) such as $(NCCH{=}CH_2)Fe(CO)_4$ are also well known.[519]

Copper, silver, and gold all form olefin complexes in the (I) (d^{10}) oxidation state. Those of silver, especially with BF_4^- as[520] anion, are particularly easily obtained, and have been extensively used to purify olefins.

C. STRUCTURES AND PHYSICAL PROPERTIES

The complexes [olefinPdCl$_2$] all appear to be dimeric; this is indicated by molecular weight studies on those which are sufficiently soluble[504, 506] (for olefins up to C_6 the complexes tend to be insoluble) and by X-ray structure determinations of [ethylenePdCl$_2$]$_2$ and [styrenePdCl$_2$]$_2$, carried out by Baenziger and his collaborators.[247, 248]

(III-3)   (III-4)

The ethylenes could not be accurately located in the former complex, (III-3), but were assumed to lie perpendicular to the Pd_2Cl_4 plane. Two types of dimer with apparently slightly differing Pd–Cl bond lengths were found, but the precision of this determination is probably not great enough to draw any conclusions.

The styrene in (III-4) was planar and perpendicular to the Pd_2Cl_4 plane, but the axis of the vinyl group was at an angle of 74° to this plane. Furthermore, the Pd_2Cl_4 plane did not bisect the vinyl group, but cut it nearer to the terminal CH_2, in the ratio 0.92:0.40. These effects are due to both steric and electronic causes, the latter arising from an asymmetric electron distribution in the vinyl group.† These authors also claimed that the bridging Pd–Cl distances were unequal, the one trans to the styrene being 2.41(2) Å, while the other was 2.32(2) Å. If this effect is indeed real and not due to optimistic estimates of standard deviations, then it represents one of the few cases yet observed where an olefin exerts such a strong trans *influence*. The terminal Pd–Cl bond length was 2.27 Å.

Older crystal structures of the olefin–platinum complexes $[C_2H_4PtCl_3]^-$ and $[C_2H_4PtBr_3]^-$ by Bokhii and Kukina,[526] $[C_2H_4PtCl_3]^-$ by Wunderlich and Mellor,[527] and of trans-$[C_2H_4PtCl_2NHMe_2]$ by Alderman, et al.,[528] while poor in detail, confirm that the complexes are square planar about the metal with the olefin occupying one coordination site.

Two recent and more accurate determinations of Zeise's salt, $K[C_2H_4 PtCl_3]\cdot H_2O$, have been reported. One, an X-ray determination by Black et al.,[529] indicated no significant lengthening of the trans-Pt–Cl [2.34(2) Å] by comparison with the cis-Pt–Cl bonds (2.29, 2.34 Å). The ethylene was at an angle of 86° to the coordination plane of the metal which passed 0.15 Å below the midpoint of the C–C bond. The carbon atoms were equidistant from the metal [Pt–C 2.15(2) Å].

$$
\begin{array}{c}
\text{C} \cdots \quad \text{Cl} \\
\quad \diagdown \;\; \diagup \\
\| \longrightarrow \text{Pt} \!\!-\!\! \text{Cl} \\
\quad \diagup \;\; \diagup \\
\text{C} \cdots \quad \text{Cl}
\end{array}
$$

The second structure determination, by neutron diffraction, reported by Hamilton[530] showed the ethylenic hydrogens [C–H 1.10(5) Å] to be coplanar and bent away from the metal. Other parameters (e.g., the Pt–C and Pt–Cl bond lengths) are in agreement with the X-ray determination, but there is a discrepancy in the ethylenic C–C bond lengths reported—1.44(4) Å (Black et al.) and 1.354(15) Å (Hamilton). The estimated standard deviations suggest this bond length is probably in the range of 1.32–1.40 Å, but this does not answer the question of whether there is a significant increase in the C–C bond

---

† A number of workers have interpreted their results on the coupling of olefinic protons with ^{195}Pt in olefinPtCl$_3^-$ and similar complexes as also due to such asymmetric bonding.[521–524] However, too little is really known about the effects of various factors on J_{PtH} for this to be regarded as reliably established.[525]

length of ethylene (normally 1.34 Å) on coordination, as has been implied by other considerations below.

A necessary consequence of the Dewar–Chatt–Duncanson picture of the bonding in olefin complexes is that by losing electron density from the bonding π orbitals and having it returned into the antibonding π^* orbitals, the olefinic C–C bond order will be decreased. This effect should be noticeable in a number of ways; the C–C bond length should increase, while the force constant and $\nu_{(C=C)}$ should decrease. Furthermore, the magnitudes of these changes should be proportional to the strength of the metal–olefin bond, as defined by the amount of $\pi \rightarrow dsp^2$ and $\pi^* \leftarrow d(p)$ interaction. The decrease of C–C bond order toward that of a normal C–C single bond might also be expected to have other indirect effects, for example on the olefinic hydrogens, and it is found that in their proton NMR spectra these hydrogens have often experienced an upfield shift compared to that of the free olefin.[531] However, this phenomenon is not universal and may well be due to other causes[522, 532] (see also Cook and Wan[533]).

Unfortunately insufficient accurate data is yet available to draw any quantitative conclusions about the coordinated C=C bond length from X-ray studies. The more readily accessible and less reactive olefin complexes are those of the heavier metals and here it becomes difficult to locate the carbon atoms sufficiently accurately in the presence of the metal atom of greater scattering power. Present indications are that the C=C bond length is increased on coordination, but the exact degree is not known. It may also be expected to vary considerably from one type of complex to another.

Perhaps the most widely used criterion of the strengths of metal–olefin interactions is the change of $\nu_{(C=C)}$ on coordination, $\Delta\nu_{(C=C)}$. Where the band is not weak owing to a symmetry-forbidden transition, and when it is not obscured by other bands, the coordinated C=C bond usually appears as a medium strength band in the infrared at 1500–1530 cm^{-1} in Pt(II) complexes, compared to a range of 1620–1680 cm^{-1} for the free olefin.[513, 521, 534–540]

However, Grogan and Nakamoto have pointed out that $\nu_{(C=C)}$ is only a qualitative measure of the strength of coordination since this vibration is coupled to other modes, for example, the $\overset{\cdot}{C}H_2$ in-plane scissoring mode.[534, 535] They suggest that a better criterion is the force constant for the M–C$_2$H$_4$ stretching mode. They observed $\nu_{(M-C_2H_4)}$ at 408 and 427 cm^{-1} for [C$_2$H$_4$PtCl$_2$]$_2$ and [C$_2$H$_4$PdCl$_2$]$_2$, respectively, and have calculated $k_{PtC_2H_4}$ as 2.25, and $k_{PdC_2H_4}$ as 2.17 mdynes/Å. The former value is very close to that for K[C$_2$H$_4$PtCl$_3$] and that for the palladium complex is similar to that for the Pd–NH$_3$ bond. However, there still appears to be disagreement over the assignment of the bands caused by M–C$_2$H$_4$ vibrations.[540]

A $\Delta\nu_{(C=C)}$ of around 120 cm^{-1} is common on bonding to Pt(II)[484, 534]; $\Delta\nu_{(C=C)}$ for bonding to Pd(II) is less, around 100 cm^{-1} [536, 537]; while for Ag(I)

TABLE III-1

Changes in Olefin Stretching Frequencies upon Coordination

$\Delta\nu_{(C=C)}$ in cm^{-1}

Complex	C$_2$H$_4$	MeCH=CH$_2$	EtCH=CH$_2$	cis-MeCH=CHMe	trans-MeCH=CHMe	Me$_2$C=CMe	CH$_2$=CMeEt
(olefin)$_2$AgBF$_4$[a]	37	53	55	59	58	70	63
K[olefinPtCl$_3$][b]	97	144	—	167	159	—	—

[a] From Quinn et al.[532]
[b] From Grogan and Nakamoto.[534]

it is only about 60 cm^{-1}.[532] These are only very approximate values and considerable differences in $\Delta\nu_{(C=C)}$ for one metal atom complexed to different olefins are observed.[534] This may well be due to different degrees of coupling between $\nu_{(C=C)}$ and other modes, as well as to the influence of the olefinic substituent.

In the absence of other, more reliable data, however, the $\Delta\nu_{(C=C)}$ for some (olefin)$_2$Ag$^+$BF$_4^-$ by Quinn and his co-workers and for K[olefinPtCl$_3$] by Grogan and Nakamoto are presented in Table III-1. No such comparable series has yet been reported for olefin–palladium complexes, but it is anticipated that they will show similar trends to those for Pt(II) and Ag(I), namely that $\Delta\nu_{(C=C)}$ increases with increasing methyl substitution at the double bond, implying a stronger bond for the more heavily methylated olefins.

Quinn and his co-workers also found an inverse linear correlation between the ionization potential (I.P.) of the olefin and $\Delta\nu_{(C=C)}$, implying that the more easily the olefin lost a π electron (low I.P.) the stronger was the bond to silver, and a linear correlation between the enthalpy of dissociation for

$$\frac{1}{n}(\text{olefin})_n\text{AgBF}_4 \rightleftharpoons \text{olefin} + \frac{1}{n}\text{AgBF}_4$$

and $\Delta\nu_{(C=C)}$.[520, 532] However, this cannot be directly correlated with the stability constants, since the entropy factor also enters into the latter. As a result, although it seems probable that increased alkylation of a double bond strengthens the overall olefin–metal bond for Pt(II) and Ag(I) this is not necessarily reflected in the equilibrium constants, as has been shown by Venanzi and his co-workers.[541] They studied the equilibria,

$$\text{olefin} + \text{PtX}_4^{2-} \rightleftharpoons \text{olefinPtX}_3^- + \text{X}^-$$

in aqueous solution for a series of allylammonium olefins (CH$_2$=CHCH$_2$ NH$_3^+$, CH$_2$=CHCMeHNH$_3^+$, trans-MeCH=CHCH$_2$NH$_2$Et$^+$) and found that K_{eqm} decreased sharply on adding an alkyl group anywhere to the allylammonium ion and particularly so if it was on the double bond. Similarly, although ΔH values and $\Delta\nu_{(C=C)}$ indicated that for (EtNH$_2$CH$_2$CH=CH$_2$) PtX$_3^-$ the stronger bond was formed when X = Br, the more stable complex was that where X = Cl.[542]

Hartley and Venanzi[543] have compared $\Delta\nu_{(C=C)}$ and stability constants for a number of allylammonium complexes of Pt(II) and Ag(I) and found that the latter metal always formed the weaker complex.

Cramer[544] has published data on the stabilities of some rhodium(I) complexes

$$\text{olefin} + \text{acacRh(C}_2\text{H}_4)_2 \xrightarrow{\;\;K\;\;} \text{acacRh(C}_2\text{H}_4)(\text{olefin}) + \text{C}_2\text{H}_4$$

and has derived approximate enthalpy and entropy data. The $\Delta H°$ values suggested that increased alkylation of the olefin, for Rh(I), decreased the Rh–

olefin bond strength, whereas halo substitution, in particular by fluorine, had the opposite effect.

In summary, however, it appears at the moment that in the absence of force-constant data on the M-olefin stretching modes, really accurate and extensive thermochemical data, or X-ray results on coordinated $C=C$ bond lengths, that the most useful indicator of the strength of a metal–olefin bond is $\Delta\nu_{(C=C)}$.

From a more practical viewpoint though, a knowledge of the equilibrium constants is of great importance, particularly in understanding the mechanisms of reactions of olefins. Table III-2 summarizes the available data for the reaction

$$\text{olefin} + \text{PdCl}_4{}^{2-} \rightleftharpoons \text{olefinPdCl}_3{}^- + \text{Cl}^-$$

TABLE III-2

Equilibrium Constants for Olefin–Palladium(II) Complexes

$$\text{olefin} + \text{PdCl}_4{}^{2-} \underset{\longleftarrow}{\overset{K_1}{\longrightarrow}} \text{olefinPdCl}_3{}^- + \text{Cl}^-$$

Olefin	K_1			
Ethylene	15.2 ± 0.7[a]	16.9[b]	13.1 ± 0.6[c]	17.4 ± 0.4[d]
Propylene	7.9 ± 0.5	7.6	—	14.5 ± 1.5
1-Butene	11.4 ± 0.7	14.3	—	11.2 ± 1.1
cis-2-Butene	—	—	—	8.7 ± 0.5
trans-2-Butene	—	—	—	4.5 ± 0.5

[a] At 20°. By Pestrikov et al.[501]
[b] At 20°. By Pestrikov et al.[502]
[c] At 25°. By Pestrikov et al.[501]
[d] At 25°. By Henry.[503]

The results of Pestrikov et al.[501, 502] and Henry[503] that are quoted are not strictly comparable to each other since they were measured under different conditions (at an ionic strength of 4 gm-ions/liter at 20°,[501] and $\mu = 2.0$, [Cl$^-$] 0.1 to 0.3 M and [H$^+$] 0.05 to 0.2 M at 25°[503]) and by different methods. However, the only serious discrepancy is in the values for propylene. The most significant conclusion is that, as seen before, alkylation of the olefin reduces K_1 quite drastically; Henry's results, in fact, show a steady decrease on increasing alkylation, and also emphasize the lower stability of the trans- as compared to the cis-2-butene complex.

Pestrikov et al.[501] have also determined the enthalpies and entropies for this process as, $\Delta H_1 = -1.5$, 0, and 0 kcal/mole and $\Delta S_1 = 0$, 4, and 5 e.u., for ethylene, propylene, and 1-butene, respectively. They also determined K_2 and the corresponding enthalpies and entropies for:

$$\text{olefin} + \text{PdCl}_4{}^{2-} + \text{H}_2\text{O} \underset{\longleftarrow}{\overset{K_2}{\longrightarrow}} \text{olefinPdCl}_2(\text{H}_2\text{O}) + 2\text{Cl}^-$$

These are presented in Table III-3.

TABLE III-3

Equilibrium Constants, Enthalpies and Entropies[a]

$$\text{olefin} + PdCl_4{}^{2-} + H_2O \underset{\longleftarrow}{\overset{K_2}{\longrightarrow}} \text{olefinPdCl}_2(H_2O) + 2Cl^-$$

Olefin	K_2 (mole/liter)	ΔH_2 (kcal/mole)	ΔS_2 (e.u.)
Ethylene	4.3 ± 0.9	-11.5	-36
Propylene	4.6 ± 0.7	0	3
1-Butene	3.4 ± 1.1	0	2.5

[a] At 20°. From Pestrikov et al.[501]

Equilibrium constants, $K_1' = 3.8 \pm 1$, and $K_2' = 1150 \pm 300$ mole/liter at 25° in acetic acid have also been determined by Moiseev et al. for the reactions[545]:

$$C_2H_4 + Na_2Pd(OAc)_4 \underset{\longleftarrow}{\overset{K_1'}{\longrightarrow}} (C_2H_4)Pd(OAc)_2 + 2NaOAc$$

$$2C_2H_4 + 2Na_2Pd(OAc)_4 \underset{\longleftarrow}{\overset{K_2'}{\longrightarrow}} (C_2H_4)_2Pd_2(OAc)_4 + 4NaOAc$$

No quantitative or even semiquantitative data of this kind are available for the metal(0)–olefin complexes. Orders of "stability" based on the isolated products from replacement reactions of the type

$$(Ph_3P)_2ML + L' \rightleftharpoons (Ph_3P)_2ML' + L$$

has been quoted by Chatt et al.[510] and Takahashi and Hagihara[62] for M = Pt. For example, L', diphenylacetylene, displaced L = 4,4'-dinitrostilbene and acenaphthalene, but not dimethyl maleate. L', maleic anhydride, displaced dimethyl fumarate, which, in turn, displaced acrylonitrile and methyl acrylate.

Very little spectroscopic data for the M(0) complexes have been reported and, in particular, no bands which are easily ascribed to coordinated C=C. This has led some workers, unnecessarily in the view of this author, to propose σ-bonded structures such as (III-5a) for the complex (III-5b).[399]

(III-5a)   (III-5b)

Five X-ray structure determinations on olefin–nickel(0) and olefin–platinum(0) but none on olefin–palladium(0) complexes have been reported. Two of these are of ethylenebis(triphenylphosphine)nickel.[546, 547]

$$\begin{matrix} Ph_3P \\ \\ Ph_3P \end{matrix} Ni— \Vert \begin{matrix} CH_2 \\ \\ CH_2 \end{matrix}$$

The two phosphorus atoms, the nickel and one carbon of the ethylene are co-planar; the second olefinic carbon is out of this plane and the C–C bond is at an angle of 12° to the plane. This may be due to nonbonded interactions of the olefinic carbons and the two phosphorus atoms.[546] The C–C bond length is given as 1.41(3)[546] and 1.46(2) Å[547] by the two groups of workers.

The structure of a tetracyanoethylene–nickel(0) complex has been reported by Stalick and Ibers.[547a] The coordination about Ni is approximately trigonal, the coordinated C=C bond (of length 1.476(5) Å) making a dihedral angle of 24° with the plane of the nickel and the isonitrile ligands. Very similar structures have been found for π-fumaronitrile- and π-tetracyanoethylenebis-(triphenylphosphine)platinum.[548]

McGinnety and Ibers reported the structure of an analogous iridium(I) complex which has approximately trigonal bipyramidal geometry if the TCNE is regarded as a monodentate axial ligand. The olefinic C–C bond length of 1.507(15) Å compares with 1.399 Å for TCNE itself.[549] In each complex the cyanides are no longer coplanar with the ethylenic carbon atoms and are bent away from the metal.

In the nickel and platinum complexes the metal–phosphine bond lengths appear to be rather short; this indicates the presence of d_π–d_π bonding.

Considerable attention has been given to the structures of acetylene complexes of platinum(II) and platinum(0), as well as to some nickel(0) and palladium(0) complexes. Accounts of crystal structure determinations of the platinum(II) and the platinum(0) acetylene complexes [p-toluidine-PtCl$_2$(Bu^{t}C$_2$Bu^t)] (III-1) and (Ph$_3$P)$_2$Pt(PhC$_2$Ph) (III-2) have been published by Davies et al.[492] and Glanville et al.[493] In the former the acetylene behaves as a monodentate ligand perpendicular to the coordination plane of the Pt(II). In the latter, as in the olefin M(0) complexes, the two phosphorus atoms, the metal, and one acetylenic carbon atom are coplanar, the coordinated C≡C making an angle of 14° with the plane. The C≡C bond length in the Pt(0) complex was reported as 1.32(9) Å and that in the Pt(II) complex as 1.24(2) Å, but the large e.s.d.'s make these values rather uncertain. They appear to be larger than the C≡C bond length in acetylene itself (1.21 Å).

Infrared data was used by Chatt and his co-workers to infer these structures before the results of the crystal structure determinations were available, and infrared and NMR data have subsequently been used to extend our knowledge of the structures of, and the bonding in, these complexes. Chatt *et al.* found that they were only able to isolate acetylene complexes of the types $K[acPtCl_3]$, $[acPtCl_2]_2$, or *trans*-$[acPtCl_2(amine)]$ for acetylenes (ac) having at least one bulky electron-releasing substituent such as Bu^t, $Me_2C(OH)$, or $Me_2C(OMe)$.[488, 490] The acetylenic $\nu_{(C≡C)}$ normally is seen in the infrared (or in the Raman spectrum for symmetrical acetylenes) in the region 2190–2260 cm^{-1}, and a decrease of about 200–240 cm^{-1} is observed on complexing to Pt(II) (Table III-4). Only one value for a palladium complex, $[Bu^tC_2Bu^tPdCl_2]_2$, is known and there $\nu_{(C≡C)}$ is not known for the parent acetylene. By comparison with the platinum complex, $\Delta\nu_{(C≡C)}$ is clearly less by a substantial amount for the Pd(II) complex, in agreement with the results for the olefin complexes.

TABLE III-4

$\nu_{(C≡C)}$ and $\Delta\nu_{(C≡C)}$ for Acetylene–Pt(II) and Acetylene–Pd(II) Complexes[a]

Complex	$\nu_{(C≡C)}$	$\Delta\nu_{(C≡C)}$
ac[b]	2224, 2248	—
$K[acPtCl_3]$[b]	2010	226
$K[acPtBr_3]$[b]	2015	221
ac[c]	2228, 2273	—
$K[acPtCl_3]$[c]	2008	—
ac[d]	2224, 2259	—
$K[acPtCl_3]$[d]	2009	232
trans-$[acPtCl_2C_5H_{11}N]$[d]	2041	200
$[Bu^tC_2Bu^tPtCl_2]_2$	2005, 2023	—
$[Bu^tC_2Bu^tPdCl_2]_2$	2050	—
$Na[Bu^tC_2MePtCl_3]$	2028	—

[a] From Chatt *et al.*[489, 490] and Hosokawa *et al.*[495].
[b] ac = $Me_2C(OH)C_2C(OH)Me_2$.
[c] ac = $Bu^tC_2CMe_2OH$.
[d] ac = $Bu^tC_2CMePhOH$.

The acetylene complexes of platinum(0), first made by Chatt, Rowe, and Williams[491] show much larger $\Delta\nu_{(C≡C)}$. The values of $\nu_{(C≡C)}$ and $\Delta\nu_{(C≡C)}$ for Ni(0), Pd(0), and Pt(0) complexes reported are listed in Table III-5. In the case of diphenylacetylene, both a Pt(0) and, very recently, a Pt(II) complex [(PhC$_2$Ph(Pt(acac)Cl] have been prepared.[549a] In the latter case a band at 1990 cm^{-1} was ascribed to $\nu_{(C\equiv C)}$, representing a change on coordination, $\Delta\nu_{(C\equiv C)}$, of 233 cm^{-1} for Pt(II).

TABLE III-5

$\nu_{(C\equiv C)}$ and $\Delta\nu_{(C\equiv C)}$ for Acetylene–Ni(0), Acetylene–Pd(0), and Acetylene–Pt(0) Complexes[a]

Complex	$\nu_{(C\equiv C)}$	$\Delta\nu_{(C\equiv C)}$[b]
(Ph$_3$P)$_2$Pt(PhC$_2$Ph)	1740, 1768	469
(Ph$_3$P)$_2$Pt(PhC$_2$Me)	1756	478
(Ph$_3$P)$_2$Pt(Me$_2$C(OH)C$_2$H)	1710 or 1681	397 or 433
(Ph$_3$P)$_2$Pt(MeOOCC$_2$COOMe)	1765(sh), 1782	474
(Ph$_3$P)$_2$Pt(CF$_3$C$_2$CF$_3$)	1775	525
(PhMe$_2$P)$_2$Pt(CF$_3$C$_2$CF$_3$)	1767	533
(Bu$^n_3$P)$_2$Pt(CF$_3$C$_2$CF$_3$)	1758	542
(Ph$_3$P)$_2$Pd(MeOOCC$_2$COOMe)	1830, 1845	410
(Ph$_3$P)$_2$Pd(CF$_3$C$_2$CF$_3$)	1811, 1838	475
(PhMe$_2$P)$_2$Pd(CF$_3$C$_2$CF$_3$)	1800, 1837	482
(Bu$^n_3$P)$_2$Pd(CF$_3$C$_2$CF$_3$)	1795, 1837	484
(Ph$_3$P)$_2$Ni(PhC$_2$Ph)	1800	423
(Ph$_3$P)$_2$Ni(PhC$_2$Me)	1795	439
(Ph$_3$P)$_2$Ni(CF$_3$C$_2$CF$_3$)	1790	510

[a] From Maitlis et al.[45] and Roundhill et al.[349]

[b] In some cases $\nu_{(C\equiv C)}$, for reasons which are not understood, appears as a doublet even in solution, and $\Delta\nu_{(C\equiv C)}$ was calculated using the mean of the two values.

An interesting consequence is that for (Ph$_3$P)$_2$M(CF$_3$C$_2$CF$_3$), the values of $\Delta\nu_{(C\equiv C)}$ decrease in the order, Pt > Ni > Pd. Since the complexes only differ in the central atom, it is reasonable to suppose that the metal–acetylene bond strengths also decrease in the same order.[45] The other ligands also play a role. Thus, Maitlis and co-workers[45] have shown that for the series (R$_3$P)$_2$M(CF$_3$C$_2$F$_3$) $\Delta\nu_{(C\equiv C)}$ increases in the order, Ph$_3$P < Ph$_2$MeP < Bu$_3^n$P, for both M = Pd and Pt. In other words, the phosphine which is the best σ donor (most highly alkylated) also gives the strongest M–acetylene bond.

In both the acetylene–Pt(II) and the acetylene–M(0) complexes $\nu_{(C\equiv C)}$, possibly modified by other interactions, appears as a medium intensity band in a region of the spectrum where few other functional groups absorb. Hence there is little ambiguity about the identification of these bands. Quite analogous results have been obtained for acetylene complexes of other metals. For hexafluoro-2-butyne, for example, a range of $\Delta\nu_{(C\equiv C)}$ from 381 [for C_5H_5Mn $(CO)_2(CF_3C_2CF_3)$] to 541 cm^{-1} [for $(Bu_3{}^nP)_2Pt(CF_3C_2CF_3)$] is known.[45]

These workers have also shown that the ^1H and ^{19}F NMR spectra of $(Ph_3P)_2M(PhC_2Me)$ (M = Ni, Pt) and $(R_3P)_2M(CF_3C_2CF_3)$ (R = Ph, M = Ni, Pd, Pt; R = Bu^n, M = Pt) are consistent only with planar or near planar structures such as (III-2). In other words, the structures in the solid state and in solution are the same.

Allen and Cook[94] have studied the kinetics and equilibria for

$$(Ph_3P)_2Ptac + ac' \rightleftharpoons (Ph_3P)_2Ptac' + ac$$

and have shown that the rate of exchange was proportional to the concentration of $[(Ph_3P)_2Ptac]$ but independent of $[ac']$. They concluded that the reaction proceeded via an intermediate, $(Ph_3P)_2Pt$, of significant stability, and that electron-withdrawing substituents on the acetylenes also stabilized the complexes. The latter conclusion had previously also been arrived at by Chatt et al.[491]

Little data is available on acetylene–palladium(0) complexes other than that described above. These compounds were difficult to isolate and also appeared rather easily autoxidized, particularly in solution. Judging by the $\Delta\nu_{(C\equiv C)}$ values (Table III-5) they also had significantly weaker metal–acetylene bonds than the platinum analogs.

Olefins and acetylenes bonded to transition metals frequently undergo exchange reactions if the complex is sufficiently labile. In addition, there is good evidence that in some cases at least, the barriers to rotation about the M–olefin or M–acetylene axis are small and rotation can occur. This phenomenon was first demonstrated by Cramer for Rh(I); he found that the ethylenes in the labile complex $[acacRh(C_2H_4)_2]$ exchanged very readily with ethylene. In contrast, no exchange was observed for π-$C_5H_5Rh(C_2H_4)_2$, but a rotation of the ethylene about the C_2H_4–Rh axis was shown to occur and the activation energy barrier for rotation was estimated by an NMR method as 15 kcal/mole.[550–552]

Similar results have been obtained by Lewis and his co-workers for some platinum–olefin complexes, (III-6),[524] who estimated the free energies of activation for rotation of the olefin (ethylene, propylene, cis- and trans-2-butene, and tetramethylethylene), as between 10.9 and 15.8 kcal/mole. From their

(III-6)

NMR data they were able to exclude all intermolecular exchange processes. They suggested that, in addition to steric effects, part of the energy barrier to rotation was electronic.

D. BONDING IN OLEFIN AND ACETYLENE COMPLEXES

As described in Section A, the best model for the bonding in olefin– and acetylene–M(II) complexes is that devised by Dewar, Chatt, and Duncanson.

This model can also be extended to the M(0) complexes. However, a number of authors have suggested that they should be regarded as square planar complexes of M(II), with the unsaturated ligand occupying two coordination sites (cis) in a three-membered metalacyclopropane or metalacyclopropene ring,

$$\begin{array}{cc} \mathrm{L}\diagdown\quad\diagup\mathrm{CR}_2 \\ \quad\mathrm{M}\diagdown|\quad \\ \mathrm{L}\diagup\quad\diagdown\mathrm{CR}_2 \end{array} \qquad \begin{array}{cc} \mathrm{L}\diagdown\quad\diagup\mathrm{CR} \\ \quad\mathrm{M}\diagup|\!| \\ \mathrm{L}\diagup\quad\diagdown\mathrm{CR} \end{array}$$

Arguments for this formulation are based on some crystal structure determinations, the rigidity of the metal(0) complexes and, in the infrared spectra of the acetylene complexes, on $\nu_{(C\equiv C)}$.

A characteristic of all these complexes is that the unsaturated ligand is coplanar, or nearly so, with the metal and the two ligand atoms, despite some nonbonding interactions.[546]

The olefinic C–Ni bond lengths in $(Bu^tNC)_2Ni(TCNE)$[547a] of 1.954(4) Å are similar to those found for some Ni^{II}–C σ bonds, for example in π-C_5H_5 $NiC_6H_5(PPh_3)$ and π-$C_5H_5NiC_6F_5(PPh_3)$ [1.919(13) and 1.914(14) Å, respectively].[354, 355] It is also shorter than many other Ni–olefinic carbon bond lengths which vary from 1.90 to 2.13 Å for both Ni(0) and Ni(II).[179, 255, 553–556] There is also evidence for lengthening of the coordinated C–C bond, but the value found in $(Ph_3P)_2Ni(C_2H_4)$, 1.41(3) Å, is not nearly as great as a normal C–C bond between two sp^3 carbons (1.54 Å).

The Pt–C bond length found for the tetracyanoethylene complex, $[(Ph_3P)_2Pt(C_2(CN)_4)]$, 2.11(3) Å,[548] is in between the range found for normal Pt(II)–olefinic carbon bonds, 2.1–2.25 Å,[199, 200, 407, 529, 530, 557] and that for normal Pt(II)–C σ bonds, 1.97–2.10 Å.[196, 199, 200, 446, 448]

In contrast, Pt–C bonds in $(Ph_3P)_2Pt(PhC_2Ph)$,[493] (III-2), are shorter, 2.01, 2.06 Å. In this complex the H_5C_6—$C\equiv C$—C_6H_5 is no longer linear; both phenyl groups are bent away from the metal and the angle CCPh is 140°. This is consistent with a change in hybridization of the acetylenic carbons toward sp^2.† In the corresponding acetylene–Pt(II) complex, [p-toluidinePtCl$_2$(Bu^{t}C≡CBu^t)], **(III-1)**, the angle is about 165°.

As indicated in Tables III-4 and III-5, there is also a substantial difference between $\Delta\nu_{(C≡C)}$ for acetylene–Pt(II) (200–240 cm^{-1}) and acetylene–Pt(0) complexes (400–530 cm^{-1}). In the latter case the decrease in the C–C stretching frequency is enough to bring it down to the values for $\nu_{(C=C)}$, in the corresponding olefins.[558]

A consideration of the bonding involved here, however, makes it clear that the metalacyclopropene and metalacyclopropane formulation is merely a different way of writing the structures.

These complexes are best considered as complexes of M(0), with trigonal coordination about the metal and the unsaturated ligand occupying one coordination site. The bonding is then similar to that in the M(II) complexes, the two components involving overlap of the olefin (or acetylene) π orbital with a vacant hybrid orbital of the metal (usually taken to be sp^2, though other combinations are possible), and of the π^* orbital with a filled metal orbital. In principle, the unsaturated ligand could be either perpendicular to the coordination plane of the metal or coplanar with it.

The two bonding situations for olefins are illustrated in Fig. III-2. The C_2 or C_{2v} axis is defined as the z direction, and the x axis is in the plane of the metal and the other ligands L. Only the coplanar arrangement [Fig. III-2(b)] is

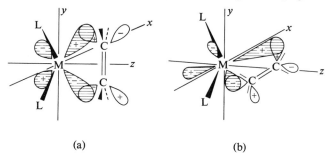

(a)   (b)

FIG. III-2. π-Bond formation in L$_2$M(0)–olefin complexes. (a) The olefin perpendicular to the coordination plane (xz); bonding between π^* and d$_{yz}$. (b) The olefin coplanar with the coordination plane (xz); bonding between π^* and d$_{xz}$.

found, and this implies the presence of strong electronic forces which keep the molecule coplanar (or nearly so). One explanation which has been put forward for this is that in the coplanar arrangement the d$_{xy}$ orbital (filled) can interact with the filled orbitals of the ligands, L. This would destabilize d$_{xy}$ if it were not also involved in bonding with the π^* orbitals. In the perpendicular arrangement [Fig. III-2(a)], the d$_{yz}$ orbital is not involved in σ-bonding to L.

† Recent determinations of $J_{13_{CH}}$ for (Ph$_3$P)$_2$Pt(CH≡CH) and (Ph$_3$P)$_2$Pt(CH$_2$=CH$_2$) also bear this out.[533]

The bonding in the planar acetylene complexes, $L_2M(0)$–acetylene (Fig. III-3), is similar to that described for the olefin complexes in Fig. III-2(b). In addition to the primary bonding involving the acetylene π_z and $\pi_z{}^*$ [Figs. III-3(a) and (b), orbitals in the xz plane], there is also the possibility of bonding involving the orthogonal orbitals π_y and $\pi_y{}^*$ [Figs. III-3(c) and (d), orbitals in the xy plane]. In this case, the metal orbitals p_y and d_{xy} have the correct symmetry for overlap with π_y and $\pi_y{}^*$, respectively.

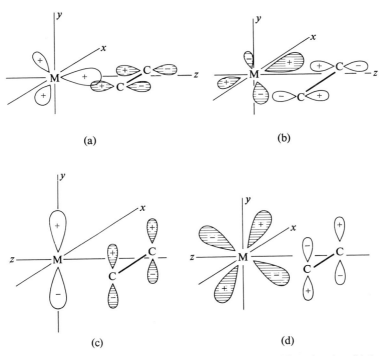

FIG. III-3. Bonding in $L_2M(0)$–acetylene complexes. Overlap of (a) sp^2 and π_z (b) d_{xz} and $\pi_z{}^*$; (c) p_y and π_y; and (d) d_{xy} and $\pi_y{}^*$.

It is not clear, however, to what extent bonding involving π_y and $\pi_y{}^*$ occurs, since overlap is probably small. Similar arguments apply to the M(II)–acetylene complexes.

The bonding situation in acetylene–M(0) and acetylene–M(II) complexes has been discussed further by Greaves, Lock, and Maitlis by considering the energies of the combining orbitals.[45] Unfortunately the energies of the combining fragments are not known and approximations have to be used. Despite this, useful information can be obtained.

If the energy of the ion and the separated electron is defined as zero, the bonding s, p, and d levels in palladium can be inferred from spectroscopic data to be approximately: Pd^0, $-24,000$ to $-67,000$ cm^{-1}; Pd^+, $-84,000$ to $-157,000$ cm^{-1}; and Pd^{2+}, $-148,000$ to $-266,000$ cm^{-1}.[559] Very similar values can be obtained for platinum. The π and π^* levels in acetylene are then at $-92,000$ and $-33,000$ cm^{-1}.†

The significant conclusion from these figures is that an increase in the formal charge on the metal results in a lowering of the energies of the orbitals involved in bonding and that the energy of an uncharged metal atom is in the same range as the π and π^* orbitals of acetylene. Accurate estimates of the energies of $L_2M(0)$ or $L'_3M(II)$ cannot be made, but it seems very likely that the former will not be too far removed from M^0 and that the energies of the latter will be lower, though not as low as for Pd^{2+}.

Using this simple picture, a number of different bonding situations between the acetylene and the metal can be investigated. Five different situations can reasonably be postulated to occur [Fig. III-4 (A–E)].

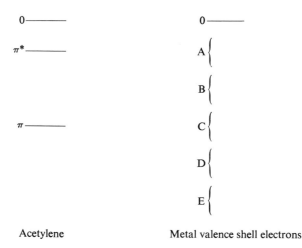

Acetylene   Metal valence shell electrons

FIG. III-4. Relative energies of acetylene and metal valence shell orbitals.

In *case A*, the metal orbitals lie as high as or even above the π^* orbitals of the acetylene, and well above the π orbitals. There is, therefore, very little interaction between the latter and the metal orbitals. However, the bonding and antibonding orbitals formed by combination of the metal and the acetylene π^* orbitals will both have large amounts of metal and acetylene character and hence "back-donation" of electron density from the metal to the acetylene will

† The first excited state of acetylene is at ca. $-50,000$ cm^{-1} and is trans bent. However, calculations have shown a cis-bent state (i.e., the same geometry as in the complex) to be at ca. $-33,000$ cm^{-1}.[560]

be large. The acetylene will resemble $C_2H_2^-$ here and will not be greatly distorted by comparison with C_2H_2 itself.[561] This situation can be depicted as

A

The orbitals of the metal in *case B* lie between π and π^* and all combinations of π and π^* with the metal orbitals will have substantial acetylene and metal character. As a result, there will be significant transfer of electron density from acetylene π orbitals into the area between the metal and the acetylene and also significant back-bonding from the metal into π^* orbitals of the acetylene. The metal–acetylene bond order is therefore high; and as electron density has effectively been transferred from the π to the π^* orbital of the acetylene, the latter now resembles the excited state of acetylene. The normally observed excited state is trans bent, but calculations show that a cis bent state is of lower energy.[560] Blizzard and Santry have shown from symmetry arguments that trans bending of the acetylene results in a considerable weakening of the metal–acetylene interactions in the complexes, and trans-bent complexes are not observed.[561] Howard and King have also shown that the spectrum of acetylene in the excited trans-bent state is consistent with a CCH angle of 120°, a $C\equiv C$ bond length of 1.39 Å, and a decrease in $\nu_{(C\equiv C)}$ of ca. 900 cm^{-1}.[560] Assuming the parameters for the unknown cis-bent state to be similar, and that these values are not greatly changed by substitution, it appears that this situation is beginning to approximate to the state of the acetylenes in the M(0) complexes,

B

In *case C* the acetylenic π orbitals are of about the same energy as the metal valence orbitals and both bonding and antibonding combinations will have considerable acetylene and metal character. However, the energy difference between metal orbitals and the π^* orbitals will minimize interaction. Hence the bonding will be largely by transfer of electron density from the acetylene to the metal.

In *case D* this trend is continued, but now the bonding combination of the π orbital with the metal orbitals will be largely metal in character, whereas the antibonding combination will be largely acetylenic. This then corresponds to the transfer of more electron density from the acetylene π orbitals to the metal,

$$M \xleftarrow{\delta-} \begin{array}{c} | \\ C^{\delta+} \\ \| \\ \| \\ C^{\delta+} \\ | \end{array}$$

D

Since interaction of the π^* orbitals with the metal orbitals will be small, the M–acetylene bond order will be less than in case B, and the C–C bond order, although less than in acetylene, will be more than in case B. This case then corresponds approximately to the M(II)–acetylene complexes.

The orbitals of the metal and the acetylene in *case E* are so disparate that no bonding now occurs.

As cases A–D represent metals with increasing positive charge, any factor which increases the charge on the metal is predicted to make the dative acetylene-to-metal bonding more important. The range of complexes known does not cover the whole range A–D, but probably only from somewhere between B and C to nearly D. Within this range increase in positive charge on the metal, e.g., M(0) to M(II), will lower the energy levels and make the dative bonding ac $\rightarrow$ M more important, while a decrease in positive charge, e.g., by a change from $(Ph_3P)_2M$ to $(Bu^n_3P)_2M$, will increase the back-bonding. The over-all effect in the latter case is to strengthen the bond.

In the limiting case A, the carbon atoms will carry a small net negative charge, and therefore complexes approaching this type will tend to be stabilized by electron-withdrawing substituents on the acetylene (CF_3, COOMe). The highest $\Delta\nu_{(C\equiv C)}$ observed for a series of $(Ph_3P)_2Pt(C_2R_2)$ complexes is for R = CF_3 (Table III-5).

On the other hand, in complexes approaching case D the acetylenic carbons will bear a small net positive charge. These should then be stabilized by electron-releasing groups. This is complicated by steric factors, but Chatt *et al.* observed that acetylene–Pt(II) complexes were only isolable when bulky, electron-releasing substituents were present.

Another consequence of the charges present on the carbon atoms in the limiting cases is that there may well be a correlation between the ease of attack of nucleophilic reagents (X^-) and case D, and of electrophilic reagents (X^+) and case A.

Since the π and π^* levels of ethylene, at $-84,000$ and $-23,200$ cm^{-1}, respectively,[562] are reasonably similar to the π and π^* levels for acetylene, it is apparent that a very similar argument to the above can be developed here too. In agreement with those conclusions, no simple olefin complexes with strongly electron-withdrawing substituents (F, CN, COOMe, CF$_3$†) are known for Pt(II), whereas Pt(0) [or Pd(0)] complexes of tetrafluoroethylene, tetracyanoethylene, maleic anhydride, etc., are well known and quite stable.

Perhaps the most important theme running throughout palladium(II) chemistry is the facility with which coordinated olefins undergo nucleophilic attack, or insert into Pd–X bonds. The ease with which these reactions proceed may well, in part, be determined by the importance of structures such as

$$Pd^{\delta-} \longleftarrow \begin{array}{c} \diagup \diagup \\ C^{\delta+} \\ \| \\ C^{\delta+} \\ \diagup \diagdown \end{array}$$

No reports of simple nucleophilic attack on olefin- or acetylene-M(0) complexes have appeared but recently a number of workers have reported that some Pt(0) complexes do react with acid (Section E,4). This agrees with the prediction that olefin–M(0) and acetylene–M(0) complexes should be susceptible to attack by electrophiles.

As pointed out by Pannatoni et al.,[548] the representation of these complexes as M(0) complexes in molecular orbital terms is equivalent to the valence-bond description as metalacyclopropane or metalacyclopropene rings. A convenient way to describe the bonding which avoids ambiguity is

$$M \begin{array}{c} \diagup C \diagup \\ \| \\ \diagdown C \diagdown \end{array} \quad or \quad M \begin{array}{c} \diagup C \diagup \\ \| \\ \diagdown C \diagdown \end{array}$$

and this can be used whenever the bonding is specifically under consideration. For most purposes, however, the conventional representation,

$$M \!-\! \begin{array}{c} \diagup C \diagup \\ \| \\ \diagdown C \diagdown \end{array} \quad or \quad M \!-\! \begin{array}{c} | \\ C \\ \| \\ C \\ | \end{array}$$

will usually suffice.

The general conclusions can be summarized for acetylene or olefin (= un) complexes of M(0) and M(II) as follows:

(a) For (un)M(0) complexes there is expected to be (i) a large distortion from linearity (or planarity) of the unsaturated ligand; (ii) a considerable

† An apparent exception is the π-acetylene complex (Me$_2$PhAs)$_2$PtMe(Cl)CF$_3$C$_2$CF$_3$, but this is rather unusual in that it is 5-coordinate and also has very soft ligands present. It also undergoes a cis-insertion reaction very readily.[483a]

increase in the coordinated C–C bond length; (iii) an increase in stability caused by electron-withdrawing substituents on the unsaturated ligand or by electron-releasing ligands on the metal; and (iv) a tendency to undergo electrophilic attack.

(b) For (un)M(II) complexes there is expected to be (i) less distortion of the unsaturated ligand and a smaller increase in the C–C bond length than for (un)M(0); (ii) an increase in stability caused by electron-withdrawing ligands and a decrease in stability caused by electron-withdrawing substituents on the unsaturated ligand; and (iii) a tendency to undergo nucleophilic attack.†

Although these conclusions should apply to all metal–olefin and metal–acetylene complexes, complications will arise and considerable caution must be exercised in carrying out any extrapolation to metals in different oxidation states, with different stereochemistries, and with different numbers of d electrons in the valence shell. For example, although Rh(I) and Ir(I) are d^8 states, the chemistry of their complexes indicates that the properties are intermediate between those of Pt(0) and Pt(II) [or Pd(0) and Pd(II)]. To give just one example, Rh(I) forms the most stable complexes with fluorinated olefins and acetylenes, but from $\Delta\nu_{(C\equiv C)}$ in $(Ph_3P)_2Rh(CF_3C_2CF_3)Cl$, 383 cm^{-1}, it appears that the metal–acetylene bond is quite weak by comparison with the palladium complex, $(Ph_3P)_2Pd(CF_3C_2CF_3)$ ($\Delta\nu_{(C\equiv C)}$, 475 cm^{-1}).

E. REACTIONS OF MONOOLEFIN COMPLEXES

Although olefin complexes, particularly of Pd(II), are implicated in many catalytic processes, comparatively little work has been reported (through 1970) on the actual chemistry of these olefin complexes. Part of the reason for this is, no doubt, the very high reactivity of these complexes which makes them somewhat difficult to work with. In this regard the olefin–platinum complexes have been studied more and are, in general, rather less reactive. This is probably due in part to the general lesser lability of Pt(II) by comparison with Pd(II). The apparent greater strength of the Pt–olefin compared to the Pd–olefin bond may well be important as well.

The olefin (and acetylene) complexes of Ni(0), Pd(0), and Pt(0) have not been known as long and little is known of their chemistry.

1. Substitution of L in trans-(Olefin)MCl₂L

One of the earliest manifestations of the trans effect was for $[C_2H_4PtCl_3]^-$ and similar complexes, and the ready labilization of the ligand trans to the

---

† Some of these generalizations may not hold where unusual stereochemistries or other special features such as chelate rings are present.

ethylene is well established. Indeed, it has been said that for $K[C_2H_4PtCl_3]$ the *trans* chlorine behaves more as ionic Cl^- than as a coordinated ligand. Leden and Chatt showed that for

$$C_2H_4PtCl_3^- + L \rightleftharpoons \textit{trans-}C_2H_4PtCl_2L + Cl^-$$

the equilibrium was established in less than 2 minutes.[563] These authors estimated K_{eqm} as 3×10^{-3} (L = H_2O), $<3 \times 10^{-2}$ (F^-), 3.4 (Br^-), 120 (I^-), and $\sim 10^5$ (NH_3). Zeise's salt is stable in dilute HCl, but slowly reacts with water under neutral or alkaline conditions. In aqueous solution the species present is probably *trans-*$C_2H_4PtCl_2OH^-$ and the acid dissociation constant for the aquo complex was estimated to be about 10^{-5}.† In ethanol $[C_2H_4PtCl_2]_2$ exists principally as *trans-*$C_2H_4PtCl_2EtOH$, and K_{eqm} for

$$C_2H_4PtCl_3^- + EtOH \rightleftharpoons C_2H_4PtCl_2EtOH + Cl^-$$

has been shown to be 5×10^{-5} at $25°$.[564]

Chesseman *et al.*[565] have measured the extent of the trans effect in the exchange reaction

$$\textit{trans-}LPtCl_2(H^{14}NEt_2) + Et_2NH \rightleftharpoons \textit{trans-}LPtCl_2(HNEt_2) + Et_2^{14}NH$$

and determined that for different ligands, L, the effect decreased in the order $C_2H_4 \sim Pr^iCH{=}CHMe \sim Me_2C(OH)C{\equiv}CC(OH)Me_2 \gg Et_3Sb > Ph_3Sb > Me_3P > Et_3P > Pr^n_3P > Ph_3P > Et_3As > Ph_3As \gg Pr^n_2S$. It should be noted that the trans effects of olefins and acetylenes are similar and much greater than stibines, arsines, and phosphines.

This high trans effect makes it easy to substitute the trans ligand; if the entering ligand is an amine, stable complexes result. The amines most commonly used are ammonia,[563] pyridine, 4-chloropyridine, 4-methylpyridine, *p*-toluidine, piperidine, and *p*-chloraniline,[484, 489, 490] since they give crystalline derivatives, as do the pyridine *N*-oxides.[521] However, sometimes the olefin ligand is also displaced, to give either L_2PtCl_2 or $[LPtCl_2]_2$.[566]

Complexes of this type are hardly known for palladium; *p*-toluidine reacted with [cyclohexene $PdCl_2]_2$ to give a black tar even at $-70°$.[566] Presumably nucleophilic attack on the olefin occurs here. Clement[567] has prepared pyridine *N*-oxide complexes from $[C_2H_4PdCl_2]_2$ and the ligand in methylene chloride at $0°$. Even in an atmosphere of ethylene these complexes slowly decomposed.

Triphenylphosphine does not usually form complexes of this type; displacement of the olefin occurs instead. However, the complexes, (**III-7**), derived

---

† Equilibrium constants K_{eqm} for

$$\textit{trans-}C_2H_4PtCl_2(H_2O) + L \rightleftharpoons \textit{trans-}C_2H_4PtCl_2L + H_2O$$

<10 (L = F^-), 3.3×10^2 (Cl^-), 1.1×10^3 (Br^-), 4×10^4 (I^-), and 3×10^7 (NH_3) have also been measured.[563]

from (1,5-cyclooctadiene)MX$_2$ and alkoxide, behave rather differently. The only ligands which give isolable adducts are triphenylphosphine, p-toluidine, pyridine, and 4-methylpyridine; triphenylphosphine apparently always gives (III-8), whereas pyridine (and p-toluidine) was reported to give only (III-9) (see Chapter II, Section C,3,d,vii).

For platinum it was also possible to prepare the very labile (C$_2$H$_4$)$_2$PtCl$_2$ (from ethylene and [C$_2$H$_4$PtCl$_2$]$_2$ in the presence of acetone at −70°),[568, 569] and the more stable, probably cis, (1-octene)$_2$PtCl$_2$.[570] Chalk has also isolated the rather reactive complexes (olefin)PtCl$_2$CO, by reaction of the dimer with CO in carbon tetrachloride, for olefin = 1-hexene, 1-octene, and styrene. Their stereochemistry is uncertain.

2. Substitution of the Olefin or Acetylene

Both in the M(II) and the M(0) complexes the coordinated olefin or acetylene is easily displaced by a wide variety of ligands, including those mentioned above when used in excess or under vigorous conditions. Cyanide, ammonia, and t-phosphines will easily replace the olefin, and Chatt and Venanzi have prepared [amine PtCl$_2$]$_2$ from [C$_2$H$_4$PtCl$_2$amine].[566]

In a number of cases the greater volatility of ethylene has been utilized in displacements by other olefins, acetylenes, etc. Some typical reactions include:

$$[C_2H_4PdCl_2]_2 + RCH{=}CHR' \rightarrow [(RCH{:}CHR')PdCl_2]_2 \text{ [281, 513, 538]}$$

$$[C_2H_4PdCl_2]_2 + Bu^tC{\equiv}CBu^t \rightarrow [Bu^tC_2Bu^tPdCl_2]_2 + C_2H_4 \text{ [495]}$$

$$(Ph_3P)_2NiC_2H_4 + RC{\equiv}CR \rightarrow (Ph_3P)_2Ni(RC_2R) \text{ [45, 494]}$$

A wide range of other replacements have also been reported, some are indicated below.[45, 62, 94, 114, 145, 510, 514]

A number of very interesting reactions have recently been reported in which $(Ph_3P)_2Pd$(maleic anhydride) is used as catalyst in the codimerization of butadiene and R–H (R = MeO, EtO, Me_3SiO, PhNH, Ph(Bu)NH, and Me_3Si) to octadienes[571–576] (see Chapter II, Volume II, Section C,3,c).

3. Nucleophilic Attack on Olefin Complexes

Some of the most important reactions of olefin–metal complexes are those where nucleophilic attack on the olefin has occurred. Paths for such a reaction involve either (a) attack by the nucleophile directly on the olefin, or (b) primary attack by the nucleophile at the metal followed by transfer onto the coordinated olefin, i.e., insertion of the coordinated olefin into the M–X bond. Evidence for path (a) has come from the reactions of diene–MX_2 complexes† and for path (b) from kinetic studies of the $PdCl_2$-catalyzed oxidation of olefins to aldehydes or ketones.[577] The reactions observed are presented here; comments on their mechanisms are deferred to later sections, but it appears that some reactions of *mono*olefins are better explained in terms of path (b) (Chapter II, Volume II).

Phillips was the first to observe that ethylene, when passed into aqueous palladium chloride, was oxidized to acetaldehyde while the palladium was reduced to metal.[578] Anderson in 1934 in his investigation on Zeise's salt, found that when $K(C_2H_4PtCl_3)$ was heated in water in a sealed tube, some ethylene was evolved, but a considerable amount of it was oxidized to acetaldehyde with concomitant reduction to platinum metal.[579]‡ The stoichiometry of this reaction is:

$$[C_2H_4PtCl_3]^- + H_2O \rightarrow CH_3CHO + 2HCl + Pt + Cl^-$$

Although no details have been published of the reaction of $[C_2H_4PdCl_2]_2$ with water, the work of Smidt and his collaborators and other workers has made it clear that this complex is much more sensitive to water, even at low temperatures, than the platinum complex.[581] Furthermore, Smidt *et al.* showed that the reaction was inhibited by acid, suggesting that attack by OH^- was occurring.

Moiseev and Vargaftik[379] reported that $[C_2H_4PdCl_2]_2$ reacted with sodium ethoxide in ether to give diethylacetal and a trace of ethylvinyl ether. Moiseev and co-workers[582] also reported the synthesis of vinyl acetate from the ethylene complex. This was confirmed by Stern and Spector who also found that $[C_2H_4PdCl_2]_2$ reacted with isopropanol in isooctane in the presence of base to give the acetal and a small amount of the vinyl ether.[583]

$$[C_2H_4PdCl_2]_2 + NaOEt \rightarrow CH_3CH(OEt)_2 + CH_2{=}CHOEt$$

$$[C_2H_4PdCl_2]_2 + Na_2HPO_4 + AcOH \rightarrow CH_2{=}CHOAc$$

$$[C_2H_4PdCl_2]_2 + Na_2HPO_4 + Pr^iOH \rightarrow CH_3CH(OPr^i)_2 + CH_2{=}CHOPr^i$$

These reactions are important industrially and are discussed further in Chapter II, Volume II.

---

† Very strong evidence that nucleophilic attack even on monoolefin complexes occurs exo, and hence directly, has come from a recent study of the reaction of diethylamine with an optically active (1-butene)platinum(II) complex by Panunzi *et al.*[576a]

‡ Halpern and Pribanic recently showed that on oxidation of $[C_2H_4PtCl_3]^-$ or $[C_2H_4$-$PtCl_2(H_2O)]$ to Pt^{IV} by $IrCl_6{}^{2-}$ the ethylene was released as such and not oxidized.[580]

The reaction of acetate ion with [cyclohexenePdCl$_2$]$_2$ has been reported by Green *et al.*[457]

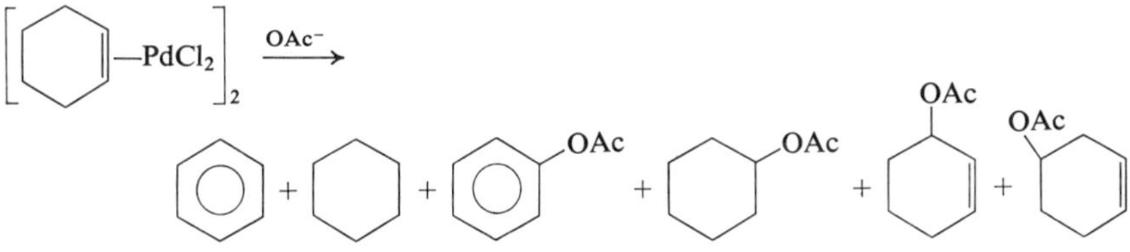

(9%)   (40%)

Similar results were obtained with [cyclopentenePdCl$_2$]$_2$,

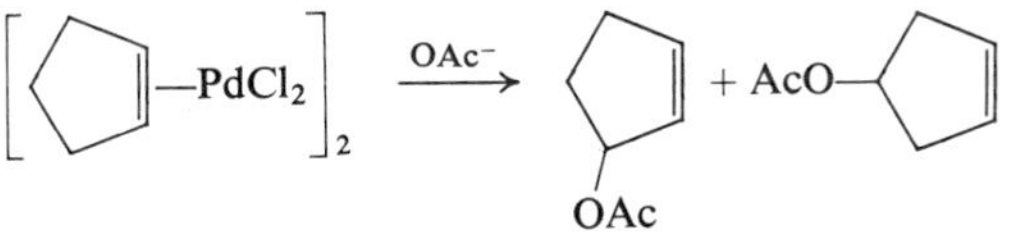

(39%)   (26%)

Reactions on the olefin complex itself do not appear very efficient, since loss of the olefin is always a strongly competing path. This has been illustrated by the study by Ketley and Fisher[584] of reactions of olefin complexes with various alcohols. They found that [C$_2$H$_4$PdCl$_2$]$_2$ reacted only with methanol and ethanol under mild conditions to give the acetals CH$_3$CH(OR)$_2$ (R = Me, Et). Even then some of the coordinated ethylene was lost; higher alcohols gave no reaction at all with the ethylene. No attack by methanol or ethanol on the olefin in propylene– or butene–palladium chloride complexes was observed.

Tsuji[585] and Medema[326] and their co-workers have studied the carbonylation of olefin–PdCl$_2$ complexes. The main products were the β-chloroacyl acid chlorides together with palladium and/or palladium carbonyl chlorides.

$$[C_2H_4PdCl_2]_2 + CO \xrightarrow{C_6H_6/20°/1\ atm} ClCH_2CH_2COCl$$

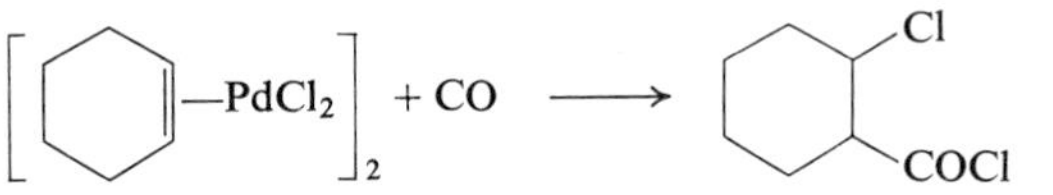

Saegusa *et al.* have reported the reaction of [(cyclohexene)PdCl$_2$]$_2$ with carbomethoxymercuric chloride, to give methyl cyclohexylacetate.[586] This may also involve a type of nucleophilic attack on the olefin, but the source of the extra H atom and the reaction path are unknown.

(10%)

The reactions of some olefin complexes with Grignard reagents have been studied by Okada and Hashimoto who obtained the *trans*-olefins from [styrenePdCl$_2$]$_2$.[587] The β-methylstyrene complex did not react, whereas the propylene complex and phenyl Grignard gave a mixture, with β-methylstyrene the major product.

Ketley and Braatz have shown that vinylcyclopropenes undergo similar reactions.[588, 589] Again, under mild conditions complexes, in which the metal is considered to be bonded to the olefin, are isolated and these rearrange to π-allylic complexes on heating (see Chapter V, Section B,3, this volume).

This type of reaction also occurs for acetylene–PdCl$_2$ complexes,[282] and is discussed more fully in Chapter II, Section 3,C,b,iv, this volume, and Chapter I, Volume II.

Borohydride and hydrogen have both been used to reduce olefin complexes to the saturated hydrocarbon.[408, 462] An early example was reported by Anderson[579]:

$$[C_2H_4PtCl_2]_2 + 2H_2 \rightarrow 4HCl + 2Pt + C_2H_6$$

No examples of such nucleophilic attack on olefins or acetylenes coordinated to metals formally in the (0) state are known (see Section D).

These reactions should be contrasted with the formation of dimethyl-(1,5-cyclooctadiene)palladium from dichloro(1,5-cyclooctadiene)palladium and methyllithium, reported by Calvin and Coates,[49]

Similar reactions are known for platinum–diene complexes.[418, 455]

It is plausible that attack of RLi or RMgX always occurs at the metal to give an (olefin)Pd–R complex. In the monoolefin complexes a cis-insertion reaction can then occur since the olefin is able to rotate so that it becomes coplanar with the other ligands and allows a four-center reaction to proceed.

$$
\begin{array}{ccc}
\overset{\displaystyle Cl}{\underset{\displaystyle R}{\overset{|}{|}\!\!-Pd-Cl}} + R'MgX & \longrightarrow & \overset{\displaystyle Cl}{\underset{\displaystyle R}{\overset{|}{|}\!\!-Pd-R'}} & \longrightarrow & \overset{\displaystyle R'}{\underset{\displaystyle R}{\overset{|}{|}\!\!-Pd-Cl}} & \longrightarrow
\end{array}
$$

$$
\underset{R}{\overset{R'}{\diagdown}}\!\!\overset{}{\underset{Pd}{\diagup}}\!\!\diagdown_{Cl} \longrightarrow \quad R\diagdown\!=\!\diagup_{R'} + HCl + Pd^0
$$

In contrast, in the 1,5-cyclooctadiene complexes, the double bonds are held rigidly and cannot rotate to give the coplanar configuration where the cis-insertion reaction can occur.[590]

Moritani et al.[591-598] have discovered some remarkable reactions in which olefins are arylated by aromatic hydrocarbons in the presence of Pd(II), particularly the acetate. This reaction also occurs with olefin–$PdCl_2$ complexes;

$$
\left[\underset{Ph}{\diagup}\!\!\diagup\!\!-PdCl_2\right]_2 + C_6H_6 \xrightarrow{\text{HOAc}} \underset{Ph}{\overset{Ph}{\diagdown}}\!=\!\diagup
$$

it and some analogous ones reported by Heck are discussed in Chapter I, Volume II.

The reaction of butadiene with $(PhCN)_2PdCl_2$ at ambient temperatures gives chloro[1-(chloromethyl)allyl]palladium dimer (**III-10**). An intermediate, (**III-11**), in this reaction has now been isolated by Donati and Conti[506] from butadiene and [(1-pentene)$PdCl_2$]$_2$ at $-40°$. The π-olefin complex (**III-11**) rearranges to the π-allylic complex (**III-10**) above $-20°$ by a reaction which probably involves "insertion" of the coordinated double bond into the Pd–Cl bond.

$$
[\{Me(CH_2)_2CH\!=\!CH_2\}PdCl_2]_2 + CH_2\!=\!CH\cdot CH\!=\!CH_2 \xrightarrow{-40°}
$$

$$
\left[\diagup\!\!\diagup\!\!-PdCl_2\right]_2 \xrightarrow{-20°} \left[\diagdown\!\!\diagup(\!-PdCl \atop CH_2Cl\right]_2
$$

$$
\text{(III-11)} \qquad\qquad\qquad \text{(III-10)}
$$

A quite different reaction which also involves nucleophilic attack, but alpha to the coordinated double bond, is the deprotonation of olefins having at least one alkyl substituent, usually to give π-allyl complexes. Reactions of this type were originally described by Hüttel in 1959, e.g.,[497,498,599]

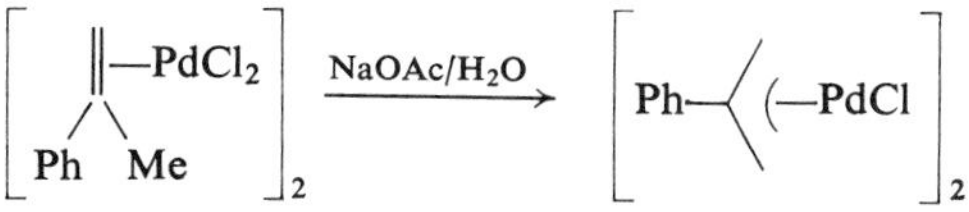

In some cases, no base was needed.

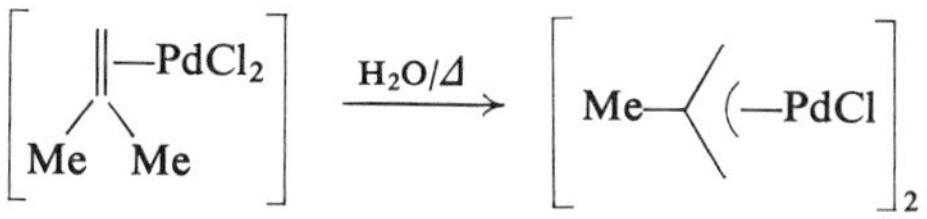

Ketley and Braatz[600] and Morelli et al.[601] have shown this to be a very general reaction. The former workers reported that [propylenePdCl$_2$]$_2$ was smoothly converted to chloro(π-allyl)palladium dimer on heating in dry chloroform in the presence of anhydrous sodium carbonate.

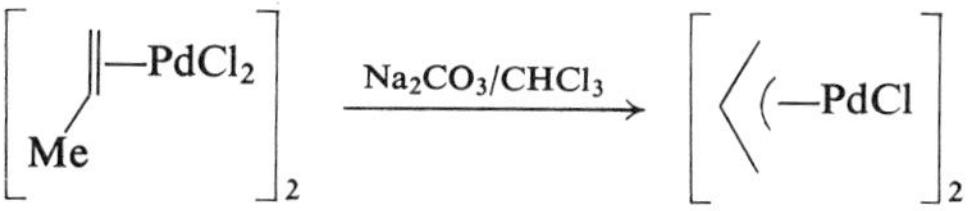

When sodium acetate was used as base the acetato complex was isolated.

Another general preparation of π-allylic complexes was from the olefin and PdCl$_2$ in DMF:

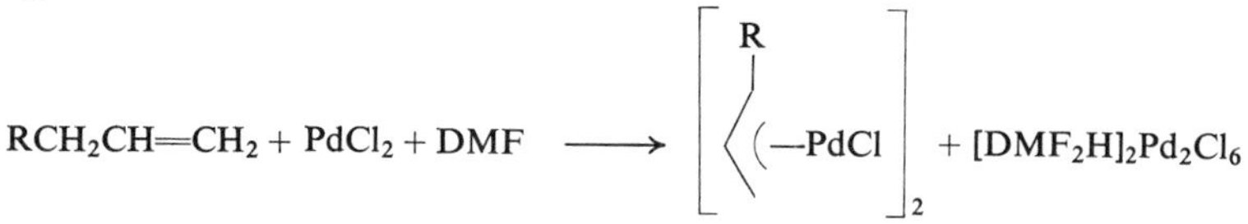

Morelli et al.[601] were able to isolate a π-olefin complex intermediate at $-80°$; this very readily transformed into the π-allylic complex. A platinum olefin complex prepared analogously was more stable and was identified as trans-[1-pentene–PtCl$_2$DMF], but this was not converted to a π-allylic complex.

In both these reactions it was not necessary to isolate the olefin complex and good yields of the π-allylic complexes were obtained by direct reaction of the free olefin and PdCl$_2$.[600] Ketley et al. have also noted that the β-methylstyrene complex was converted to a π-allylic complex by isobutene in chloroform.[507]

4. Electrophilic Attack on Olefin and Acetylene Complexes

Recently a number of examples of reactions in which attack by an electrophile, H$^+$, on the coordinated olefin or acetylene occurs, have been reported.

They are only known for complexes of Ni(0) and Pt(0) (see Section D). Barlex *et al.* found that fluoroolefin and fluoroacetylene complexes of Pt(0) reacted with trifluoroacetic acid to give σ-vinyl or σ-alkyl complexes.[602]

$$(Ph_3P)_2Pt\!-\!\overset{CCF_3}{\underset{CCF_3}{|||}} + CF_3COOH \longrightarrow (Ph_3P)_2Pt\!\!\begin{array}{c} \overset{H}{\underset{}{}}\!\!\diagdown\!\!\overset{}{\underset{}{C}}\!\!\diagup\!\!\overset{CF_3}{\underset{}{}} \\ \| \\ C\!-\!CF_3 \\ \diagdown OCOCF_3 \end{array}$$

$$(Ph_3P)_2Pt\!-\!\overset{CF_2}{\underset{CF_2}{|||}} + CF_3COOH \longrightarrow (Ph_3P)_2Pt\!\!\begin{array}{c} \diagup CF_2CF_2H \\ \diagdown OCOCF_3 \end{array}$$

A diphenylacetylene complex gave *trans*-stilbene and the bistrifluoroacetate,

$$(Ph_3P)_2Pt\!-\!\overset{CPh}{\underset{CPh}{|||}} + CF_3COOH \rightarrow (Ph_3P)_2Pt(OCOCF_3)_2 + PhCH\!\!=\!\!CHPh$$

Wilke *et al.* have reported the reaction of bis(1,5-cyclooctadiene)nickel with acetylacetone, which is apparently a strong enough acid to protonate one double bond.[603]

Many other examples of such reactions will undoubtedly now be found.†

5. Insertion of Coordinated Olefin into a Pd–C Bond

A number of reactions are now known where a coordinated olefin is inserted into a Pd–C σ bond. Perhaps the best example of this is the reaction described by Coulson.[466] (See also Ref. 201a).

† See for example, Tripathy and Roundhill,[602a] or Mann *et al.*[602b]; the latter workers confirmed that overall cis- addition to the π-complexed acetylene occurs to give the σ-vinyl Pt(II) compounds.

This is analogous to the reaction of norbornadiene with mercuric chloride in methanol; however, an exo orientation for the metal in the product has been proposed by Alexander et al.[604]

In the above palladium reaction and in a number of others, an intermediate which can be described as a type of homoallylic π complex may be involved.[462, 466]

Similar reactions of complexes derived from 1,5-cyclooctadiene and of $[Cl(MeC_2Me)_3PdCl]_2$ are discussed in Section C,3,d,v and in Chapter I, Volume II.

An unusual reaction which probably involves both insertion into Pd–Cl and Pd–C bonds has been described by Hosokawa et al.,[495] who obtained the complex (III-12) from reaction of t-butyl(phenyl)acetylene and $[C_2H_4PdCl_2]_2$; no ethylene was evolved. $(PhCN)_2PdCl_2$ reacted with the same acetylene to give a cyclobutadiene–$PdCl_2$ complex.[605]

$$[C_2H_4PdCl_2]_2 + PhC{\equiv}CBu^t \longrightarrow \left[Ph{-}C \begin{matrix} CH_3 \\ | \\ CH \\ \diagup \\ ({-}PdCl \\ \diagdown \\ C{-}Cl \\ | \\ Bu^t \end{matrix} \right]_2$$

(III-12)

The reaction mechanism is discussed in Chapter I, Volume II, Section C,3,d. Very similar reaction paths can be postulated to occur in the formation of 1,4-disubstituted 2,3-diphenylbutadienes from α-olefins, diphenylacetylene, and $(PhCN)_2PdCl_2$, reported by Mushak and Battiste.[606]

$$PhC{\equiv}CPh + 2RCH{=}CH_2 \xrightarrow{(PhCN)_2PdCl_2/C_6H_6}$$

6. Thermal Decomposition of Olefin—Metal Complexes and the Oligomerization of Olefins

The thermal decomposition of $[C_2H_4PtCl_2]_2$ was studied by Gow and Heinemann,[607] who found that the primary step was dissociation of C_2H_4, followed by a complex series of addition and substitution reactions to give a variety of chlorinated C_2 compounds (e.g., EtCl, CH_2=CHCl, CH_2CHCl_2, etc.). By contrast, Van Gemert and Wilkinson[608] found that $[C_2H_4PdCl_2]_2$ was much less thermally stable, especially in solution. In the solid 87 % could be recovered after 91 hours at 50°, but in solution after 113 hours at 50° it had completely decomposed to palladium and the following organic products,

$$[C_2H_4PdCl_2]_2 \xrightarrow{\Delta}$$

10	:	65	:	25

Solvents: dioxane 39%   17%   13%

benzene 63%   6%   trace

The most interesting feature was the formation of butenes, particularly in benzene, and the authors noted that this could be carried out catalytically by passing ethylene into a solution of $[C_2H_4PdCl_2]_2$.

Ketley *et al.*[507] have investigated this reaction more fully and have found that in chloroform containing 0.03 M ethanol the ethylene complex absorbed ethylene to give a red solution which was the active catalyst for the dimerization. They found that the product was a mixture of *cis*- and *trans*-2-butenes (47 and 52 %) together with 1 % 1-butene. Traces of other alcohols or water also acted as cocatalysts, unless the concentration was higher than that of the $PdCl_2$ in which case palladium metal was formed and the catalyst was deactivated. Other solvents such as CH_2Cl_2, $EtNO_2$, and $PhNO_2$ caused dimerization without a cocatalyst, but many solvents (CCl_4, saturated hydrocarbons) were quite inactive. The authors suggested that the solvent-dependence of the reaction is only in the step in which the $[C_2H_4PdCl_2]_2$ forms $(C_2H_4)_2PdCl_2$.†

They found that $(C_2H_4)_2PdCl_2$ at $-40°$ in $CHCl_3$ gave the red catalytic solution without needing a cocatalyst. The red solution gave red crystals which had very similar spectroscopic properties to, but greater solubility than, the trans dimeric 2-butene complex (III-13), and the authors suggested that it might be the cis dimeric 2-butene complex (III-14) (the configuration of the olefin in the complex is not known). On standing at 20° the complex (III-14)

† This complex was stable only under ethylene pressure.

$$\underset{Cl}{\overset{MeCH=CHMe}{\diagdown}}Pd\underset{Cl}{\overset{Cl}{\diagup}}\underset{Cl}{\overset{Cl}{\diagdown}}Pd\underset{MeCH=CHMe}{\overset{Cl}{\diagup}}$$

(III-13)

$$\underset{Cl}{\overset{MeCH=CHMe}{\diagdown}}Pd\underset{Cl}{\overset{Cl}{\diagup}}\underset{Cl}{\overset{MeCH=CHMe}{\diagdown}}Pd\underset{Cl}{\overset{}{\diagdown}}$$

(III-14)

isomerized to **(III-13)** and this could be partially reversed. The red solution obtained from $(C_2H_4)_2PdCl_2$ under ethylene pressure at $-40°$ for 4 hours gave another set of red crystals, again with spectroscopic properties similar to those of **(III-13)** and **(III-14)**. The authors concluded that this was yet another 2-butene complex to which they assigned the structure **(III-15)**; this complex did not isomerize to **(III-13)** on standing.

$$\underset{MeCH=CHMe}{\overset{MeCH=CHMe}{\diagdown}}Pd\underset{Cl}{\overset{Cl}{\diagup}}\underset{Cl}{\overset{Cl}{\diagdown}}Pd\underset{Cl}{\overset{Cl}{\diagup}}$$

(III-15)

$$Me\diagdown\overset{\diagup}{\underset{Me\diagdown}{\Big(}}\!\!\!\Big(-Pd\underset{Cl}{\overset{Cl}{\diagup}}Pd-\Big)\!\!\Big\rangle\underset{\underset{Me}{\diagdown}}{\overset{Me}{\diagup}}$$

$$CHClMe$$

(III-16)

After some days the red solutions lost their catalytic activity and an allylic complex, formulated as **(III-16)**, was isolated. It was found that addition of either hydride (NaBH₄) or H_2 or acid (anhydrous HCl) inhibited the dimerization reaction. The authors proposed a reaction scheme to account for these observations, and suggested that a transient Pd(IV) hydride species may be an intermediate, e.g.,

$$\underset{H_2C}{\overset{H_2C}{\diagup}}\!\!\overset{\overset{CH_2}{\diagdown}}{\diagdown}Pd\underset{Cl}{\overset{Cl}{\diagup}}\quad\longrightarrow\quad \Big\langle\!\!-\overset{CH_2-CH_2\ H}{\underset{Cl}{\diagdown Pd-Cl}}\quad\xrightarrow{\ C_2H_4\ }\quad \underset{\diagdown\diagdown}{\overset{H_2C}{\diagdown}}\!\!\overset{\overset{CH_2}{\diagup}}{\diagup}Pd\underset{Cl}{\overset{Cl}{\diagup}}$$

It is not clear, however, that a Pd(II) hydride intermediate is ruled out by this work (see also Chapter I, Volume II, Section C,1).

Propylene was also dimerized (to a mixture of 65% 2- and 3-hexenes and 35% 2-methyl- and 4-methylpentenes) by $PdCl_2$ in chloroform or methylene chloride. Butenes were not dimerized by $PdCl_2$, and both *cis*- and *trans*-2-butene and 1-butene gave the same mixture of isomeric [butenePdCl₂]₂ complexes.

The formulations (**III-13**), (**III-14**), and (**III-15**) for the butene complexes in solution should be treated with some caution; it is not clear that the organic ligand is exactly the same in all of them as the butenes are so readily isomerized. Furthermore, the molecular sizes of the complexes are not known and their differences in solubility may also be explained on the basis of different-sized aggregates, involving trimers or tetramers.

Kawamoto *et al.*[609] have isolated a complex from the reaction of ethylene and acetylacetone with $[(C_2H_4)PdCl_2]_2$ in methylene chloride to which they assigned the structure (**III-17**) on the basis of the NMR spectrum at $-45°$. At higher temperatures, decomposition, to a mixture of butenes, palladium acetylacetonate, and metal occurred. These authors also noted that the formation of butenes was accelerated by β-diketonates. The reaction of ethylene, acetylacetone, and $[C_2H_4PdCl_2]_2$ in acetic acid gave ethylideneacetylacetone.

$$[C_2H_4PdCl_2]_2 + C_2H_4 + MeCOCH_2COMe \xrightarrow{CH_2Cl_2}$$

$$\downarrow HOAc$$

$$Pd + CH_3CH{=}C\overset{COMe}{\underset{COMe}{<}}$$

(III-17)

$$Pd(acac)_2 + Pd + \quad + \quad + \quad$$

$$5 \quad : \quad 68 \quad : \quad 27$$

7. Miscellaneous Reactions

Fitton and McKeon[88] observed that tetrakis- or bis(triphenylphosphine)-palladium reacted with a number of chloroolefins to give σ-vinyl complexes (see also Mukhedkar *et al.*[67])

$$(Ph_3P)_4Pd + trans\text{-}CHCl{=}CHCl \longrightarrow$$

Very similar reactions involving $(Ph_3P)_4Pt$ have been reported by Kemmitt and Stone and their co-workers,[121, 133, 134, 135, 399, 610] except that here the intermediate olefin–Pt(0) π complexes were isolated. The mechanisms of the reactions involving platinum have been discussed in Chapter I, Section B,2,d, this volume, and it seems probable that the palladium complexes react analogously.

F. OLEFIN COMPLEXES OF THE NEIGHBORING ELEMENTS

Olefin complexes are known for Cu, Ag, and Au in the (I) oxidation state (d^{10}) but probably not in the (II) or (III) states. They are also known for Co(0), Co(I), Rh(I), and Ir(I). Of the group Ib metals, silver forms the most stable olefin complexes, but none appear known where the olefin has a strongly electron-withdrawing substituent such as F. It appears that these complexes bear a closer resemblance to those of Pd(II) and Pt(II) (d^{8}) than to the zero-valent complexes. Although many are known, the chemistry of silver–olefin complexes is not yet well developed, although silver undoubtedly displays interesting catalytic properties.

In contrast, there has been considerable activity in the field of olefin complexes of Rh(I) and Ir(I), which to some extent at least, resemble those of Pd(0) and Pt(0); for example, the strongest olefin complexes are formed with fluoro-olefins. The lower stability of the hydrocarbon olefin complexes has led, in particular for Rh(I), to a number of very important catalytic reactions involving hydrogenation and oligomerization such as those described by Cramer and Wilkinson and their co-workers. These are discussed in Chapters I and III, Volume II.

The significant feature appears to be that olefins are readily inserted into Rh–H and Rh–C bonds but, in contrast to palladium(II) chemistry, there appears to be little evidence for "insertion" into Rh–O or Rh–halide, or for attack by an external nucleophile on the coordinated olefin.

Chapter IV
Diene Complexes

A. INTRODUCTION

The complexes considered here are those formed by 1,3-, 1,4-, and 1,5-diolefins in which four carbon atoms are bound to the metal. Palladium(II) undergoes reaction with 1,3-dienes usually to form π-allylic complexes in which only three carbon atoms are coordinated to the metal; these are considered in this volume, Chapter V. Pt(II) again has a low tendency to complex with 1,3-dienes, the best known example is the 1,3-butadiene complex in which each double bond coordinates to a different metal atom,[535]

$$\left[Cl_3Pt-\| \quad \|-PtCl_3 \right]^{2-}$$

The most important exceptions are the tetraphenylcyclobutadiene–metal

$$\left[\begin{array}{c} R \quad R \\ R \quad R \\ MX_2 \end{array} \right]_2$$

(structure with Me substituents and MCl₂)

complexes which are known for all three metals[611-614] (the tetramethylcyclo-butadiene complex is also known for nickel[615]) and the pentamethylcyclo-pentadiene complexes, known for Pd(II) and Pt(II).[616]

The most common, and most thermodynamically stable, diene complexes are those of 1,5-dienes, in particular, 1,5-cyclooctadiene and dicyclopenta-diene. The stability of these complexes arises by virtue of the good chelating qualities of the bidentate 1,5-diene.

Similar considerations apply to a number of 1,4-dienes with special geometry such as norbornadiene [bicyclo[2.2.1]heptadiene] and hexamethyl-(Dewar benzene) [hexamethylbicyclo[2.2.0]hexadiene] and complexes of these ligands with both Pd(II) and Pt(II) are known. A report has also appeared claiming the isolation of complexes of 3-alkyl-1,4-cyclooctadienes with these metals.[617]

One diene complex of Pd(0) has been prepared[458a]; they are, however, well known for Ni(0),[603] and one example, (1,5-cyclooctadiene)$_2$Pt, has also been reported for Pt(0).[455]

In addition, some unusual 1,5-cyclooctadiene (1,5-COD) complexes, one apparently of Ni(I), [1,5-CODNiI],[152] and palladium–tin and platinum–tin clusters have been described.[618,619] A preliminary crystal structure of the platinum–tin cluster complex has been published,[618] and the palladium complex, (1,5-CODPd)$_3$Sn$_2$Cl$_6$, is probably isostructural.[619]

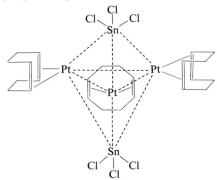

B. STRUCTURES OF THE DIENE COMPLEXES

The Pd(II) and Pt(II) complexes of 1,5-dienes which have had their structures reported include (1,5-hexadiene)PdCl$_2$[209] (**IV-1**), (cyclooctatetraene)PdCl$_2$[227] (**IV-2**), (1,3,5,7-tetramethyl-2,6,9-trioxabicyclo[3.3.1]nonadiene)PtCl$_2$[557] (**IV-3**), and (dipentene)PtCl$_2$[620] (**IV-4**). 1,5-COD complexes of Rh(I)[621] (**IV-5**), Cu(I)[622] (**IV-6**), and Ni(0) (**IV-7**) and (**IV-8**)[555,623] have had their structures determined.

In the d^8 complexes, **(IV-1)**, **(IV-2)**, **(IV-3)**, **(IV-4)**, and **(IV-5)**, the plane of the metal and the two chlorines attached to it is usually very nearly the bisector of the double bonds, and the coordination about the metal is therefore square planar. The d^{10} complexes **(IV-6)**, **(IV-7)**, and **(IV-8)** all have approximately tetrahedral stereochemistry about the metal. The angle subtended by the diene at the metal is 90° or greater in all cases. These 1,5-dienes therefore act as very effective chelating ligands.

(IV-1)   **(IV-2)**

Coordination of the olefin to the metal probably increases the C–C bond length, and values of 1.40 Å have been reported. This is not always the case,[209] but as in the monoolefin complexes, the estimated standard deviations are too large to allow any conclusions to be drawn from the reported bond lengths. Both **(IV-3)** and **(IV-4)** show some unusual features; in the former the $PtCl_2$ plane does not bisect the C=C bonds, but cuts them further from the =CMe

(IV-3)   **(IV-4)**

(IV-5)   **(IV-6)**

(IV-7)   **(IV-8)**

group. In **(IV-4)** the exocyclic double bond is not perpendicular to the $PtCl_2$ plane but at an angle of 62° to it. See also the structure of [styrene $PdCl_2]_2$, **(III-4)**, pp. 112–113.

An example of the structure of a complex involving a 1,4-diene is (nor-bornadiene)$PdCl_2$,

Here the angle subtended by the diene at the metal is only 74° and the angle ClPdCl is now opened out to 94°.[226] A very slight lengthening of the double bonds [to 1.366(10) Å] by comparison with norbornadiene itself (1.333 Å) was reported. An example of a 1,4-diene bonding to Pt(II) is the rather unusual complex derived from hexamethyl(Dewar benzene)$PtCl_2$, **(I-39)**,[199] already discussed on p. 40 and p. 202.

The only detailed information on 1,3-diene complexes is for tetramethyl-cyclobutadienenickel chloride dimer.[624]

The C_4 ring is square and planar (C–C, 1.43 Å) and the four carbons are equidistant from the metal. However, the coordination is not square planar about the metal, but rather 5-coordinate (if the C_4Me_4 is taken to be a biden-tate ligand) or distorted tetrahedral, if the C_4Me_4 is regarded as a monodentate ligand. The tetraphenylcyclobutadienepalladium(II) complexes are dimeric and probably have the same structures.[625] A reason for the increased co-ordination number about the metal here may be the small angle (45°) which the cyclobutadiene subtends at the metal.[626]

In all these complexes the metal–carbon and metal–halogen bond lengths appear to be quite normal. A preliminary report of the structure of $C_4Ph_4PdB_9C_2Me_2H_9$ **(IV-12)** has shown this to contain a tetraphenylcyclo-butadienepalladium coordinated to a carboranyl cage.[627]

C. BONDING IN DIENE COMPLEXES

The bonding situation in complexes of 1,4- and 1,5-dienes, where there is no great interaction between the olefins can be regarded as similar to that in the monoolefin complexes, and is therefore quite straightforward (Chapter III, Section D, this volume).

However, in 1,3-dienes and other systems where there is an appreciable interaction between the two olefinic bonds (e.g., in p-benzoquinones), complications arise.

The molecular orbitals for square planar cyclobutadiene and butadiene, respectively, are given in Fig. IV-1.

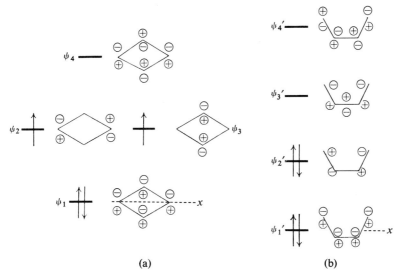

(a)   (b)

Fig. IV-1. Energy levels and representations of the molecular orbitals of (a) cyclobutadiene and (b) *syn*-butadiene. The z axis is perpendicular to the plane of the molecule.

The ground state of cyclobutadiene, if square and planar, would be expected to be a triplet as shown. Neither cyclobutadiene nor any simple substituted derivatives thereof have yet been isolated and studied spectroscopically, but the complexes are well known and contain square planar cyclobutadiene.[628] Butadiene, on the other hand, is well known though it usually exists in the anti rather than the syn (cisoid) form. The bonding between both of these ligands and metals has been discussed by various authors,[359, 626, 629] Possible combinations of suitable symmetry are: ψ_1 or ψ_1' and s, p_z or d_{z^2}; ψ_2 or ψ_2' and p_x or d_{xz}; ψ_3 or ψ_3' and p_y or d_{yz}; ψ_4 and $d_{x^2-y^2}$ and ψ_4' and d_{xy}.

The relative energies of the various molecular orbitals of the C_4 ligand and the metal atomic orbitals now become important in deciding which particular combination is the more significant in the bonding. However, it is qualitatively clear that there must be an electron flow both from the ligand (occupied orbitals) to the metal as well as back-bonding from the metal orbitals to the ligand. Cotton has given a possible orbital correlation diagram for

$[C_4Me_4NiCl_2]_2$ based on reasonable assumptions about the relative energies of the orbitals[629] (Fig. IV-2).

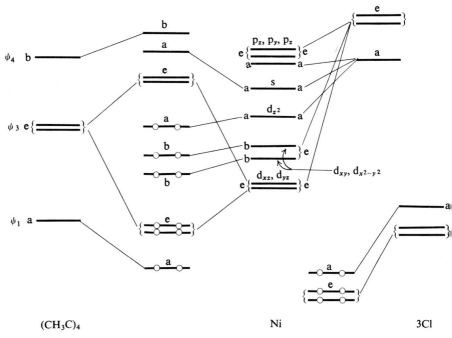

FIG. IV-2. Orbital correlation diagram for $[C_4Me_4NiCl_2]_2$. Reprinted, with permission, from Cotton.[629]

The main bonding here is then between the metal s orbital and the lowest (occupied) orbital of the C_4 ring (ψ_1) and between the metal d_{xz}, d_{yz}, and the ψ_2 and ψ_3 orbitals of the ring. Modifications to give more directional character to the metal orbitals, and hence better overlap with the C_4 orbitals, by hybridizing them have also been suggested.[626]

D. PREPARATION OF DIENE COMPLEXES

1. 1,4- and 1,5-Diene Complexes of Pd(II)

(Dicyclopentadiene)PdCl$_2$ and (1,5-COD)PdCl$_2$ were formed more readily than the platinum analogs by reaction of the diene with sodium tetrachloropalladate in acetone.

The COD complex could also be prepared in water or ethanol,[365,421] but dicyclopentadiene gave the σ-bonded ethoxy derivative in the latter solvent (p. 74). This illustrates the lower reactivity of these complexes by comparison with the monoolefin complexes. Similar reactions have been reported for phenylcyclooctatetraene,[630] a tricyclodecadiene,[419] norbornadiene,[604,631] but not for dipentene, which only gave palladium metal.[365]

The usual variant, using $(PhCN)_2PdCl_2$ and the diolefin in an inert solvent such as CH_2Cl_2, benzene, etc., has been successfully employed by numerous workers for norbornadiene[604,631-633] a tricyclodecadiene,[419] cyclooctatetraene[634] [and $(COT)PdBr_2$[635]], 1,6-cyclodecadiene,[636] unsubstituted- and hexamethyl(Dewar benzene)†[637-639] (See also Ref. 639a):

$$\text{diene} + (PhCN)_2PdCl_2 \rightarrow \text{(diene)}PdCl_2$$

Another variant which has been widely used is the displacement of an olefin, particularly by 1,5-COD, from a monoolefin complex,[505,506] e.g.,

$$(R = CH_2{=}CH, Et)$$

Fischer and Werner[337] have prepared the norbornadiene and 1,5-COD complexes from "$PdCOCl_2$".

$$\text{diene} + \text{"}PdCOCl_2\text{"} \rightarrow \text{(diene)}PdCl_2$$

Tsuji et al.[643] obtained some $(1,5\text{-COD})PdCl_2$ from the reaction of 1,5-COD and palladium black (metal) in ethanol in the presence of HCl under pressure at 110°.

Another general reaction is that of the σ-bonded alkoxy derivatives with acids, e.g.,[365,462]

† Dewar benzene complexes are accessible only by this route; they do *not* arise[640] from Pd(II)-catalyzed cyclotrimerization of acetylenes as has been claimed.[641,642]

This reaction is also known for –HC(COMe)$_2$ in place of OR; cleavage of the C–C bond occurs with HCl or Ph$_3$C$^+$BF$_4^-$.[463]

Some very intriguing syntheses of 1,5-diene complexes in which the initial diene has undergone isomerization to give the complexed diene have been reported. Trebellas *et al.*[636] found that *cis,trans*-1,5-cyclodecadiene reacted with (PhCN)$_2$PdCl$_2$ to give (*cis*-1,2-divinylcyclohexane)PdCl$_2$ (**IV-9**). This could also be obtained from *cis*-1,2-divinylcyclohexane, which was itself liberated on treatment with aqueous cyanide.

(IV-9)

cis,trans-1,6-Cyclodecadiene underwent a similar but slower reaction with Na$_2$PtCl$_4$ in propanol to give the platinum analog of (**IV-9**). It is probable these are examples of a metal-catalyzed Cope rearrangement since Heimbach[644] has observed that a *cis,trans*-1,5-cyclodecadiene could also be thermally transformed into a *cis*-1,2-divinylcyclohexane. A hydride shift is not a necessary step in the formation of the complex (**IV-9**), although it may occur.

The reaction of 4-vinylcyclohexene with (PhCN)$_2$PdCl$_2$ gave a brown uncharacterized complex, (C$_8$H$_{12}$)$_2$PdCl$_2$, which, on standing, changed into yellow needles of (1,5-COD)PdCl$_2$.[645]

This rearrangement did not occur for platinum or for 4-alkyl-4-vinylcyclohexenes.[646] However, Paiaro *et al.*[409] have reported the isolation of (4-vinylcyclohexene)PdCl$_2$ and some reactions of it (Section F,1,c).

Another reaction of this type has been reported by Vedejs[647] who found that bullvalene reacted with (PhCN)$_2$PdCl$_2$ to give a dark amorphous complex which, on treatment with pyridine, gave primarily 2,5,7,9-bicyclo[4.2.2] decatriene. The nature of the dark amorphous solid is not known.

The author suggested a stepwise path involving addition of Pd–Cl to the bullvalene; an uncharacterized orange solid could be isolated if the reaction was carried out at −40°.

Zakharova *et al.*[209] have described a remarkable synthesis of (1,5-hexa-diene)PdCl$_2$. It was obtained (60% yield) by low temperature reaction of (PhCN)$_2$PdCl$_2$ with dry allyl chloride in the absence of solvent. Characterization of the complex included a complete X-ray structure determination.

$$CH_2{=}CH \cdot CH_2Cl + (PhCN)_2PdCl_2 \longrightarrow$$

In the presence of water, or using allyl alcohol and palladium chloride in benzene, chloro(π-allyl)palladium dimer was always obtained. Although they are probably mechanistically different, this recalls the reaction described by Powell and Shaw,[648] who found that [Rh(CO)$_2$Cl]$_2$ reacted with methylallyl chloride in two ways: at 0° to give the bis-π-allylic complex and at 60° to give the (3,4-dimethyl-1,5-hexadiene)RhCl complex.

A protonation step occurs in the formation of the dimethylhexadiene complex here.

2. Complexes of Pd(II) with Other Dienes

1,3-Dienes, in general, react with (PhCN)$_2$PdCl$_2$ or Na$_2$PdCl$_4$ to give π-allylic complexes.[634, 649, 650] In two cases intermediates in which the Pd is complexed to one double bond of the diene have been isolated, for example, [(butadiene)PdCl$_2$]$_2$; above −20° this rearranges to chloro[1-(chloromethyl)-allyl]palladium dimer[506] (Chapter III, Section E,3 and Chapter V, Section B,2,a, this volume).

Pentamethylcyclopentadiene is an unusual ligand in that it forms stable complexes with both Pd(II) and Pt(II) in which both double bonds are coordinated.[616]

Only the isomer with H endo to the metal is obtained. The reason why only pentamethylcyclopentadiene forms readily isolable 1,3-diene complexes is probably partly steric, in that the diene is forced into a syn–cis configuration, and partly electronic, in that a deformation of the organic ligand towards a "homocyclobutadiene" structure appears to be occurring.

Lukas and Kramer[651] have briefly reported obtaining cationic complexes of cisoid butadiene by reaction of the [1-(chloromethyl)allyl]palladium chloride complexes with SbF_5 in SO_2ClF.

A Pd(II) complex of *cis,cis*-1,6-cyclodecadiene has been reported by Trebellas *et al.*[636]; the analogous Pt(II) complex could not be isolated. The structure of the palladium complex is not certain since it is very insoluble; it may be polymeric with bridging dienes and $PdCl_2$ groups.

3. Cyclobutadiene-Pd(II) Complexes

The preparation of tetraphenylcyclobutadienepalladium dihalide complexes from diphenylacetylene was first described by Malatesta, Santarella, Vallarino, and Zingales[611] and later confirmed and discussed in greater detail by Maitlis and co-workers,[612,625] Hüttel and Neugebauer,[652] and Vallarino and Santarella.[653]

Several reactions are possible, depending on the conditions. In ethanol diphenylacetylene reacted to give the *endo*-ethoxycyclobutenyl complex virtually quantitatively; other alcohols gave analogous complexes.[653] $(PhCN)_2PdCl_2$, Na_2PdCl_4, and H_2PdCl_4 were all used as sources of palladium. On treatment of this alkoxy complex with hydrogen halides, the tetraphenylcyclobutadienepalladium dihalide complexes (X = Cl, Br, I) were obtained in high yields.

These reacted with alcohols to give the *exo*-alkoxycyclobutenyl complexes; this reaction was again reversed on treatment with hydrogen halides.

Very similar reactions have been shown to occur with 4,4'-ditolylacetylene and bis(4-chlorophenyl)acetylene[625] (see Chapter I, Volume II, Section C,3,d).

More complex products in general arise from other acetylenes and Pd(II), but Hosokawa and Moritani[605] reported that $PhC{\equiv}CBu^t$ gave 1,2-di-*t*-butyl-3,4-diphenylcyclobutadiene dichloride by the following route. Only one isomer was obtained; see also Avram *et al.*[654]

In aprotic solvents (or solvents containing little alcohol) the reaction of the diphenylacetylene with $(PhCN)_2PdCl_2$ takes a different course. The major products here are the hexaphenylbenzene and a red cyclobutadiene complex,

$$PhC\equiv CPh + (PhCN)_2PdCl_2 \xrightarrow{C_6H_6}$$

$$\downarrow DMF; HX$$

$$[Ph_4C_4PdX_2]_2$$

The composition of this cyclobutadiene complex varies rather with the method of preparation and usually contains between 2 and 3 $PdCl_2$ units per cyclobutadiene. The structure is not known, but it has been proposed that it consists of a chain of $PdCl_2$ units bridging two rings as shown.[625, 652] This complex is easily converted to the normal cyclobutadiene complex either by treatment of a suspension in methylene chloride with HBr or, better, by dissolving the complex in DMF and precipitating $[Ph_4C_4PdX_2]_2$ by addition of the concentrated hydrohalic acid. The mechanisms of these reactions are discussed in Chapter I, Volume II.

Acetylenes with substituents other than phenyl do not usually appear to react in this way. Müller *et al.*[655] have described the reaction of *o*-bis(phenylethynyl)-benzene with $(PhCN)_2PdCl_2$. A complex of two moles of the acetylene to one of $PdCl_2$, which readily lost part of its halogen above 20°, was obtained. The structure (IV-10) was suggested, but this appears rather unlikely since the palladium would then have an effective atomic number corresponding to (Xe + 2), and a formal coordination number of 6, both of which are highly unusual. A more likely formulation would be as a diene complex, but clarification will have to await a crystal structure determination. A labile complex, $C_{45}H_{32}OPdCl_2$, was obtained from the reaction in methanol.

(IV-10)

The complex (IV-10) reacted with triphenylphosphine to give $(Ph_3P)_2PdCl_2$ and unidentified products, and with hot DMF to give metal and a mixture of five red hydrocarbons, some of which may be dibenzopentalenes.

A tetraphenylcyclobutadiene complex of Pt(II) has recently been reported to be formed on reaction of diphenylacetylene with $Pt(CO)_2Cl_2$ in ether[614]:

$$Pt(CO)_2Cl_2 + PhC_2Ph \xrightarrow{Et_2O/\Delta} [Ph_4C_4PtCl_2]_n$$

On reaction with sodium iodide, $[Ph_4C_4PtI_2]_2$ was obtained, the molecular weight of which showed it to be a dimer.

Quite a large number of cyclobutadiene complexes of Ni(II) are known, but these are not as easily accessible as the palladium ones. Routes by which these have been prepared directly include:

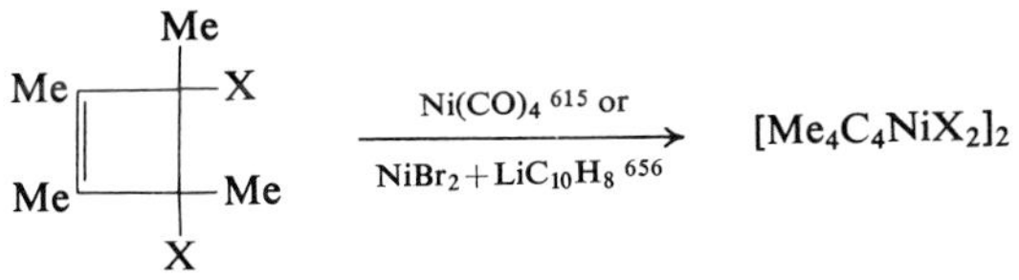

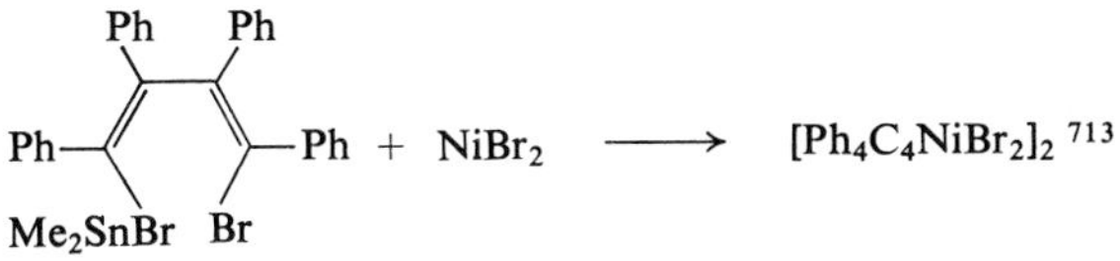

Wittig and Fritze[657] found that cyclooctyne reacted with $NiBr_2$ or NiI_2 in THF containing a trace of water to give the cyclobutadiene complexes (in 9 and 7% yields, respectively). The major product, and the only one under anhydrous conditions, was the organic trimer.

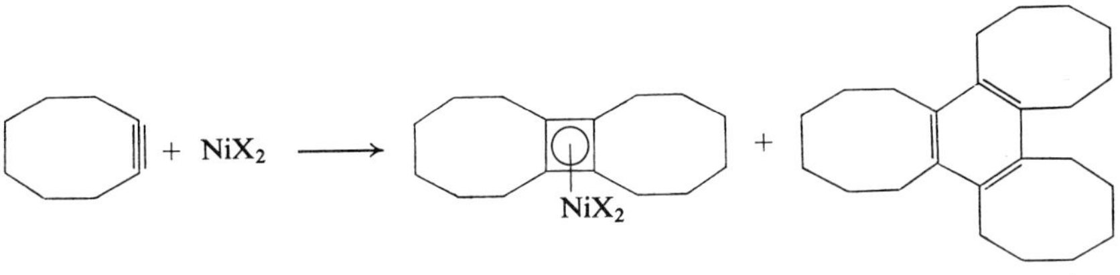

A very useful route to the tetraphenyl- or tetrakis(p-substituted-phenyl)-cyclobutadiene–metal complexes is by a ligand-exchange reaction with the appropriate palladium complex. These reactions are discussed in Section F,2.

4. Diene Complexes of M(0) and M(I)

A diene complex of Pd(0), bis(dibenzylideneacetone)palladium, has recently been prepared,[485a] but its exact structure is unknown. As it is very easily made and relatively stable, it is a very useful complex for the preparation of a wide variety of other Pd(0) and Pd(II) complexes.

$$PhCH\!=\!CHCOCH\!=\!CHPh + Na_2PdCl_4 + NaOAc \xrightarrow{MeOH/60°}$$

$$(PhCH\!=\!CHCOCH\!=\!CHPh)_2Pd°$$

Müller and Göser[455] have reported that when diisopropyl(1,5-cyclooctadiene)platinum(II) [from dichloro(1,5-cyclooctadiene)platinum and isopropyl Grignard at −50°] was photolyzed in 1,5-cyclooctadiene, $(1,5\text{-COD})_2Pt$ was obtained.

$(1,5\text{-COD})_2Ni$ was obtained in a similar fashion, but bis(norbornadiene)-platinum could not be prepared in this way.

Several routes to $(1,5\text{-COD})_2Ni$ have been reported; the original one by Wilke and co-workers involved treating nickel acetylacetonate with diethylaluminum ethoxide or triethylaluminum in the presence of 1,5-COD.[511, 603]

$$Ni(acac) + 1,5\text{-COD} + (Et_2AlOEt)_2 \xrightarrow[0°-5°]{C_6H_6} (1,5\text{-COD})_2Ni$$

Cyclooctatetraene nickel complexes, $(COT)_2Ni$, $(COTNi)_n$, and all-*trans*-(1,5,9-cyclododecatriene)nickel were all prepared similarly. Otsuka and Rossi[658] prepared $(1,5\text{-COD})_2Ni$ from 1,5-COD, nickel chloride, and sodium in THF and pyridine.

One cyclooctadiene in $(1,5\text{-COD})_2Ni$ is easily replaced by various ligands, including duroquinone [to give (**IV-7**)], and hexafluoroacetone.[623, 659]

(**IV-7**)

Bis(1,5-cyclooctadiene)nickel has also found use as a catalyst for oligomerization of butadiene and other olefins[90, 660, 661] as well as for the preparation of π-allylnickel complexes.[91]

Reaction of (π-allyl)nickel iodide dimer with 1,5-COD in the presence of norbornene (bicyclo[2.2.1]heptene) gave a complex, apparently of Ni(I).[152]

$$\left[\left\langle \!\! \left\langle \!\!-\text{NiI} \right. \right]_2 + \bigcirc \xrightarrow{\quad /80° \quad} \left[\quad \text{NiI} \right]_n$$

(1,5-COD)PdCl$_2$ reacted with an excess of stannous chloride in methanol to give the dark red complex (1,5-CODPd)$_3$Sn$_2$Cl$_6$[619]; it is presumed to have the same structure as the platinum analog.[618] Formally the Pd (or Pt) appears to be in the +2/3 state here (p. 146).

E. PROPERTIES OF THE DIENE COMPLEXES

All the diene–palladium(II) complexes, with the sole exception of the tetraphenylcyclobutadienepalladium halides, are monomers with an approximately square planar arrangement of ligands about the metal. Similar considerations apply to the diene–platinum complexes.

Comparatively little spectroscopic data are available on these complexes. Where it is observed, the coordinated double bonds show a C=C stretching mode at around 1500 cm^{-1} [403, 636, 637, 662, 663] [compared to $\nu_{(C=C)}$ in excess of 1650 cm^{-1} for the free ligand], the exact position depending on the metal and the diene. Fritz and Sellmann have assigned a bond at 478 cm^{-1} to an M–olefin mode in (1,5-COD)PdCl$_2$ and at 472 cm^{-1} in 1,5-CODPtCl$_2$.[664] Analyses of the infrared spectra of some 1,5-hexadiene and 1,5-cyclooctadiene complexes have also been reported.[662]

In the NMR spectrum (cyclooctatetraene)PdCl$_2$, (IV-2), only showed a single resonance,[663, 665] but this was due to the accidental identity of the chemical shifts of the two types of proton. The NMR spectra of many of these complexes have been quoted,[665] but few useful generalizations can be made.

Fritz[666] has measured the infrared spectra of some cyclobutadiene–metal complexes including some of the palladium and nickel complexes. A band at ca. 1380 cm^{-1} was intense in these complexes, but its origin is not clear. Since all these complexes are heavily substituted on the C$_4$ ring, neither NMR nor infrared spectra reveal much.

Two Dewar benzene (bicyclo[2.2.0]hexadiene) complexes of Pd(II) are known (IV-11a,b)[637-639]; the platinum analog (IV-11c) is also known.[616] The hexamethyl complexes have the expected NMR spectra (the platinum complex shows coupling of both types of methyl to ^{195}Pt) but that of (IV-11a) appears

(IV-11a) R = H, M = Pd, X = Cl
(IV-11b) R = Me, M = Pd, X = Cl, Br
(IV-11c) R = Me, M = Pt, X = Cl

as a broad singlet at τ 5.60, probably owing to accidental identity of the chemical shifts. A $\Delta\nu_{(C=C)}$ on coordination of about 150 cm^{-1} for the palladium complexes (IV-11b) was observed; that for the platinum complex was 165 cm^{-1}.

In general, the palladium complexes of all these dienes are more easily prepared, more reactive, and more labile than the platinum analogs. 1,5-Cyclooctadiene followed by *endo*-dicyclopentadiene forms the least labile complexes; dipentene forms a very labile complex with platinum (which has two isomers),[403] but palladium is reduced to the metal by this ligand and no complex is formed.[365] An order of thermal stability for (diene)MX$_2$ complexes, M = Pt > Pd, X = Cl > Br > I has been proposed.[365] In DMSO all the palladium complexes are immediately solvolyzed, but the solvolysis rates of the platinum complexes are much less and depend on the X-ligand, in the order of decreasing rate, X = tolyl > I > Br > Cl.[665]

One of the most interesting examples of instability is that of the Dewar benzene–Pd(II) complex (IV-11b) (X = Cl). This undergoes a PdCl$_2$-*catalyzed* decomposition to hexamethylbenzene and PdCl$_2$.[637] The mechanism of this is not known. The bromo complex decomposes more slowly, and the platinum analog (IV-11c) appears stable to this type of decomposition (see also Chapter I, Volume II, Section F,2).

The tetraphenylcyclobutadienepalladium halides all have exceptional thermal and oxidative stability. Probably part of this, at least, is due to the steric hindrance exerted by the phenyl groups. These complexes do not appear to be as labile as the 1,5- or 1,4-diene complexes, though quantitative data is not available. Although the cyclobutadiene complexes are dimeric, no authenticated example of reaction with a ligand to give the monomeric species R$_4$C$_4$PdX$_2$L has been reported, but several examples are known for the nickel complexes.[628] The (pentamethylcyclopentadiene)PtCl$_2$ complex is more labile than the platinum complex (IV-11c).[616]

F. REACTIONS OF DIENE COMPLEXES

1. Nucleophilic Attack on Diene–Pd(II) Complexes

By far the most characteristic reactions of diene–Pd(II) complexes (and of their platinum analogs) are those with nucleophiles of various types. These can be divided into three types: (1) the nucleophile displaces the diene; (2) the nucleophile attacks the metal, does not displace the diene, but rather the other, anionic, ligands; and (3) the nucleophile attacks at an olefinic carbon of the diene. Although the actual product obtained is probably largely governed by the relative thermodynamic stabilities of the various possible products, some tentative rules and their exceptions can be formulated.

Anionic nucleophiles such as halides, $B_9C_2Me_2H_9^{2-}$, and $C_5H_5^-$ will usually displace the anionic ligands. However, anionic nucleophiles with a high tendency to attack at carbon will cause reactions of type (3) to occur. Examples are $(RCOCHCOR')^-$ ($R = R' = Me$, OEt; $R = Me$, $R' = OMe$) and especially alkoxide, but strangely enough, Grignard (and organolithium) reagents give products where attack at the metal appears to have been favored. Neutral nucleophiles, especially "soft" ones with a high tendency to complex with the metal will displace the ligands, for example, t-phosphines, pyridine, DMSO. Cyanide has a very high tendency to form $M(CN)_4^{2-}$ complexes with Pd(II) and Pt(II) and displaces all the ligands.

a. Nucleophilic Displacement of the Diene

Some typical examples where this occurs without rearrangement are as follows. The platinum complexes usually react more slowly.[665]

$$MCl_2 + 2DMSO \longrightarrow \quad + (DMSO)_2MCl_2 \quad [665]$$

$$+ 2L \longrightarrow \quad + L_2MCl_4 \quad [637-639]$$

$$(R = H, Me; L = PPh_3, py)$$

The tetraphenylcyclobutadienepalladium complexes react smoothly on heating with various t-phosphines and t-phosphites to give the cyclo-octatetraenes.[625, 667–669]

(R = Ph, p-tolyl, p-anisyl, p-ClC$_6$H$_4$; R′ = Ph, n-Bu; X = Cl, Br)

In the cold, [Ph$_4$C$_4$PdCl$_2$]$_2$ reacts with phosphines to give a green solution characterized by a complex ESR spectrum. Several possible formulations for this material have been proposed,[626] including

by Sandel and Freedman.[670] The cyclooctatetraene must arise by dimerization of two cyclobutadienes to a tricyclooctadiene, which then isomerizes,

Experiments in which the intermediate was trapped by reaction with a dienophile have been reported,[668] but it is not clear whether a free cyclobutadiene or a weakly complexed one is the reactive species. These trapping experiments were not invariably successful.[626, 660] (see also Chapter I, Volume II, Section C,3,a).

In the presence of oxygen, little of the cyclooctatetraene was formed; the main product was tetraphenylfuran.[667] An analogous reaction occurred with [Ph$_2$(Bu^t)$_2$C$_4$PdCl$_2$].[605] Hüttel and Neugebauer[652] have also reported the reactions of [Ph$_4$C$_4$PdCl$_2$]$_2$ with pyridine, to give tetraphenylfuran, and with DMSO to give cis-dibenzoylstilbene. However, it is possible that an oxidation which has nothing to do with the ligand displacement occurs here.

No information is yet available on [Ph$_4$C$_4$PtCl$_2$]$_2$[614]; the nickel complexes [Me$_4$C$_4$NiCl$_2$]$_2$ and [Ph$_4$C$_4$NiBr$_2$]$_2$ appear to be more strongly bonded and give adducts with PPh$_3$ and o-phenanthroline.[626]

b. *Nucleophilic Displacement of the Anionic Ligands*

Typical reactions of this type include

The last two reactions are of interest since here the cyclopentadienyl entering ligand is acting as a weak nucleophile. When $[Me_4C_4NiCl_2]_2$ was reacted with sodium cyclopentadienide, the product obtained arose by attack of $C_5H_5^-$ both at the cyclobutadiene and the metal.[673-675]

The $B_9C_2H_{11}^{2-}$ group has many properties comparable to those of $C_5H_5^-$ and Hawthorne and his co-workers[627, 669] have prepared the complexes **(IV-12)** (R = H, Me); an X-ray study of the latter has shown the structure to be correct (see p. 164).

c. *Nucleophilic Attack at an Olefinic Carbon*

Hofmann and von Narbutt in 1908 showed that dicyclopentadiene reacted slowly in alcohols with K_2PtCl_4.[402] Two types of product were isolated, $ROC_{10}H_{12}PtCl$ from methanol and ethanol (R = Me, Et) and $C_{10}H_{12}PtCl_2$ from propanol. These authors suggested that the first product arose by addition of RO– and –PtCl to a double bond. Chatt, Vallarino, and Venanzi in

$$[Ph_4C_4PdCl_2]_2 + B_9C_2H_9R_2{}^{2-} \longrightarrow$$

(IV-12)

1957 modified and extended this work to other dienes.[403] They showed that the diene–$PtCl_2$ complexes reacted with alkoxides to give the σ- and π-bonded alkoxy complexes, the dicyclopentadiene complex reacting most readily.

Two X-ray studies, one by Powell and co-workers[200] on the methoxy-*endo*-dicyclopentadiene complex (II-24) (p. 75) and one by Pannatoni *et al.*[407] on the pyridine adduct of the methoxycyclooctenyl complex (II-26), show that the attack was exo to the metal on a diene carbon. It is assumed that all such reactions involve attack exo to the metal and hence that metal–alkoxy complexes are not intermediates. For further details, see Chapter II, Section C,3,b,iii, this volume.

Panunzi *et al.* have resolved the *S*-α-methylbenzylamine adduct of (II-24) by fractional crystallization. They then treated this material with HCl to convert it to the optically active *endo*-dicyclopentadiene–$PtCl_2$ complex which, on treatment with cyanide, gave the optically active *endo*-dicyclopentadiene.[677]

The tetraphenylcyclobutadiene–metal complexes also undergo reaction with alkoxide in a similar manner[612, 652, 653, 672] to give alkoxycyclobutenyl complexes. Again here the alkoxide group is exo to the metal.[206]

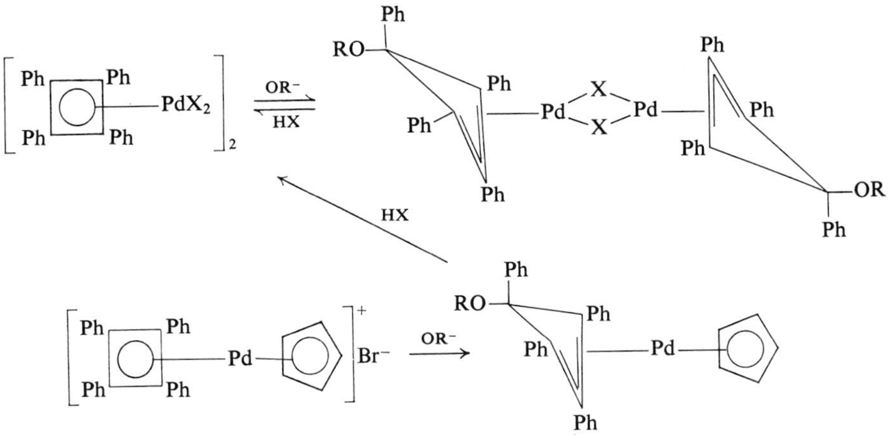

These reactions are reversed on treatment with acids to give back the cyclobutadiene complex. In the latter reaction, cleavage of the Pd–C_5H_5 bond also occurs. However, the nickel complex, exo-$ROC_4Ph_4NiC_5H_5$, gave $[Ph_4C_4NiC_5H_5]^+Br^-$ with HBr.[672]

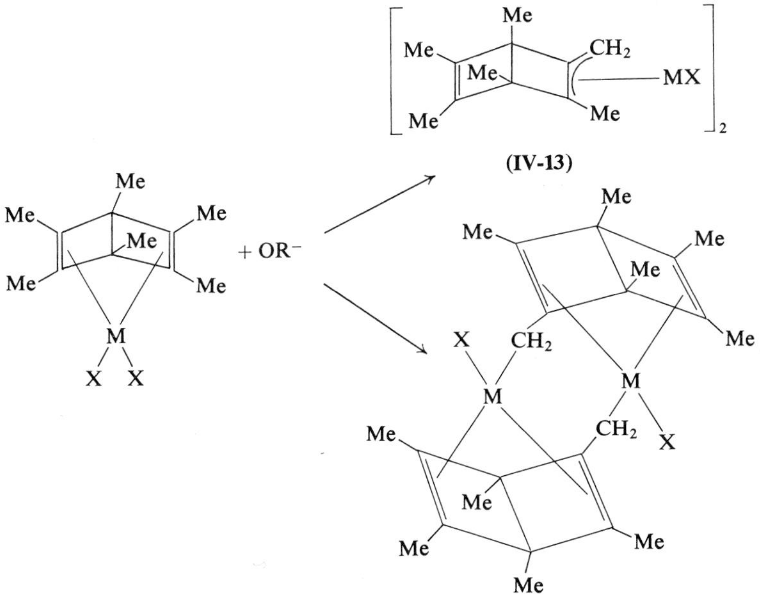

(IV-13)

(IV-14)

The hexamethyl(Dewar benzene)–palladium and –platinum complexes undergo rather unusual reactions with alkoxide. The favored reaction path here is deprotonation rather than addition and two types of product (**IV-13**) and (**IV-14**), can be obtained.[199, 221, 639]

In (**IV-13**) one double bond is uncoordinated and the metal is π-allylically bonded; this occurs for M = Pd, X = halogen and for M = Pt, X = I. In the other cases (M = Pt; X = Cl, Br) the metal atom is π-bonded to the two double bonds of one diene and σ-bonded to a methylene group of the other, (**IV-14**). Interestingly, these reactions are not reversed by acid under mild conditions.[639]

Another example of deprotonation, which is probably a fairly common reaction, was mentioned by Takahashi and Tsuji[414] who found that malonate did not add to (*endo*-dicyclopentadiene)PdCl$_2$. The product obtained, formulated by these workers as

was also isolated from reaction of the diene complex with sodium carbonate in ether–acetone.

A number of π-allylic complexes of this general type can also be formed directly from the diene, Na$_2$PdCl$_4$, and an alcohol. In some cases, with 1,3-dienes, this is the only route to the resultant π-allylic complexes since 1,3-diene complexes are generally not known. Robinson and Shaw[634] have shown in some cases (e.g., for 1,3-cycloheptadiene and 1,3-cyclooctadiene) that at low temperatures an alkoxy adduct results, whereas at higher temperatures the deprotonated product is obtained.

This makes it likely that deprotonation occurs via primary addition of OR$^-$ followed by elimination of ROH, but the causal relationship has not been established.

Ammonia, primary and secondary amines, and azide ion will also add to an olefinic bond of some 1,5- (and 1,4-) diene complexes.[416, 417]

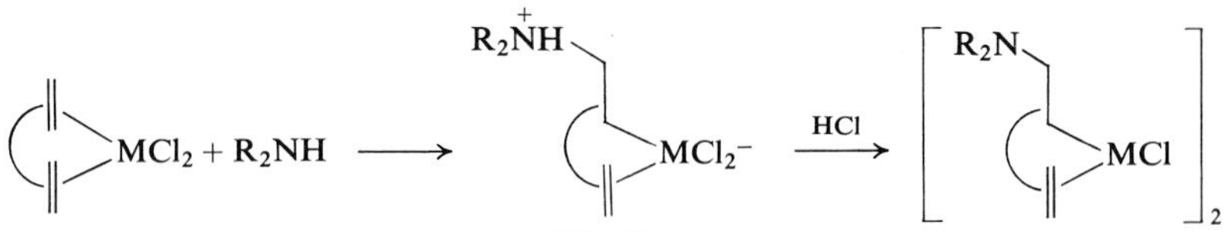

(IV-15)

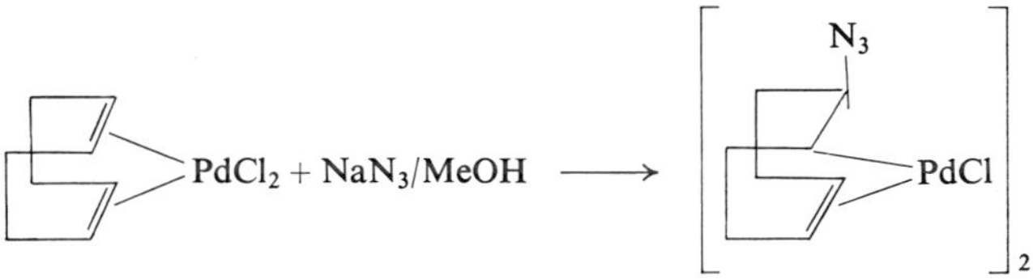

The intermediate (IV-15) could in some cases [e.g., for (1,5-hexadiene)-PtCl$_2$] be isolated. In the reaction of (1,5-hexadiene)PtCl$_2$ Palumbo *et al.* showed that attack could occur either at the 1- or at the 2-position of the co-ordinated olefin. The latter was favored for amines of low steric requirements (ammonia, ethylamine).[416]

Takahashi and Tsuji[412–414] have found that malonates, acetylacetonates, and acetoacetates react similarly to alkoxides with 1,5-cyclooctadiene–palladium and 1,5-cyclooctadiene–platinum halides. Similar reactions have been described for (*endo*-dicyclopentadiene)PtCl$_2$ by Stille and Fox[678] and Lewis and co-workers for the 1,5-cyclooctadiene complexes with thallium(I) β-diketonates.[415] In the case where the attacking group is malonate (R = R′ =

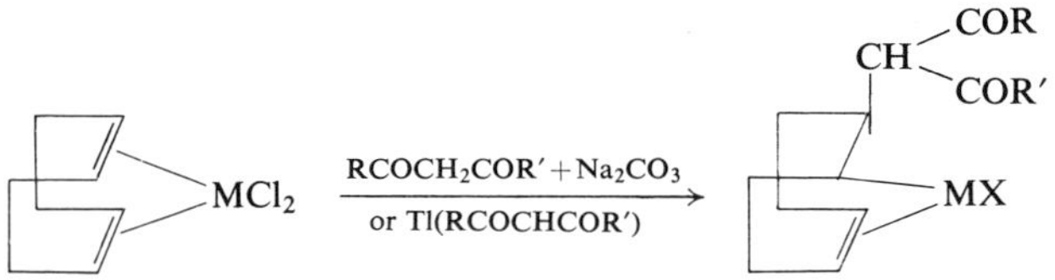

(IV-16)

OMe, OEt) or acetoacetate (R = Me, R′ = OMe) the product is the halide dimer (IV-16) (X = Cl, Br), but when β-diketonate or thallous acetylacetonate is used the complex (IV-16) (R = R′ = Me, Ph) is isolated as the β-diketonate (=X). Reaction of the latter with HCl first cleaves off the metal-bonded β-diketonate to give the halide dimer; the C-bonded β-diketone is removed by excess acid.[463]

The tetraphenylcyclobutadienepalladium dichloride also reacts with malonate to give, presumably, the cyclobutenyl complex (IV-17). This complex was incorrectly formulated by the authors.[414]

All these products arise under very mild conditions; typically the diene complex is stirred with the β-diketone, malonate, etc., in ether in the presence of anhydrous sodium carbonate. Under more drastic conditions further reactions occur which have already been described in Chapter II, Sections C,3,d,iii and v, this volume.

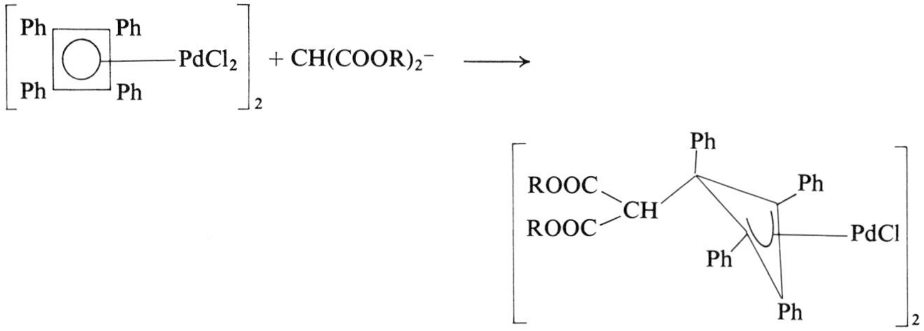

(IV-17)

Tsuji *et al.* have also reported that the carbonylation of (1,5-cyclooctadiene)-palladium dichloride in ethanol gave the following products[643]:

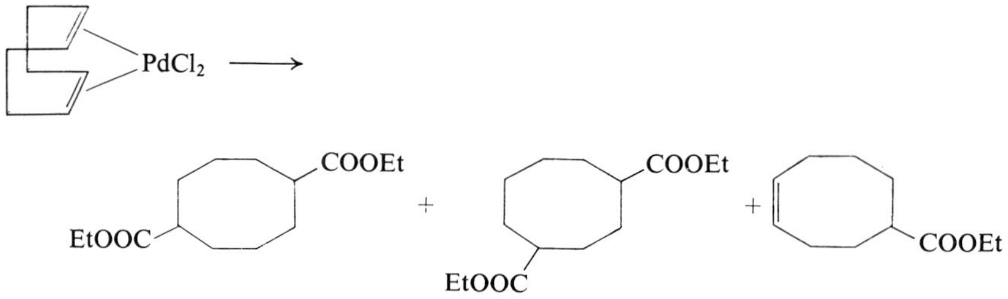

2. Reactions of Cyclobutadiene–Palladium(II) Complexes

The tetraphenylcyclobutadienepalladium complexes, because of the novelty of the ligand and the simplicity of their syntheses, have been studied to a greater extent than some other diene complexes. This work has been reviewed[626, 628] and only a brief summary will be given here. See also Sections F,1,a–c for other reactions.

Some of the reactions described here, proceed in unusual ways owing to the instability of the cyclobutadiene, but others, for example, the oxidation, reduction, and ligand-transfer reactions may well also be possible with the complexes of more conventional dienes.

The thermal decomposition of $[Ph_4C_4PdCl_2]_2$ at above $350°$ *in vacuo* gave two isomers of 1,4-dichlorotetraphenylbutadiene and a small amount of the diphenylindenoindene,[612, 668]†

$$ClCPh{=}CPh \cdot CPh{=}CPhCl \ + \qquad\qquad\qquad\qquad + \ Pd$$

By contrast, the nickel complex $[Ph_4C_4NiCl_2]_2$ did not lose halogen to the organic ligand and the major product of thermal decomposition was the octaphenylcyclooctatetraene,[679]

The thermal decomposition of $[Me_4C_4NiCl_2]_2$ gave a mixture of hydrocarbons, their nature and amount depending on the exact conditions.[680]

Vallarino and Santarella[653] showed that $[Ph_4C_4PdCl_2]_2$ reacted with CO at $70°$ and 200 atm to give the tetraphenylcyclopentenone,

Maitlis and Games[681] observed that tetraphenylcyclopentadienone was formed in quite high yield under relatively mild conditions ($80°$) in the reaction of $[Ph_4C_4PdBr_2]_2$ with nickel carbonyl in benzene. At lower reaction temperatures this product was hardly formed, the major product being the nickel complex $[Ph_4C_4NiBr_2]_2$, p. 172,

† The thermal decomposition of $[Ph_2Bu_2{}^tC_4PdCl_2]_2$ to give a homotetrahedrane [2-*t*-butyl-1,5-diphenyl-3,3,4-trimethyltricyclo[2.1.0.0^{2,5}]pentane] has been reported by Hosokawa and Moritani.[668a]

[Ph$_4$C$_4$PcCl$_2$]$_2$ was reduced by LiAlH$_4$ or NaBH$_4$ to *cis,cis*-tetraphenyl-butadiene,[652, 653, 682]

However, Hosokawa and Moritani[605] observed reduction of [Ph$_2$(Bu^t)$_2$C$_4$ PdCl$_2$], without ring-opening, to give the cyclobutene,

The nickel bromide complex [Ph$_4$C$_4$NiBr$_2$]$_2$ gave *cis*-tetraphenylcyclobutene (which isomerized to *cis-trans*-tetraphenylbutadiene above 50°) on treatment with LiAlH$_4$,[683]

The tetramethylcyclobutadienenickel complex underwent catalytic hydrogenation (or reduction with Zn–HCl) to give the all-*cis*-tetramethylcyclobutane,[680]

Oxidation of this complex with aqueous sodium nitrite gave *cis*-3,4-dihydroxy-1,2,3,4-tetramethylcyclobutene.[680] The tetraphenylcyclobutadiene-nickel complex, in contrast, gave tetraphenylfuran under similar conditions.[613]

$[Ph_4C_4PdCl_2]_2$ gave *cis*-dibenzoylstilbene with a more vigorous oxidizer (e.g., HNO_3–THF)[652, 653] (see also Chapter II, Volume II, Section I,2).

Perhaps the most interesting and unusual reactions of the cyclobutadiene-palladium and cyclobutadienenickel halide complexes are the ligand transfer reactions. These have allowed the synthesis of a large number of otherwise inaccessible cyclobutadiene–metal complexes. Two types of reaction have been observed, the first being the ligand-exchange reactions between the cyclobutadienepalladium halides and bis(*t*-phosphine)nickel dihalide complexes,[669]

$(R = Ph, p\text{-}ClC_6H_4, p\text{-}MeC_6H_4, p\text{-}MeOC_6H_4; X = Cl, Br)$

This reaction utilizes the known lability of the bis(phosphine)nickel complexes which dissociate appreciably to give the free phosphine. The mechanism of this reaction is not known, but some suggestions have been made.[626]

The other variant, of more general utility, is the ligand-transfer reaction, in which the palladium complex is reacted with a metal carbonyl, or a substituted metal carbonyl in some cases.[625, 672, 681, 684–687] Some typical examples are described in Chart (IV-1), p. 172; this reaction also proceeds with both the tetraphenyl- and tetramethylcyclobutadienenickel halide complexes.[628]

3. Reactions of Diene Complexes of Nickel(0)

Very little work has yet appeared on the chemistry of diene–metal complexes of the nickel triad in which the metal is not in the (II) state. The one exception is for Ni where the only diene–Ni(II) complexes known are the cyclobutadiene complexes. The chemistry of diene–Ni(0) complexes has been mainly investigated by Schrauzer[517] and Wilke[603] and their co-workers.

Schrauzer has prepared a series of complexes based on the unstable bis-(duroquinone)nickel (**IV-18**). On heating *in vacuo* this loses one mole of duroquinone to give a pyrophoric monoduroquinonenickel complex. In solutions of cyclooctatetraene or other dienes one duroquinone is exchanged for the diene to give (**IV-19**). These can also be prepared directly. The diene ligand in

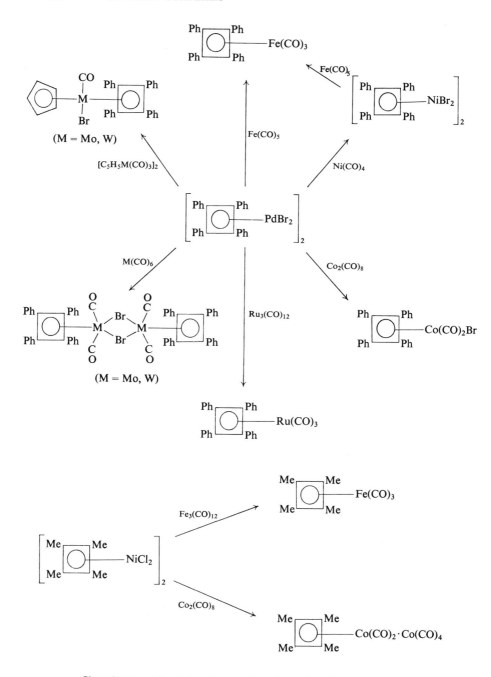

Chart (IV-1).   Ligand-transfer reactions of cyclobutadiene-metal complexes.

(IV-19) can be exchanged, and is released first on thermal decomposition. On reaction with CO both **(IV-18)** and **(IV-19)** are converted to $Ni(CO)_4$.[517, 688]

Bis(1,5-cyclooctadiene)nickel is a particularly useful starting material for syntheses of various organo complexes of Ni(II) and Ni(0).[255, 401, 603, 659] Some typical reactions are summarized in Section D,4.

Another complex with similar properties is all-*trans*-1,5,9-cyclododeca-trienenickel. Both this and (1,5-COD)$_2$Ni are very useful polymerization catalysts; these reactions will be further discussed in Chapter I, Volume II.

G. DIENE COMPLEXES OF OTHER METALS

Diene complexes are known for all the group IB triad of metals; however, in many cases only one double bond is complexed. One exception is μ-dichloro-bis(1,5-cyclooctadiene)dicopper, **(IV-6)**[622]; some silver and gold complexes of dienes have a 1:2 ratio of diene to metal and presumably each double bond is complexed to a different metal atom. Olefin complexes of these metals are all in the +I (d^{10}) state.

The cobalt triad forms a vast number of complexes of dienes; in contrast to the nickel subgroup, both chelating and conjugated dienes appear to form complexes equally easily. Diene complexes are known for cobalt in the (0) state and for Co, Rh, and Ir in the +I states. Some iridium(III) diene complexes have also been described.[689]

Exchange reactions of chelating diene complexes of Rh(I) and Ir(I) have been studied by Vrieze and Volger[690-692] and have been found to involve 5-coordinate intermediates. A similar conclusion has been reached by Catalini et al.[693] in a study of the replacement of the stibine by amine in

Perhaps the most significant feature of the chemistry of diene complexes in the (I) oxidation state here is that nucleophilic attack on the diene is again almost unknown. Examples of electrophilic attack are more common,[694-697]

This difference in reactivity may well be a consequence of the change in formal oxidation state. This characteristic is even more pronounced in the d^8 [M(0)] complexes of the iron triad; for example,[698, 699]

The chemistry of diene–iron tricarbonyl complexes is well developed and has been reviewed by Emerson and Pettit.[700] Of particular interest is that no evidence has been obtained for nucleophilic attack in tricarbonylcyclobutadiene-iron; on the contrary, electrophilic attack here is very easy.[626] No completely unsubstituted cyclobutadiene–metal(II) complexes are known, but indications are that they would be very sensitive to nucleophilic attack.[628]

Chapter V

π-Allylic Complexes

A. INTRODUCTION

In addition to π complexes derived from simple mono- and diolefins where the free ligand is usually stable, numerous metal π complexes of organic ligands, which are themselves not stable in a conventional sense, are known. Some of the simplest of these are derived from the allyl radical ($CH_2\!\!=\!\!CH\!\!-\!\!CH_2\cdot$), in which three carbon atoms are bonded, more or less equivalently, to the metal.

The allyl group can also be σ-bonded to the metal ($CH_2\!\!=\!\!CHCH_2\!\!-\!\!M$); here it occupies one coordination site, whereas a π-allyl group is usually thought of as a bidentate ligand. π-Allylic complexes of Ni, Pd, and Pt are normally 4-coordinate, but some examples of 5-coordination are also known.

For electron "bookkeeping" purposes, π-allylic complexes are best regarded as made up of an allylic *anion* and M^+. The π-allylic ligand is considered to "donate" four electrons to the metal. In a typical complex such as

the metal is square planar Pd(II), d^8, and since the two Cl^- ligands also "donate" four electrons to each Pd(II), the effective atomic number of the metal is two less than that for the next noble gas, xenon. This formalism has no real meaning in terms of the bonding involved, but is a convenient reference point for a discussion of the chemistry, and is true for all four-coordinate d^8 complexes. The σ-allyl ligand (CH_2=$CHCH_2$–) is a normal 2-electron monodentate ligand.

The first π-allylic complex was prepared by Prichard[701] in 1952 from butadiene and $HCo(CO)_4$; Slade and Jonassen[702] obtained an allylic palladium complex in 1957. However, the true natures of these complexes were not established until somewhat later.[649, 703]

Palladium forms a large number of allylic complexes in the (II) oxidation state and these have been more extensively studied, to date, than those of other metals. This is because the complexes of palladium are easy to prepare and to handle and because of their interest as intermediates in a number of catalytic reactions.

Nickel(II) forms a range of π-allylic complexes, probably at least as extensive as those of palladium, but these are usually much more reactive. π-Allylic complexes of Pt(II) are also known, but their number is not so extensive as platinum appears to prefer to σ-bond to carbon.

B. PREPARATION OF π-ALLYLIC PALLADIUM COMPLEXES

The main routes are those starting from olefins, dienes, and allyl halides and alcohols, respectively.

1. From Monoolefins (Substituted Propenes)

Many olefins react with palladium chloride in protic solvents to give ketones (or aldehydes) or organic coupling products. In certain cases, however, for example, where attack on the double bond is hindered, π-allylic complexes are obtained in satisfactory yield. Hüttel et al. have investigated the reactions of alkyl-substituted propenes with palladium chloride in 50% aqueous acetic acid.[497, 498, 599, 704, 705] There does not seem to be any simple correlation between the structure of the olefin and the resulting complex. Hydrogen can be (formally) removed from a CH_3, CH_2, or CH adjacent to the double bond, and the structure and yield of the resultant π-allylic complex would appear to be determined more by thermodynamic than by kinetic factors. Volger[706] has suggested that in the reactions of highly branched olefins with Na_2PdCl_4 in 100% acetic acid containing some sodium acetate, the hydrogen is almost always abstracted from a secondary rather than a primary carbon. Examples of the reaction described by Hüttel *et al.* are:

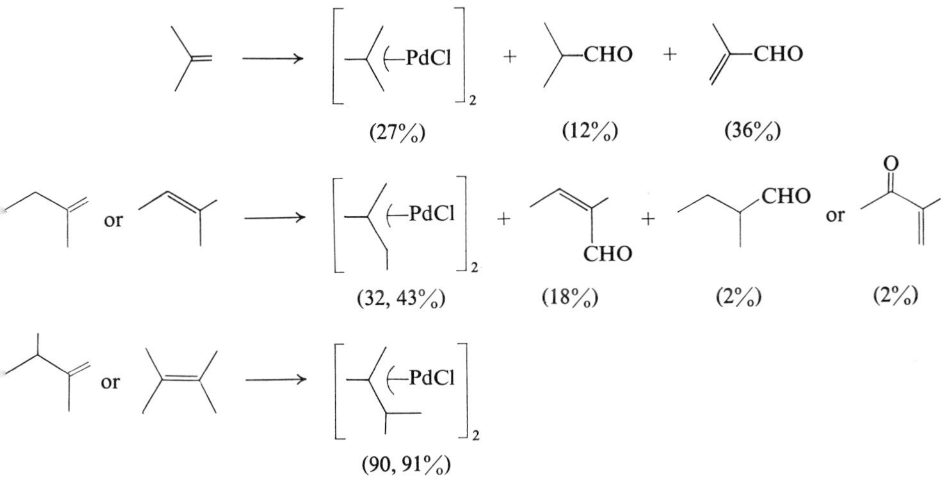

In addition some palladium was precipitated (more from the reactions which gave the lower yields), and various aldehydes and ketones were detected as by-products. Hüttel and Christ[705] have suggested that these products arise from oxidation (by $PdCl_2$) of the π-allylic complexes. Thus chloro(1,2-dimethylallyl)palladium dimer gave 57% 2-methyl-2-buten-1-al and 60% metal. See Section F,7.

Since it seemed very likely that the intermediates in the formation of both the π-allylic complexes and the carbonyl compounds were π-olefin complexes, Hüttel and Christ suggested that the products obtained depended on the relative rates of the two decomposition routes of the π-olefin complex. The route (a), which was effectively the Wacker reaction, competed with (b), the formation of the π-allylic complex. Whereas the former route yielded Pd and a saturated carbonyl the latter product (π-allylic complex) reacted with $PdCl_2$ in the presence of water to give unsaturated carbonyls and metal.

$$\text{olefin} + PdCl_2 \longrightarrow \text{olefin}PdCl_2 \xrightarrow[H_2O]{(a)} \text{saturated carbonyl} + Pd + HCl$$

$$\text{olefin}PdCl_2 \xrightarrow{(b)\ -HCl} (\pi\text{-allylic }PdCl)_2$$

$$(\pi\text{-allylic }PdCl)_2 \xrightarrow{PdCl_2/H_2O} \text{unsaturated carbonyl} + Pd + HCl$$

Path (a) was more important for unbranched and (b) for branched olefins. Addition of Cu(II) salts [to reoxidize Pd(0)] increased the yield of the π-allylic complexes (see also Section E,7).

The intermediacy of π-olefin complexes has been demonstrated by Hüttel, Kratzer, and Bechter[599] and other workers. For example, the isobutene complex on brief heating with water yielded the 2-methyllalyl complex in 23% yield.

$$\left[\!\!\!\!\!\!\text{—PdCl}_2\right]_2 \xrightarrow{\text{H}_2\text{O}/\varDelta} \left[\!\!\!\!\!\!\text{—PdCl}\right]_2$$

Ketley and Braatz[600] also noted that the π-isobutene complex reacted with sodium carbonate in chloroform under anhydrous conditions to give a quantitative yield of the π-2-methylallyl complex.

$$\left[\!\!\!\!\!\!\text{—PdCl}_2\right]_2 \xrightarrow{\text{Na}_2\text{CO}_3/\text{CHCl}_3} \left[\!\!\!\!\!\!\text{—PdCl}\right]_2$$

NaHCO$_3$ and Na$_2$HPO$_4$ were also used successfully both in this reaction and in the synthesis of allylic complexes directly from the olefins without isolation of the π-olefin complexes. However, LiCO$_3$, CaCO$_3$, CaHPO$_4$, and K$_2$CO$_3$ were inactive.

Another intriguing example of the base-assisted conversions of olefin–PdCl$_2$ complexes into π-allylic complexes is[639]:

$$\text{PdX}_2 \xrightarrow{\text{NaOMe/MeOH}} \left[\begin{array}{c}\text{CH}_3 \\ \text{XPd—} \\ \text{CH}_2\end{array}\right]_2$$

Volger[706] has published a modification of the Hüttel reaction which gives higher yields of π-allylic complexes and lower amounts of by-products. This involves using glacial acetic acid as solvent (i.e., no water is present) and tetrachloropalladate (rather than PdCl$_2$) as source of palladium, in the presence of stoichiometric amounts of sodium acetate. The reaction occurs at 85°, can be adapted to the preparation of the π-allylic palladium bromide complexes, and usually gives yields in the range 60–98%. The completion of the reaction is indicated by a color change from red to yellow. Using this as criterion the relative rates for reaction of three olefins are CH$_2$=CMeCH$_2$CO·CH$_3$ > CH$_2$=CMe(CH$_2$)$_2$CH$_3$ > CH$_2$=CPh·Me. The reaction is only applicable to 2-alkyl- or 2-aryl-substituted olefins.

No reaction occurs under these conditions in the absence of sodium acetate and Volger has described the system in terms of two equilibria, K_1 and K_2.

$$\text{—} + \text{PdCl}_4{}^{2-} \underset{}{\overset{k_1}{\rightleftharpoons}} \text{—PdCl}_3{}^- + \text{Cl}^- \underset{}{\overset{k_2}{\rightleftharpoons}} \left[\!\!\!\!\!\!\text{—PdCl}\right]_2 + \text{HCl}$$

In the absence of base the equilibrium described by K_2 is strongly to the left-hand side, but base causes a shift to the right. The rate of formation of the

π-allylic complex is probably given by $k_1 k_2$[olefin][$PdCl_4^{2-}$]. Hence the reactivity of an olefin is not determined solely by k_2, but also by k_1. Evidence for this is obtained from deuteration studies. Deuterium (from CH_3COOD) is incorporated only slowly and the solution containing the olefin and $PdCl_4^{2-}$ needs to stand for some time before the acetate (base) is added if deuterium-incorporation is to occur. Not only terminal allylic hydrogens† are replaced, but the others usually exchange more slowly. Thus in chloro(π-2-neopentylallyl)-palladium dimer (V-1) incorporation of deuterium at A is much faster than at B. This is explained if k_3 is much smaller than k_2. Since the deuteration studies

(V-1)

(V-2)

indicate the formation of the isomeric complex (V-2), its absence in the product must clearly be due to the greater thermodynamic stability of (V-1). However, under other conditions this isomer is formed (see below and Section F,10).

The reversibility of the π-olefin $\rightleftharpoons$ π-allyl step has been a subject of some dispute,[599, 600] but this may be due to the high reactivity of the π-olefin complex and the consequent difficulty in obtaining it pure. Volger has suggested that the hydrogen is usually lost from the π-olefin complex as H^- unless activating groups (e.g., the carbonyl in CH_2=$CMe \cdot CH_2COMe$) which make this hydrogen acidic are present. In that case, the hydrogen is probably removed as H^+ by base. However this point is not yet clear.

Volger[706] has also shown that cis-3-methyl-3-pentene isomerized first to the trans-olefin complex (possibly by addition and elimination of Pd–Cl to the double bond) which then reacted rapidly to the 1,2,3-trimethylallyl complex.

† The term "allylic hydrogen" is used here to apply to any hydrogen attached to a carbon ("allylic carbon") which is π-bonded to the metal.

Another variation has been described by Brown *et al.*,[291] who discovered that 1-hexene reacted slowly with palladium acetate in acetic acid to give 1-propyl-allylpalladium acetate (52%) together with other products. In the presence of perchloric acid, however, a fast reaction, leading to 1-ethyl-3-methylallyl-palladium acetate (75%) (and other organic products) was observed. The increase in rate in the presence of acid was ascribed to protonation of the palladium acetate with formation of a more strongly electrophilic species. The overall reaction was interpreted in terms of a series of reversible equilibria between olefin and allylic complexes:

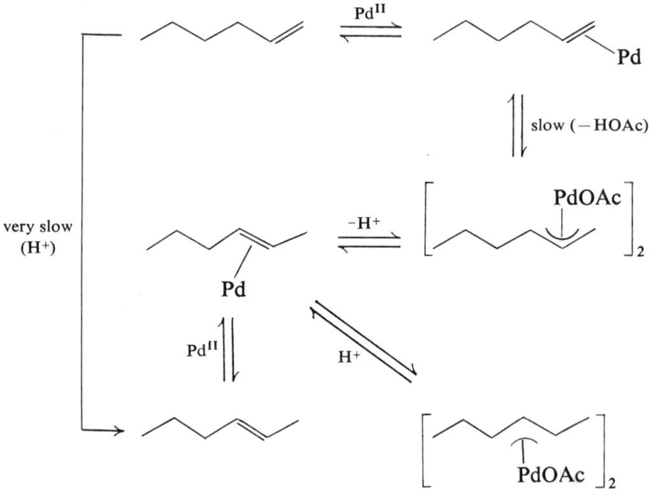

Cyclic olefins also underwent the Hüttel reaction. The best yields were obtained from cycloheptene[499] and cyclodecene.[707] The π-cyclohexenyl complex was obtained in low yield from the π-cyclohexene complex; a better route was from 1,3-cyclohexadiene.

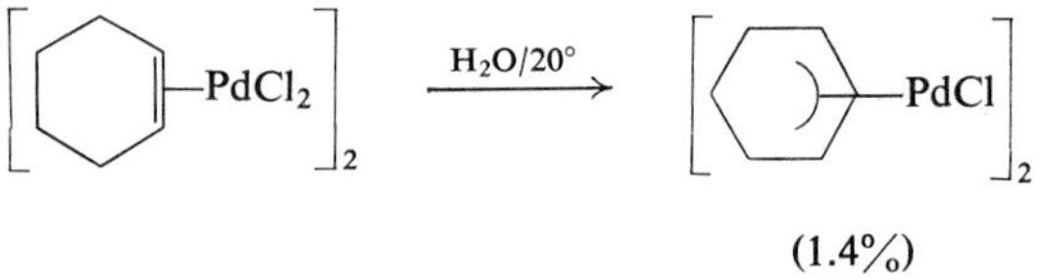

(1.4%)

1-Methylcycloheptene and 1-methylcyclooctene readily formed the exocyclic π-allyls, (**V-2a**).

Morelli *et al.*[601] reported another variant on the formation of π-allylic complexes from 1-olefins, by heating them with $PdCl_2$ in DMF. Two complexes, one the desired π-allylic complex (commonly isolated in 30–45% yield) and the other, identified as $[DMF_2H]_2{}^{2+}Pd_2Cl_6{}^{2-}$, were obtained. Again π-olefin complexes were shown to be intermediates.

(V-2a) $(n = 5$ or $6)$

Ketley et al.[507] noted that isobutylene will also act as a hydrogen acceptor in the formation of π-allylic from olefin complexes.

This type of reaction may also explain the formation of π-allylic complexes in the reactions in which ethylene is dimerized by $PdCl_2$ in aprotic solvents, e.g., **(V-3)**.

Donati and Conti[708] have shown that l-octene reacts with palladium chloride itself at 100° to give metal, HCl, and a solution from which two allylic complexes (both mixtures of isomers) could be obtained.

$$CH_2\!\!=\!\!CH(CH_2)_5CH_3 + PdCl_2 \xrightarrow{100°} [C_8H_{15}PdCl]_2 + [(C_8H_{15})_2Pd_3Cl_4]_n$$

The nature of the latter complex is discussed on p. 229.

Tsuji et al. observed that olefins with electron-withdrawing β-substituents, especially carbonyl, will very readily form π-allylic complexes.[709,710]

$(L = PhCN, Cl^-; R = H, COOEt)$

Electron-withdrawing substituents alpha to the double bond[710-718] or even on the double bond[220] have a similar effect.

$$\text{(cyclohexanone=CHPh)} + Na_2PdCl_4 \longrightarrow \left[\begin{array}{c} \text{PdCl} \\ \text{C—H} \\ \text{Ph} \end{array} \right]_2$$

$$CH_3CH=\overset{+}{C}HPPh_3Cl^- + PdCl_2 \longrightarrow \left(\hspace{-4pt}-PdCl_2^- \atop \overset{+}{P}Ph_3 \right) \text{ or } \left[\left(\hspace{-4pt}-PdCl \atop \overset{+}{P}Ph_3 \right) \right]_2 2Cl^-$$

$$MeCOCH_2COOEt \rightleftharpoons MeC(OH)=CHCOOEt \longrightarrow \left[HO-\hspace{-4pt}\left(\hspace{-4pt}-PdCl \atop COOEt \right) \right]_2$$

$$\underset{O}{\overset{CH_2}{\square}} \xrightarrow{Na_2PdCl_4/EtOH}$$

Hüttel and Schmid[714] have examined the effect of structure of some unsaturated acids on the yield and nature of the π-allylic complex produced. In some cases decarboxylation occurred:

$$\underset{CH_2}{\overset{Me}{>}}C-CH\underset{COOH}{\overset{Me}{<}} \xrightarrow[(3\%)]{90°} \left[Me-C\underset{COOH}{\overset{CH_2}{\underset{CMe}{(}}}-PdCl \right]_2 \xleftarrow[(13\%)]{100°} Me_2C=CMe\cdot COOH$$

$$\underset{CH_2}{\overset{Me}{>}}C-\overset{Me}{\underset{COOH}{C}}-COOH \xrightarrow[(97\%)]{25°}$$

$$\text{(cyclohexenyl)}-CH(COOH)_2 \longrightarrow \left[\text{(cyclohexenyl)}-C\underset{COOH}{\overset{COOH}{<}} \atop PdCl \right]_2$$

$$\text{(cyclohexenyl)}-CMe(COOEt)_2 \longrightarrow \left[\text{(cyclohexenyl)}-CMe(COOEt)_2 \atop PdCl \right]_2$$

Some simple olefins will couple to give allylic complexes containing dimers or even trimers of the olefin; for example, Moiseev and his collaborators have reported,[719-723]

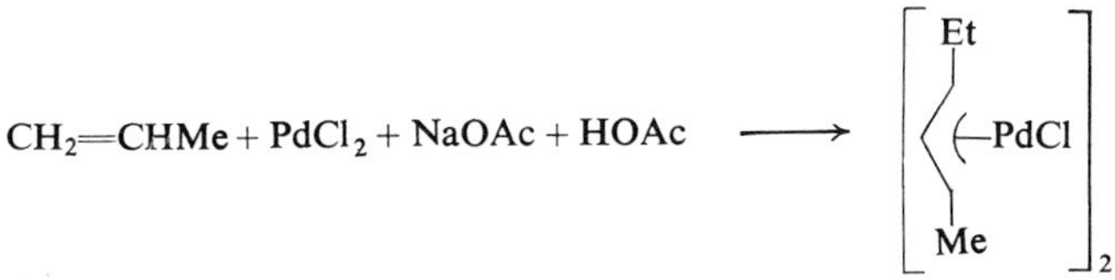

$$CH_2{=}CHMe + PdCl_2 + NaOAc + HOAc \longrightarrow \left[\begin{array}{c} Et \\ \diagup\!\!\!\!\diagdown\!(\!\!-PdCl \\ Me \end{array} \right]_2$$

Ketley et al.[507] have reported the isolation of a complex, which they formulated as (V-3), as a terminating step in the reaction of ethylene with $PdCl_2$ in $CHCl_3$. However this material *may* be a mixture.

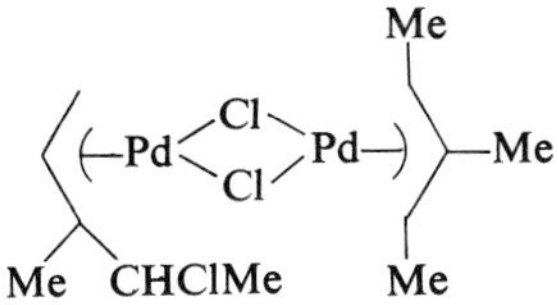

(V-3)

In general (see below), steric repulsions are minimized if substituents at the 1- or 3-position in π-allylic complexes are syn and this is the configuration normally observed.† However, Lukas et al.[724] find that 2,4,4-trimethyl-2-pentene gives a mixture of the syn and the anti complexes, (V-4) and (V-5). The anti

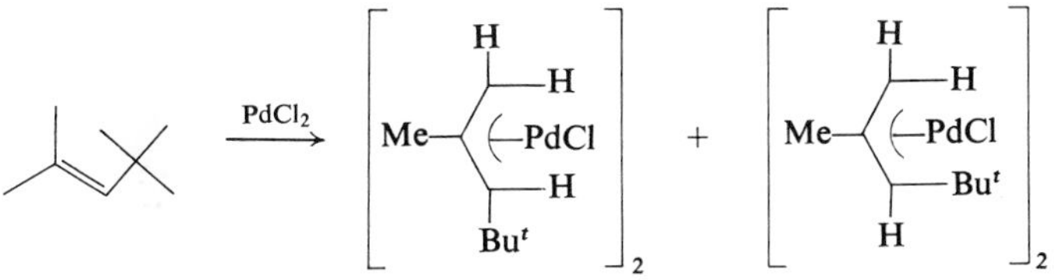

(V-4)   (V-5)

complex, (V-5), is kinetically favored and is the major product using short reaction times (in HOAc/NaOAc). Longer reaction times give (V-4). They explain the initial formation of the anti isomer in terms of steric repulsion between the Cl of the metal and the *t*-butyl group in the olefin π-complex intermediate. Twisting to relieve this interaction brings the *cis*-methyl close to the chlorine and results in the formation of (V-5) (see Section F, 1,d, and 10).

π-Allylic complexes derived from some steroids and terpenes have been reported.[715]

In contrast to the facility with which olefin–Pd(II) π complexes are converted to π-allylic complexes, olefin–Pt(II) complexes do not lose hydrogen as readily

---

† The generally accepted nomenclature is that substituents at carbon 1-(or 3-) on the same side as the 2-substituent are *syn*, while those on the opposite side are *anti*.

and attempts to prepare π-allylic platinum complexes by a number of these routes have failed.[601,711,712] π-Allylic nickel complexes have not been obtained by these routes either, which is not surprising since olefin–Ni(II) complexes are virtually unknown.

2. π-Allylic Complexes from Dienes

a. 1,3-*Dienes*

Slade and Jonassen[702,725] first isolated the complex $[C_4H_6PdCl_2]_2$ from reaction of butadiene and $(PhCN)_2PdCl_2$ in benzene. They formulated it as a butadiene complex, but this was subsequently disproved by Shaw[649] who showed it to be chloro(2-chloromethylallyl)palladium dimer (V-6) (X = Cl).

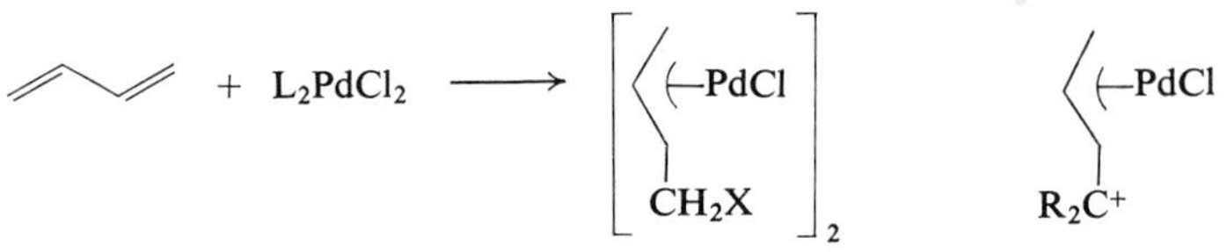

(V-6)   (V-7)

He also showed that if the reaction was carried out in methanol the product was (V-6) (X = OMe). Other 1,3-dienes and other alcohols reacted similarly.[650] The alkoxy complexes (V-6) (X = OMe, OEt) were easily interconverted in the presence of 10^{-2} M HCl; this was ascribed to the high stability of the intermediate carbonium ion (V-7), particularly when R = alkyl. In acetic acid and in the presence of sodium acetate the chloro(1-acetoxyallyl)palladium complex (V-6) (X = OAc) was obtained by Rowe and White.[726] These authors also observed that 1,3-pentadiene reacted with sodium chloropalladate in methanol to give two products, (V-8) at 0°–10° and (V-9) at 50°–60°.

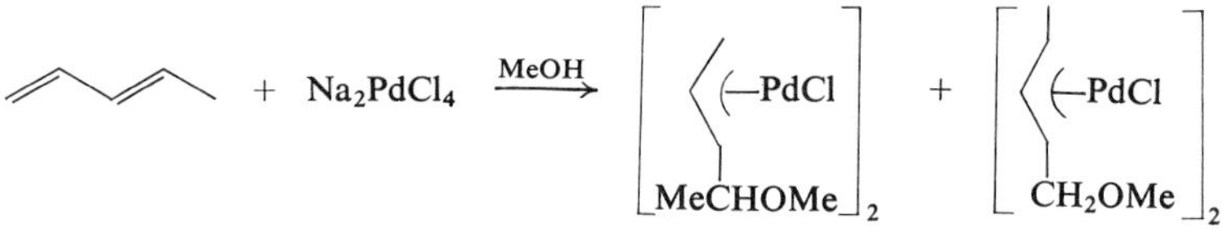

(V-8)   (V-9)

Isopropanol did not always give the isopropoxy analog of (V-6); with 2,4-dimethyl-2,4-hexadiene a complex formulated as (V-10) was obtained. This reaction can be viewed in terms of a "normal" reaction of the diene, followed by deprotonation and rearrangement[650]; or else the diene can be regarded as made up of two non-interacting dimethyl-substituted olefins, one of which reacts as isobutene does.

$$
\text{(structure)} + \text{Na}_2\text{PdCl}_4 \xrightarrow{\text{Pr}^i\text{OH}} \left[\underset{\text{Me}}{\overset{\text{Me}}{>}}\text{C=CHCH}\underset{\text{PdCl}}{\overset{\text{Me}}{\overset{|}{-\text{C}-}}}\text{CH}_2 \right]_2
$$

(V-10)

Methyl sorbate gave two products[727]:

$$
\text{(structure)}\text{COOMe} + \text{Na}_2\text{PdCl}_4 \xrightarrow{\text{MeOH}} \left[\underset{\text{MeCHOMe}}{\overset{\text{COOMe}}{(\text{--PdCl})}} \right]_2 + \left[\underset{\text{COOMe}}{\overset{(\text{--PdCl})}{\text{H--C}\underset{\text{C--H}}{}}} \right]_2
$$

Cyclic dienes react analogously, as has been shown by Hüttel,[499] Shaw,[634,728] Fischer,[339] and Wilkinson[338] and their co-workers. Cyclopentadiene did not give an isolable pure product.[634] 1,3-Cycloheptadiene and 1,3-cyclooctadiene gave an adduct analogous to (V-6) at low temperatures, whereas at higher temperatures the unsaturated cyclic allyl was formed.[499,634,729]

$$
\text{(cycloheptadiene)} + \text{Na}_2\text{PdCl}_4 + \text{MeOH}
$$

$$
\xrightarrow{-15°} \left[\underset{\text{OMe}}{\text{(ring)}}\text{--PdCl} \right]_2
$$

$$
\xrightarrow{60°} \left[\text{(ring)}\text{--PdCl} \right]_2
$$

The double bond is not conjugated with the π-allylic group in either the C_7 or the C_8 complex, though some interaction is possible in the former case, p. 202.[207,499,634]

1,3-Cyclohexadiene reacted rather differently; at −20° the methoxy adduct was obtained, but on heating reaction occurred to give the cyclohexenyl complex.[499,634] The latter complex was also obtained from cyclohexadiene and the palladium carbonyl chloride formulated as PdCOCl_2.[338,339]

There have been no reports of the conversion of the lower temperature alkoxy–allylic complex to the higher temperature dienyl or enyl complex and it may be that the former is not intermediate in the formation of the latter but that both derive from a common intermediate.

The mechanisms by which these products arise are interesting. Donati and Conti[506] showed that at low temperatures (−40° and 20°, respectively) butadiene and 1,3-cyclooctadiene gave the π-olefin complexes (V-11) and (V-12), from which the diene could be recovered unchanged. The complex (V-12) was

stable up to 110°, but the butadiene complex (V-11) was quantitatively transformed to the π-allylic complex (V-6) above −20°.

This result suggests that the first step in the reaction is coordination of one double bond which is then followed by addition of Pd–X to the double bond to give the π-allylic complex (Scheme V-1). Loss of X⁻, followed by loss of H⁺,

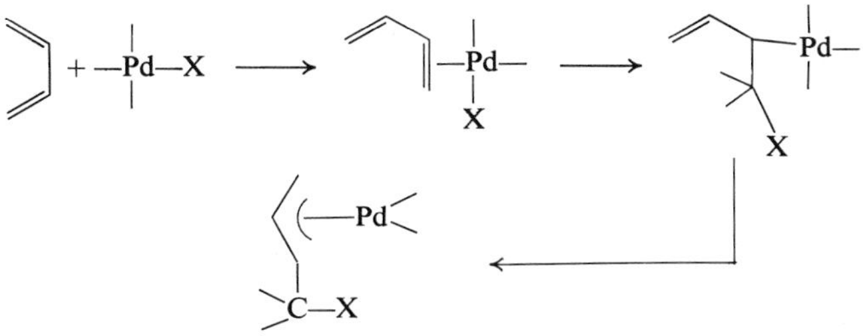

SCHEME V-1.

from the resultant carbonium ion (accompanied by a rearrangement in some cases) leads to the other products. As the various allylic intermediates involved are labile and as it is not certain just how Pd–X adds to the diene (if indeed it adds in the same sense in each case), further mechanistic speculation is unwarranted at this time. Evidence in support of this scheme comes from some work of Heck.[386,730] He has shown that a phenylpalladium complex (formed *in situ* from PhHgCl or Ph₂Hg and LiPdCl₃ in acetonitrile, but not isolated) will react very readily with 1,3-dienes to give phenyl-substituted π-allylic complexes, e.g.,

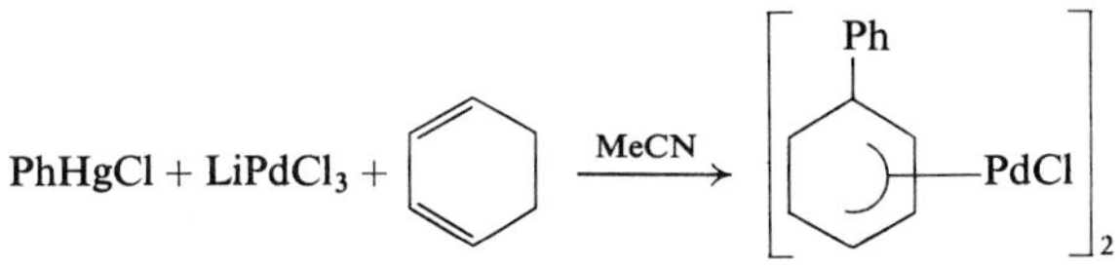

Analogous reactions occur with other organomercury compounds.

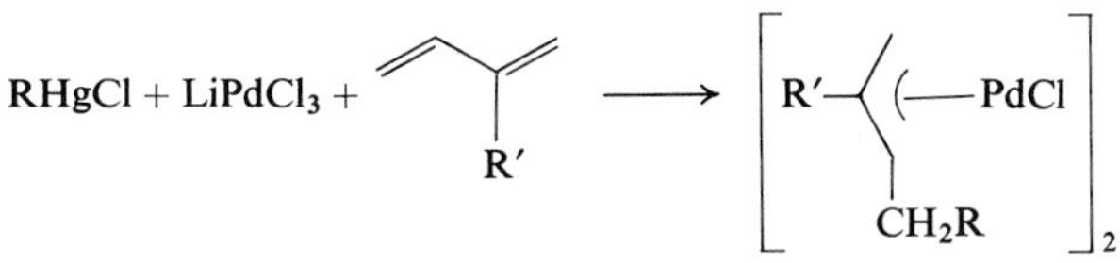

(R = Ph, Me, *p*-anisyl, benzyl; R′ = H or Me)

The yields are not high, but can be as much as 56 %. Organotin compounds have also been used in place of organomercurials. Obviously therefore, X in Scheme V-1 can be either Cl, alkoxy, alkyl, or aryl. Since all the reactions are carried out in the presence of chloride as ligand, one possible explanation for the isolation of π-allyls with OR and R rather than Cl on the organic ligand is that Pd–OR and Pd–R add more readily to a diene than Pd–Cl. The alternative suggestion, that Pd–Cl always adds first, followed by replacement of the reactive Cl by R or OR, is also possible. The addition of Pd–Cl in these systems has recently been shown to occur stereospecifically.[730a]

A rather interesting case of a 1,3-diene already coordinated to the metal adding nucleophiles to give π-allylic complexes is that of the cyclobutadiene–metal complexes. Several variants of this reaction are known, see Chapter IV, Section F,1,c, this volume.[206, 612, 625, 652, 653, 672–675, 731, 732]

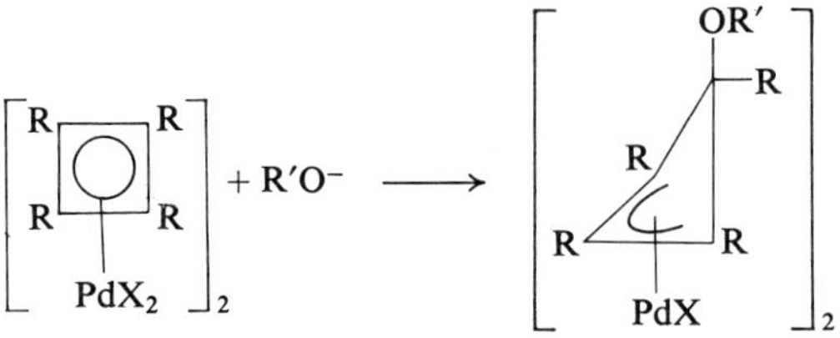

(R = phenyl or p-substituted phenyl)

In each case the nucleophilic attack on the cyclobutadiene ring occurs exo to the metal and a π-cyclobutenyl complex results. These reactions suggest another possible path for the formation of complexes such as (V-6), by formation of a π-diene complex, followed by attack on a terminal olefinic carbon by X⁻, not coordinated.

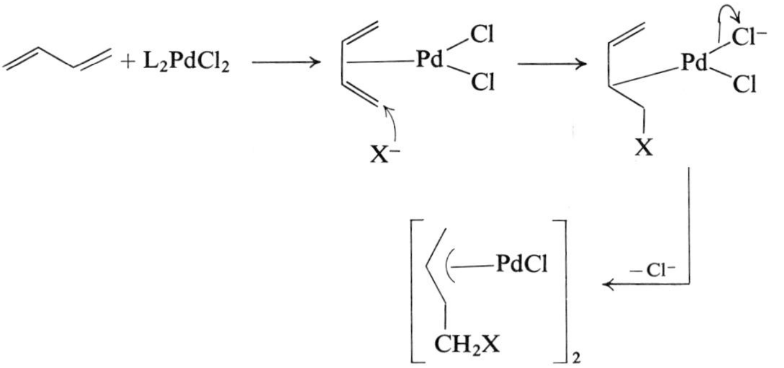

b. 1,2-Dienes (Allenes)

The reactions of allenes with palladium chloride have been investigated by Schultz[733, 734] and Lupin, Powell, and Shaw.[735, 736] Two types of π-allylic

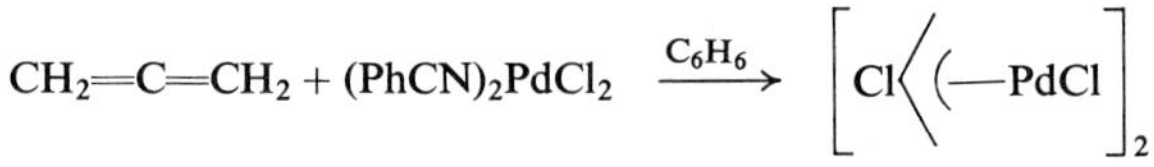

(V-13)

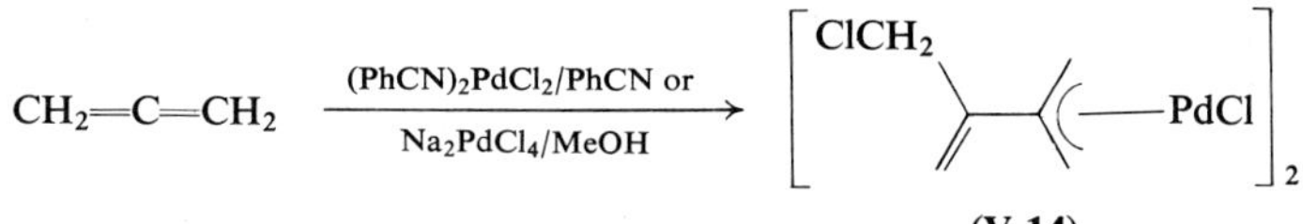

(V-14)

complex, (V-13) and (V-14) ,were isolated from allene itself. The former was the main product in nonpolar solvents (e.g., benzene), whereas the latter was isolated from reactions in methanol, benzonitrile, or methylene chloride. Schultz[733] also reported the reaction of allene with $(PhCN)_2PdCl_2$ in methanol gave (V-15) in addition to (V-14).

$$
\left[\quad \underset{MeOCH_2}{} \diagup\kern-0.4em\diagdown \kern-0.8em \diagup\kern-0.9em= \kern-0.3em\diagdown (\!-\!PdCl \right]_2
$$

(V-15)

The methyl-substituted allenes only gave complexes analogous to (V-13),

$$
\left[\quad Cl\!-\!\!\diagdown (\!\!\begin{array}{c} R' \\ \diagup\!-\!R' \\ \hline \\ \diagdown\!-\!R \\ Me \end{array}\!\!-\!PdCl \right]_2
$$

(R = R′ = H; R = R′ = Me; R = Me, R′ = H)

The mechanisms of these reactions have been considered.[733, 736] The first step involves coordination to allene, followed by insertion of C=C into the Pd–Cl bond. Two intermediates, (V-16a) and (V-16b), can be postulated to be formed initially; the former then rearranges to give (V-13), while the latter reacts with more allene to give, finally, (V-14).

An analogous mechanism, involving insertion of a coordinated allene double bond into Pd–OMe can be written to explain the formation of (V-15).

Evidence for a stepwise mechanism has been presented by Hughes and Powell from the reaction of allene with allylpalladium acetylacetonate.[737] Here a complex with a σ-allyl and π-bonded allene can be postulated as an intermediate in the formation of the 2-propenylallylpalladium acetylacetonate. The formation of bis-2,2′-[π-allylylpalladium acetylacetonate] and 1,5-hexadiene is harder to explain.

$$
\diagup\kern-0.4em\diagdown(\!-\!Pd\ acac + CH_2\!=\!C\!=\!CH_2 \quad \longrightarrow
$$

$$
\left[CH_2\!=\!CH\!\cdot\!CH_2\!-\!\!\diagup\kern-0.4em\diagdown(\!-\!Pd\ acac \right] + \left[acac\ Pd\!-\!\!\diagdown\!\!\diagup\kern-0.6em=\kern-0.3em\diagdown(\!-\!Pd\ acac \right] + CH_2\!=\!CH(CH_2)_2CH\!=\!CH_2
$$

The reaction of allene with palladium acetate and with bis-2,2′-[π-allylyl-palladium acetate] has been studied by Okamoto et al.[737a]

A recent paper by Stevens and Shier[738] sheds a very interesting light on these insertion reactions; trans-$(Et_3P)_2PdRBr$ is normally inert to olefins, but in the presence of $AgBF_4$ a very rapid reaction occurs with allene to form a π-allylic complex.

The halogen is abstracted from $(Et_3P)_2PdRBr$ by $AgBF_4$, presumably to leave a coordinatively unsaturated complex, $[(Et_3P)_2PtR]^+$, which can then coordinate allene and react further. Under normal conditions bis-t-phosphine

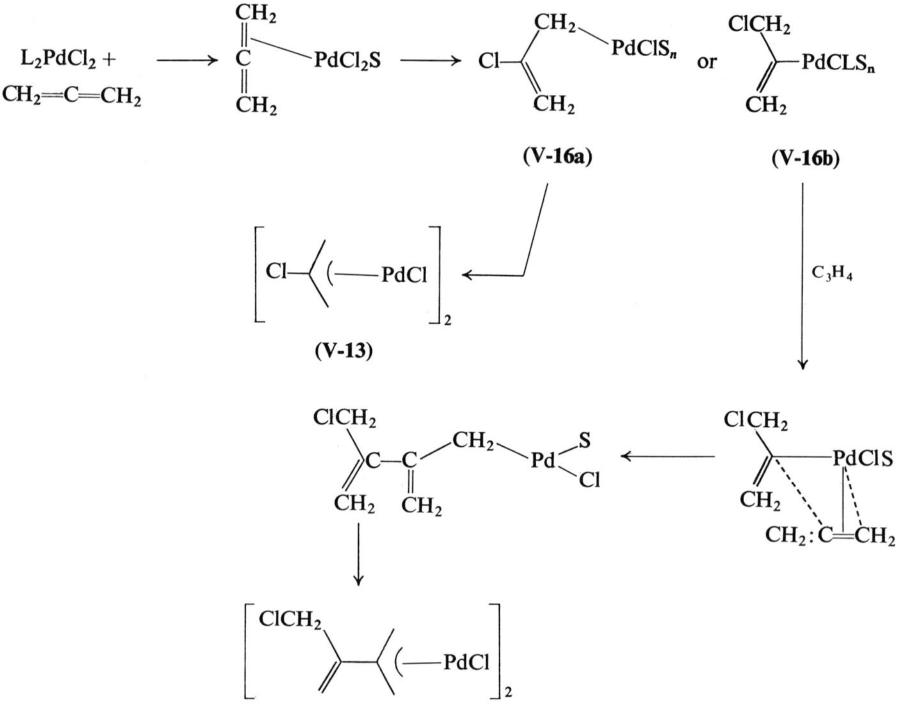

(V-16a)   (V-16b)

(V-13)   (V-14)   (S = solvent)

complexes of Pd(II) are very inert in reactions with unsaturated compounds; from this work it appears that this is only because of the absence of a vacant coordination site. Presumably other olefins and acetylenes will undergo reactions similar to the above.

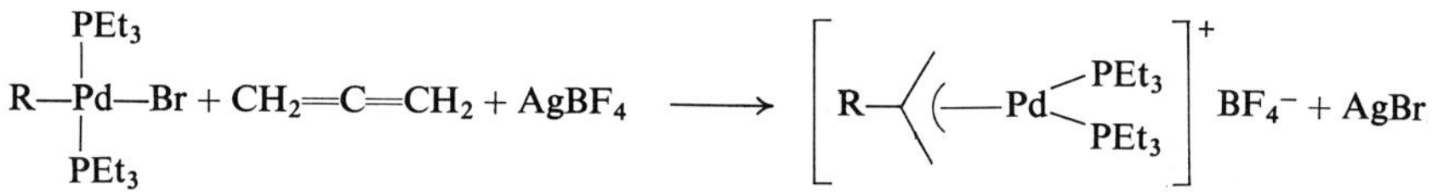

(R = Me, Ph)

When R = benzyl, no reaction with allene occurs, but $AgBF_4$ promotes a rearrangement to a π-benzyl complex, which exhibits fluxional behavior.

$$PhCH_2\underset{\underset{PEt_2}{|}}{\overset{\overset{PEt_2}{|}}{Pd}}\!\!-\!\!Br + AgBF_4 \longrightarrow \left[\begin{array}{c} \bigcirc \\ \diagdown\!\!-\!\!Pd\!\!\overset{PEt_3}{\underset{PEt_3}{\diagdown}} \end{array} \right]^+ BF_4^- + AgBr$$

3. From Cyclopropanes and Related Compounds

The chemistry of cyclopropane shows many analogies to that of olefins, and Walsh[739] has discussed the bonding in terms of each carbon being sp^2 hybridized. These chemical similarities have led to some interest in the reactions of cyclopropanes with palladium chloride. Ketley et al.[588, 589, 740] have examined the reactions of various cyclopropanes. Bromocyclopropane reacted in benzene at 80° to give allylpalladium chloride dimer,

$$Br\!-\!\triangleleft + [C_2H_4PdCl_2]_2 \longrightarrow \left[\diagdown\!\!\!\diagup\!\!(\!-\!PdCl \right]_2$$

but cyclopropane did not appear to give a π-allylic complex. Tsuji et al.[741] have reported that cyclopropane was isomerized to a small extent to propylene by palladium chloride in an autoclave. This should be contrasted with the Pt(IV) complex isolated from reaction of cyclopropane with H_2PtCl_6.[464, 465] (see also Brown[741a]).

$$\triangleleft + H_2PtCl_6 \longrightarrow \left[\triangleleft\!\!\!\square PtCl_2 \right]_4 \overset{py}{\longrightarrow} \triangleleft\!\!\!\square \underset{\underset{py}{|}}{\overset{\overset{py}{|}}{Pt}}\overset{Cl}{\underset{Cl}{\diagdown}}$$

A crystal structure of the bispyridine adduct confirms this[448]; p. 88.

Vinylcyclopropanes react with ethylenepalladium chloride dimer to give olefin complexes, presumably (V-17), which on heating undergo rearrangement to give π-allylic complexes of various types.[589] Substitution on either the cyclopropane or the olefin considerably slows the rate at which the π-allylic complexes are formed. A mechanism involving the intermediate production of a cyclopropylcarbinyl cation has been proposed.[589]

The same mixture of π-allylic complexes for R = H was obtained by reacting 1,3-pentadiene and $[C_2H_4PdCl_2]_2$. Similar results have been obtained by Shono et al.[742]

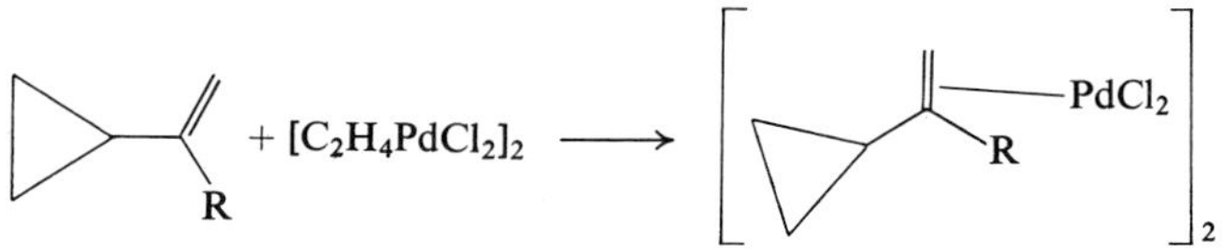

(V-17)

(R = H, Me, Ph)

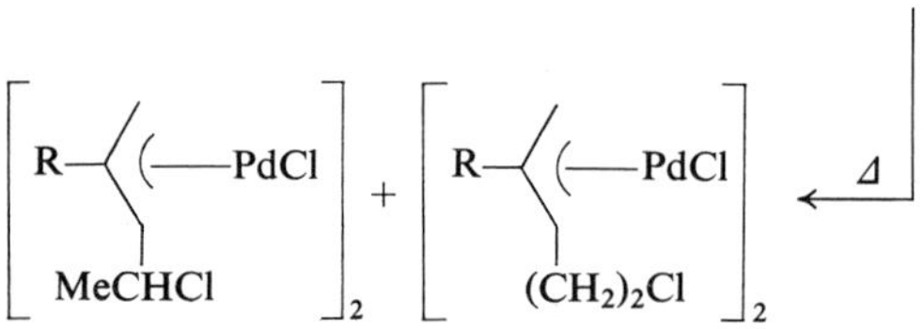

(R = H)   (R = H, Me)

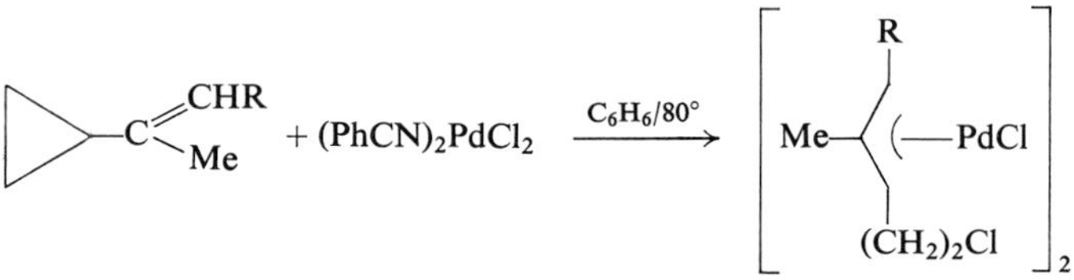

(R = Me, Ph, COOEt)

Noyori and Takaya[743] have obtained π-allylic complexes from methylene-cyclopropanes. The direction in which the cyclopropane ring opened and hence the type of π-allylic complex obtained was strongly dependent on the substituents.

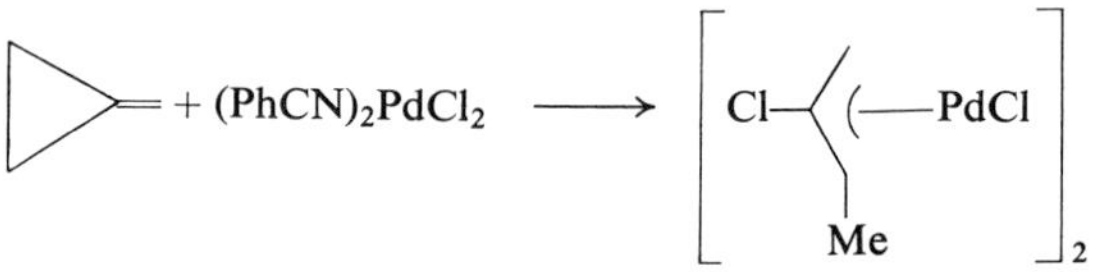

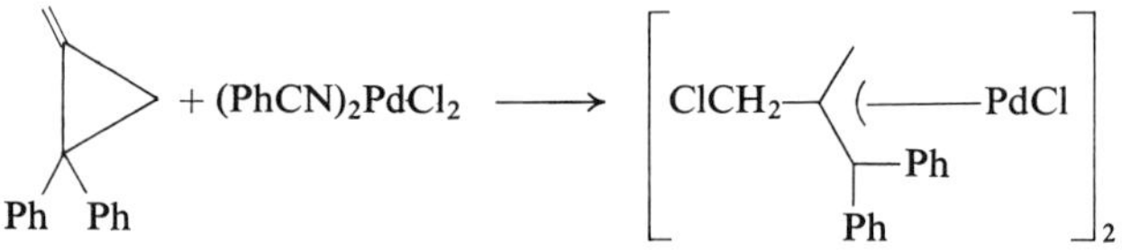

Spiropentane (but not spiro-2,4-heptane or spiro-2,5-octane) reacted readily with ethylenepalladium chloride dimer to give the 2-(β-chloroethyl)allyl-palladium chloride complex, quantitatively.[589]

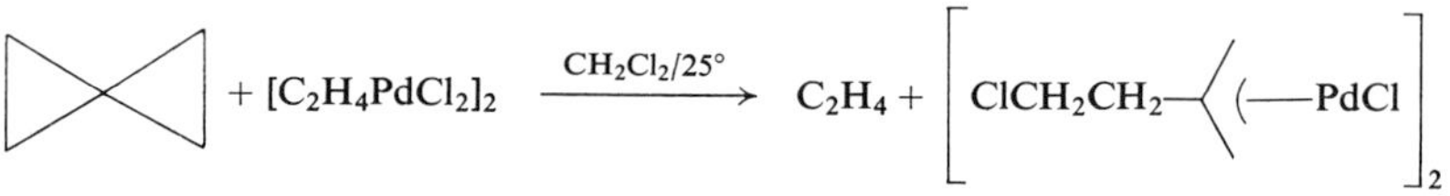

Dicyclopropyl and dicyclopropylmethane also reacted readily to give π-allylic complexes.[589]

Mechanisms for these reactions have been proposed. The ring-opening reactions are also discussed in Chapter I, Volume II, Section F,1.

Triphenylcyclopropene also underwent a ring-opening reaction with $(PhCN)_2PdCl_2$ in a variety of solvents, to give *anti*-1-chloro-*syn*-1,2,3-triphenylallylpalladium chloride.[744]

A mechanism, involving coordination of the olefin, cis addition of Pd–Cl to the olefin, and finally a stereospecific ring opening of the σ-cyclopropylpalladium intermediate was proposed.

However, other cyclopropenes underwent di- or oligomerization reactions.[744a]

4. From Allylic Halides, Alcohols, and Related Compounds

Smidt and Hafner[745] and Moiseev et al.[746] in 1959 showed that allyl alcohol reacted with palladium chloride either without solvent or in aqueous acid to give, besides organic products and metal, allylpalladium chloride dimer.

The mechanisms of these reactions have been investigated by a number of workers.[747,748,748a] The amount of propylene formed depended on the amount of palladium chloride used, but the amount of the cyclic alcohols was independent of this. The formation of metal could be avoided by working at lower temperatures. Shaw and collaborators[749] have extended this reaction to the formation of substituted π-allylic complexes from substituted allyl alcohols and found that acid promotes this reaction, presumably by protonation of the –OH group.

Zakharova and Moiseev[750] reported the reaction of allyl alcohol with $(PhCN)_2PdCl_2$ in benzene at 20° to give allylpalladium chloride and a π-olefin complex.

These workers also obtained a complex, $[C_3H_6Pd_2Cl_4]_n$, from allyl chloride and $(PhCN)_2PdCl_2$ in benzene. They suggested a structure (V-18), but (V-19) would appear a more reasonable alternative if there is an error in the number of protons present. However, see p. 153.

(V-18)   (V-19)

Hüttel and Kratzer[498,599] first reported the synthesis of π-allylic complexes from allylic chlorides and palladium chloride in 50% aqueous acetic acid at 30°–60°.

The yields in this reaction seldom exceed 50% and a modification due to Dent, Long, and Wilkinson[328, 751] is preferred for synthetic purposes. In this, the allylic chloride is reacted with sodium tetrachloropalladate in methanol in the presence of carbon monoxide. It is also found that traces of water are essential for this reaction to proceed.

Nicholson, Powell, and Shaw[752] have discussed the mechanism of this reaction. They suggest an intermediate hydroxo(carbonyl)-π-olefin complex, **(V-21)**, to be formed. This arises from a species such as **(V-20)** by loss of a pro-

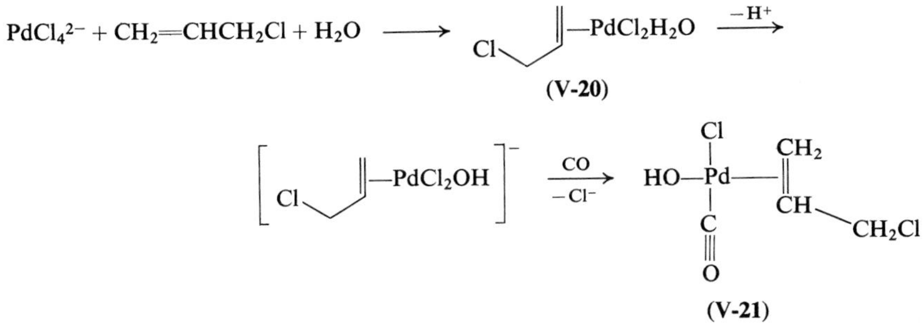

$PdCl_4^{2-} + CH_2{=}CHCH_2Cl + H_2O$
Cl
—$PdCl_2H_2O$
$-H^+$
(V-20)
Cl
—$PdCl_2OH$
CO
$-Cl^-$
Cl
HO—Pd
C
O
CH_2
CH
CH_2Cl
(V-21)

ton [water coordinated to Pt(II) has an acid dissociation constant of $\sim 10^{-5}$ [563] and hence is appreciably acidic; the same is usually presumed to be true for Pd(II)] followed by addition of CO. The next step proposed is an intramolecular attack of coordinated OH onto the carbonyl carbon† (to give a Pd–carboxylic acid) followed by rearrangement to give the π-allyl complex, CO_2, and acid. One mole of CO_2 is found per mole of allyl chloride converted.

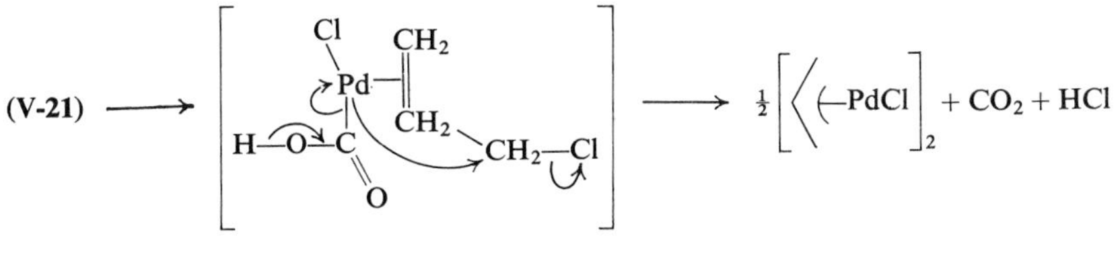

(V-21)
Cl
Pd
CH_2
CH_2
H—O—C
O
CH_2—Cl
$\frac{1}{2}$
PdCl
2
+ CO_2 + HCl

† Nucleophilic attack on the carbon of a metal carbonyl has been shown to be a common reaction,[754, 755] e.g.,

$$[(Ph_3P)_2M(CO)_4]^+ + OMe^- \rightarrow (Ph_3P)_2M(CO)_3(COOMe) \qquad (M = Mn, Re)$$

$$M(CO)_6 + RLi \longrightarrow \left[M(CO)_5C{<}^{R}_{O^-}\right] Li^+ \xrightarrow{H^+,\ CH_2N_2} \left[M(CO)_5{\cdot}C{<}^{R}_{OMe}\right]$$

(M = Cr, W)

A similar type of process in which the reaction is carried out in the presence of 1 mole of $SnCl_2$, in place of carbon monoxide, has also been described by Sakakibara et al.[753] Yields of up to 98% of π-allyl complexes were obtained.

$$R \diagdown \overset{R'}{\diagup} X \ + \ Na_2PdCl_4 + SnCl_2 \ \xrightarrow[H_2O]{MeOH} \ \left[R' \diagdown \diagup (\!\!\leftarrow PdX) \atop R \right]_2$$

$$(R = R' = H; R = H, R' = Me; R' = H, R = Me, Ph)$$

A Shell patent[756] describes a reaction again apparently similar to the above in which ethylene in used in place of CO or $SnCl_2$. It seems that a good π-acceptor ligand is required for the reaction to proceed. Another similar reaction was also described by Moiseev et al. in 1964,[757] in which a complex suggested to have the structure (V-22), with a π-cyclopropenyl ring coordinated to the metal, was prepared.

$$\underset{Cl}{\overset{Ph}{\diagdown}}\!\!\overset{Ph}{\triangle}\!\!\overset{Ph}{\diagup} \ + \ PdCl_2 + C_2H_4 \ \xrightarrow{MeCN/H_2O}$$

$$Ph\!\!-\!\!\bigcirc\!\!-\!\!\underset{Ph}{\overset{Ph}{Pd}}\!\!\underset{Cl}{\overset{Cl}{\diagup}}\!\!Pd\!\!\underset{Cl}{\overset{Cl}{\diagup}}\!\!\underset{Ph}{\overset{Ph}{Pd}}\!\!-\!\!\bigcirc\!\!-\!\!Ph$$

(V-22)

Another interesting variation, due to Tsuji and Iwamoto, involves the use of primary amines in place of water as the nucleophile in the reaction in the presence of CO.[758] The initial by-product here is the isocyanate which can then react either with methanol or with more amine to give the carbamate or urea, depending on which is used as solvent. A very similar mechanism to that proposed by Nicholson et al., involving attack on coordinated CO by coordinated RNH_2, has been suggested.

$$RNH_2 + Na_2PdCl_4 \ + \ \diagup\!\!\diagdown\!\!\diagup^{Cl} \ + \ CO \ \longrightarrow \ \left[\diagup\!\!\!\!(\!\!\leftarrow PdCl) \right]_2 + RNCO$$

$$RNCO + MeOH \ \rightarrow \ RNHCOOMe$$

$$RNCO + RNH_2 \ \rightarrow \ (RNH)_2CO$$

Allyl halides will also oxidatively add to Pd(0); Fischer and Burger[759] showed that finely divided metallic palladium slowly reacted with allyl bromide on heating. An analogous reaction with allyl iodide gave a red solution, but metal was precipitated before the complex could be isolated.

$$\text{allyl–Br} + \text{Pd} \longrightarrow \left[\text{allyl}(\text{—PdBr}) \right]_2$$

Fitton et al.[117] showed that methallyl chloride added to tetrakis(triphenylphosphine)palladium(0) to give the σ-methallyl complex (V-23). This had an

$$(\text{Ph}_3\text{P})_4\text{Pd} + \text{methallyl chloride} \longrightarrow \begin{array}{c} \text{Me} \\ | \\ \text{Ph}_3\text{P} \diagdown \quad \diagup \text{CH}_2\text{C}{=}\text{CH}_2 \\ \quad \text{Pd} \\ \text{Cl} \diagup \quad \diagdown \text{PPh}_3 \end{array}$$

$$\textbf{(V-23)}$$

$$\left[\text{Me}{-}(\text{—PdCl}) \right]_2 + 4\text{PPh}_3 \quad \longrightarrow$$

identical NMR spectrum to that obtained from the product of the reaction of four moles of triphenylphosphine with one mole of π-methallylpalladium chloride dimer. Powell and Shaw[51] reported that an excess of allyl chloride reacted with $(\text{Ph}_3\text{P})_4\text{Pd}$ to give $(\pi\text{-allylPdCl}\cdot\text{PPh}_3)$.

Allylic Grignard reagents have also been used to prepare π-allylic complexes from palladium chloride. Wilke[760] in a patent described the synthesis of bis(π-allyl)palladium by this route.

$$\text{allyl–MgCl} + \text{PdCl}_2 \xrightarrow{\text{Et}_2\text{O}} (\text{—Pd—})$$

Klimenko et al.[761] and Powell and Shaw[51] have prepared other complexes in this way.

$$\text{allyl–MgCl} + (\text{PhCN})_2\text{PdCl}_2 \longrightarrow \left(\text{—Pd} \diagdown \begin{array}{c} \text{Cl} \\ \text{NCPh} \end{array}\right)$$

$$\text{methallyl–MgCl} + (\text{Ph}_3\text{P})_2\text{PdCl}_2 \xrightarrow{\text{Et}_2\text{O, H}_2\text{O}} \left[(\text{—Pd(PPh}_3)_2) \right]^+ \text{Cl}^-\cdot 2\text{H}_2\text{O}$$

5. From Acetylenes

A number of diphenylacetylenes react with palladium chloride in alcoholic solution to give π-cyclobutenyl complexes (V-24).[611, 612, 625, 652, 653, 731] Dahl and Oberhansli[206] have shown that the alkoxy group is endo with respect to the metal in the complex where R = Ph, R′ = Et; this reaction should be compared with that of the tetraphenylcyclobutadienepalladium halides and alkoxide, which gave the *exo-* alkoxy complex (p. 188). The latter reaction is reversible.

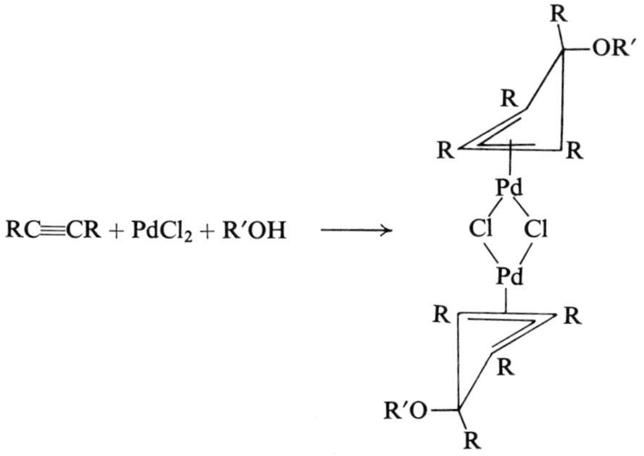

$$RC \equiv CR + PdCl_2 + R'OH \longrightarrow$$

(V-24)

(R = C₆H₅, *p*-ClC₆H₄)

Hosokawa *et al.*[495] showed that di-*t*-butylacetylene reacted with [C₂H₄Pd Cl₂]₂ to displace the ethylene and to form a π-acetylene complex. However, *t*-butyl(phenyl)acetylene in the same reaction did not displace ethylene; the product was identified as the π-allylic complex (V-25). Mechanisms of these

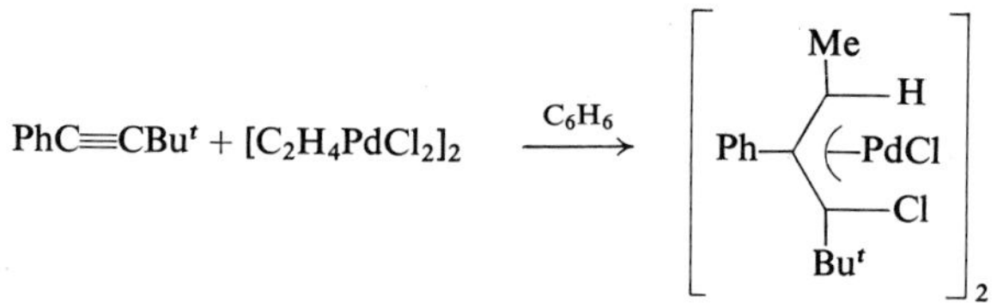

$$PhC \equiv CBu^t + [C_2H_4PdCl_2]_2 \xrightarrow{C_6H_6}$$

(V-25)

reactions are discussed in Chapter I, Volume II.

C. STRUCTURES AND SPECTRA OF "SYMMETRICALLY" π-BONDED ALLYLIC COMPLEXES

A relatively large number of X-ray structure determinations have been carried out on π-allylic complexes of palladium, including four on allyl-palladium chloride dimer itself. Two of these were only two-dimensional analyses,[202, 203] one was a three-dimensional analysis at room temperature by Oberhansli and Dahl,[205] and the last, and most accurate, by Smith[208] was carried out at −140° to minimize thermal vibrations. The overall conclusions are similar and the data of Smith are summarized here.

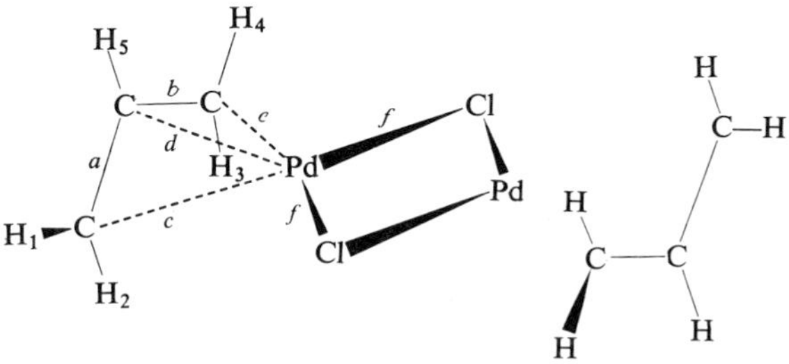

FIG. V-I. The structure of chloro(π-allyl)palladium dimer. After Smith.[208]

As shown in Fig. V-1, the allyl group is π-bonded to the metal and effectively occupies two coordination sites; the remaining two are occupied by the bridging chlorines. The molecule has C_i symmetry. The C–C bond lengths are equal [a, 1.357(15); b, 1.395(15) Å] and the central carbon is very close to sp^2 hybridized (<CCC 119.8 ± 0.9°). All three carbon atoms are approximately equidistant from the metal [c, 2.132(7); d, 2.108(9); e, 2.121(7) Å], but there is some evidence which suggests that the central carbon may be slightly closer to the palladium than the other two. The Pd_2Cl_2 is very nearly a square plane (<ClPdCl, 88°; <PdClPd, 92°). However, contrary to the preliminary results,[202, 203] the plane of the allyl group is *not* perpendicular to the Pd_2Cl_2 plane, but at an angle of 111.5 ± 0.9°. This curious feature is common to all the complexes so far examined. Although the hydrogens could not be clearly observed even in the low-temperature determination, Smith has calculated that two of them, the *anti*-hydrogens, H_2 and H_3, are significantly closer to the metal (~2.1 Å) and hence more shielded by it than the *syn*-hydrogens, H_1 and H_4, and H_5 which are all at ca. 2.96 Å from the palladium. This result has been used to interpret the NMR spectra of the complexes.

Further information has been obtained from the structures of some substituted π-allylic palladium chlorides in two papers by Mason and Wheeler[216, 217] and preliminary communications by other authors.[212, 213, 220]

(V-26)

Mason and Wheeler have shown that in the 2-methylallyl complex **(V-26)** (R_1–R_4 = H, R_5 = Me) the methyl group is not in the plane of the other carbons, but is bent toward the metal. Similar observations have been made for 2-MeC$_3$H$_4$PdCl·PPh$_3$,[211] (2-MeC$_3$H$_4$)$_2$Ni,[553] the 5-coordinate complex 2-MeC$_3$H$_4$NiBr(diphos),[179] and for the 6-coordinate Rh^{III} complex, 2-MeC$_3$H$_4$ RhCl$_2$(AsPh$_3$)$_2$.[762] In the palladium complex, the angle at the central carbon is now 112.4 ± 1.6°, but the dihedral angle between the plane of the allylic carbons and Pd$_2$Cl$_2$ remains at 111.6°.

The tetramethylallyl complex, **(V-26)** (R_1–R_4 = Me, R_5 = H),[217] was shown to have a similar geometry but with all the four methyls displaced out of the plane of the allyl group (R_2 and R_3 toward, and R_1 and R_4 away, from the metal). The dihedral angle is 121.5° here and the angle at the central carbon is 118.5°.

Less detailed analyses of a complex derived from ethyl acetoacetate, **(V-26)** (R_1, R_2, R_3 = H, R_4 = COOEt, and R_5 = OH),[220] and the 2-neopentyl complex, **(V-26)** (R_1–R_4 = H, R_5 = Bu^{t}CH$_2$),[212] have appeared. The dihedral angle in the former is 108°. Davies *et al.* in a preliminary report on the structure of the 1,3-dimethylallyl complex, **(V-26)** (R_1, R_4 = Me; R_2, R_3, R_5 = H) showed that in this complex the Pd$_2$Cl$_2$ group is not planar but bent at the Cl–Cl axis (150°); this probably arises from crystal packing forces. The dihedral angle was 123° (and 127°) here. Again the methyls were not coplanar with the allylic carbons.[213]

A more spectacular bending of the bridging group was observed by Churchill and Mason[204] in allylpalladium acetate dimer **(V-27)**, where the allyl groups are now adjacent. The Pd–Pd distance in this complex (2.94 Å), while less than that in allylpalladium chloride dimer, (3.48 Å), still implies very little direct interaction between the metals. Dihedral angles between the allyl and PdO$_2$ planes of 110° and 125° were observed.

The ease with which the dimeric molecules bend at the bridging group, while still retaining the square-planar coordination about the metal, explains the unexpectedly high dipole moments which have been observed for these complexes in solution (2.2–2.3 D).[498, 746, 759]

Although few details are given, Minasyants and Struchkov[214] have reported that in allyl(cyclopentadienyl)palladium the π-allyl and π-cyclopentadienyl rings are not parallel. The cyclopentadienyl carbons also appear to be significantly further from the metal (2.25 Å) than the allylic carbons (2.05 Å). Again, within each ligand, the C–C and Pd–C bonds are sensibly constant.

(V-27)

A number of complexes in which the π-allylic group is part of a ring have also been examined. Dahl and Oberhansli[206] studied the *endo-* and *exo*-1-ethoxy-1,2,3,4-tetraphenylcyclobutenyl complexes, (V-28a) and (V-28b), respectively. The geometry of both were very similar, a dihedral angle of 95° being observed in each case. A value of 91° for the internal angle at the central allylic carbon

(V-28a)   (R = Ph, R' = OEt)
(V-28b)   (R = OEt, R' = Ph)

was found. A similar structure was observed for a substituted tetramethyl-cyclobutenylnickel complex.[675]

Kilbourn *et al.*[219] have reported the structure of cycloheptenylpalladium bromide. The dihedral angle (C_3 to Pd_2Br_2) was 118°, and again the Pd_2Br_2 bridge was bent 41° away from the planar. As expected, the substituents on the π-allyl (i.e., the tetramethylene chain) were anti and not syn.

Similar features were also observed by Churchill in his determination of cyclooctadienylpalladium acetylacetonate,[207, 210] where the dihedral angle was 121.4°. The interesting feature here was that the double bond did not appear to be conjugated with the π-allyl group and, in fact, was twisted away from it so that conjugation was not practicable. Nonetheless, the $v_{C=C}$ of the uncoordinated double bond only showed up very weakly at 1625 cm^{-1}, in contrast to other systems containing a double bond adjacent to a π-allylic group where the $v_{C=C}$ (uncoordinated) was much stronger.[634]

Two interesting complexes derived from hexamethylbicyclo[2.2.0]hexadiene (hexamethyl Dewar benzene) have had their structures determined. In the palladium complex (V-29a) the CH$_2$⋯C⋯C—Me was π-allyllically bonded and the metal did not interact with the other double bond.[221] Platinum, by contrast, has a much lower tendency to form π-allylic complexes† and in this case had the unusual coordination depicted in (V-29b), with each metal atom π-bonded to the diene of one C$_{12}$H$_{17}$ unit and σ-bonded to the CH$_2$ of the other.[199] It was presumed (from their spectroscopic properties) that the chloride and bromide analogs of the palladium complex (V-29a) also have the π-allylic bonding. Mason et al. also showed that the iodide analog of (V-29b) appeared to be π-allylically bonded,[199] i.e., similar to (V-29a); this has not yet been confirmed by a crystal structure determination.

In addition to these complexes, in which the π-allylic group is approximately symmetrically bound to the metal, a number of structure determinations of complexes (π-all)PdXY, where the allylic group is asymmetrically π-bonded to the metal, have been carried out. These are considered in Section E.

Because of their utility in structure determination and for other reasons discussed below, the NMR spectra of π-allylic complexes have been very thoroughly examined, and substantial agreement has now been reached on the assignment of chemical shifts and coupling constants.

---

† The structure of allylplatinum chloride has recently been determined.[762a] The molecule is a tetramer with two Pt$_2$Cl$_2$ and two Pt$_2$(C$_3$H$_5$)$_2$ bridges. The allyl groups are σ-bonded to one Pt atom and π-bonded to the other. Each half of the dimer can be represented thus,

Allylplatinum acetylacetonate is also a dimer with bridging σ- and π-allyl groups but the 2-methyllallyl ligand in [2-MeC$_3$H$_4$PtCl]$_2$ and in 2-MeC$_3$H$_4$Pt(acac) is symmetrically π-bonded.[762b]

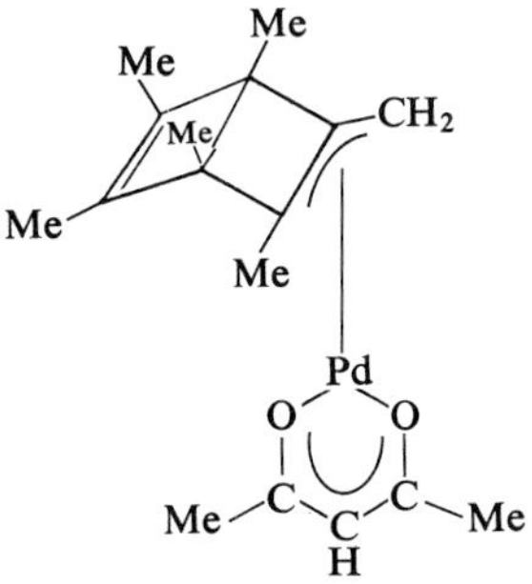

(V-29a) 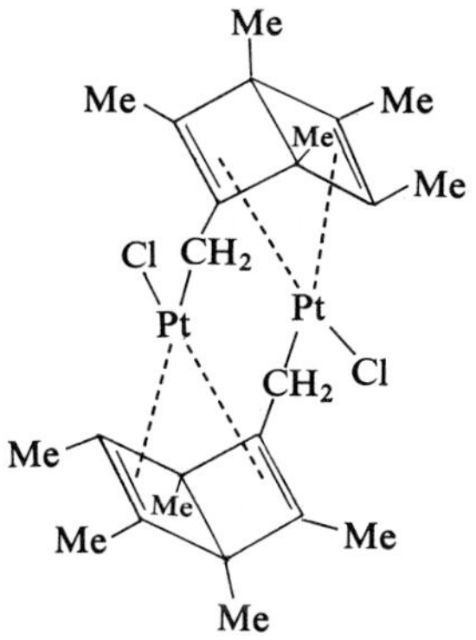   (V-29b)

For [C_3H_5PdCl]₂ in $CDCl_3$, H_5 is seen as a multiplet at lowest field,† τ 4.58, coupled to the syn protons, H_1 and H_4, equivalent, with $J_{5,1(4)} = 6.8$ Hz, in the range expected for two protons cis on an olefin. The protons H_1 (H_4) appear as a doublet at higher field (τ 5.89) and are only (significantly) coupled to H_5. The coupling to the geminal H_2 (H_3) is rarely seen (<1.0 Hz). The anti protons H_2 and H_3 are closest to the metal,[208] and therefore most shielded, and come at higher field (τ 6.95), again as a doublet owing to coupling to H_5 ($J_{5,2(3)} = 11.8$ Hz). The magnitude of $J_{5,2(3)}$ is that expected for two protons situated trans on an olefin. The intensity ratios observed are 1(H_5):

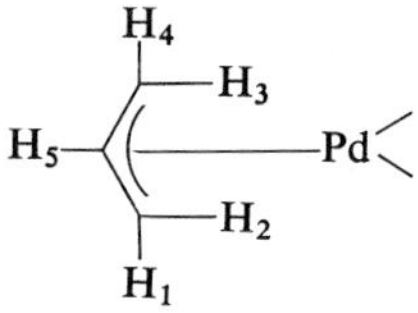

2($H_2 + H_4$):2($H_2 + H_3$).[650,764] These values of J and, to a lesser extent, of chemical shift have allowed the assignment of stereochemistry to a large number of such complexes.[649,650,709,765] In general, substituents on the terminal carbons are found to be in the syn rather than the anti positions. The main exceptions are when there are bulky groups adjacent on the allylic ligand[724] and for allyls in smaller cyclic molecules. (However, see Ref. 709.)

The tendency for substituents at C-1 and C-3 to be syn is well illustrated by the three isomers seen by Hüttel and Dietl[707] in the NMR spectra of cyclododecylpalladium chloride dimer [($C_{12}H_{21}PdCl_2$)₂] and which were assigned to the configurations a, b, and.c in a 1:5:4 ratio.

† The danger in relying on chemical shift arguments for assignment is illustrated by cyclo-heptenylpalladium chloride dimer[219] and bis(allyl)platinum.[763] In both cases the proton on the central carbon of the allylic group is not at lowest field and, in the latter case, its position is solvent-dependent.

It seems likely that the energy barriers to this type of twisting are not very high, and the thermodynamically favored configuration (to reduce nonbonded interactions) is usually adopted.

As long as the π-allylic complex has a plane of symmetry the above arguments hold, and relatively simple NMR spectra are observed. This applies also to complexes in which the halide has been replaced by other groups, for example, acetate,[765] acetylacetonate,[634, 650] cyclopentadienyl,[634, 649, 650, 766] and so forth, but not for mixed halides or thiocyanate[767] (see p. 223).

The NMR spectra of bis-π-allylic complexes are also of interest. The crystal structure of bis(2-methylallyl)nickel shows it to have the trans form (V-30a) in the solid.[553] In solution, however, an equilibrium exists between the favored

(V-30a)	(M = Ni; R = Me)	(V-30b)	(M = Ni; R = Me)
(V-31a)	(M = Ni; R = H)	(V-31b)	(M = Ni; R = H)
(V-32a)	(M = Pd; R = H)	(V-32b)	(M = Pd; R = H)
(V-33a)	(M = Pt, R = H, Me)	(V-33b)	(M = Pt; R = H, Me)

trans form, (a), and the cis form, (b), for both the bis(π-allyl), (V-31), and bis-(π-2-methylallyl), (V-30), complexes as shown by their NMR spectra by Bönnemann et al.[768] Here again, the chemical shifts of the syn and anti protons in each isomer were assigned on the basis of their (assumed) proximity to the metal.

Bönnemann et al. also showed that once equilibrium had been established, the relative amounts of the two isomers were temperature invariant from −70° to 30°. This was confirmed for bis(π-allyl)nickel, (V-31), by Becconsall and O'Brien,[53, 769] who found a broadening of the resonances at 60° in benzene. The latter workers also observed the same phenomena in the spectra of bis(π-allyl)palladium (V-32), where the ratio of trans:cis was 3:1 and temperature invariant below 0°. At room temperature and above, the doublets due to the hydrogens on the terminal carbons broadened. This was interpreted as due to two processes—one, an exchange between the syn and anti protons, the other involving an increase in the rate of interconversion of the trans and cis forms (a) and (b). If both processes go via a common intermediate, this may well be σ-allylic.[769]

The same phenomenon has been observed for bis(π-allyl)- and bis(2-methyl)allylplatinum (**V-33**) in a 220 MHz study.[763]

Shaw and co-workers have reported palladium–chlorine and palladium–bromine stretching frequencies for a number of π-allylic palladium halide dimers.[327, 639, 770] Two strong bands are usually observed at around 250 cm^{-1} for ν_{PdCl}. Fritz[771] has also discussed the infrared spectra of some π-allylic complexes; a band at ca. 1455 cm^{-1} (asym. ω_{CC}) appears characteristic for the π-allylic group in the absence of interfering vibrations. A band in the region 360–405 cm^{-1} has been assigned by Grogan[535] to the Pd–allyl group frequency (see also Shobatake and Nakamoto[772]).

Ultraviolet and visible spectra of a number of π-allylic complexes of palladium have been reported.[499, 634, 650, 773, 774] Two or three bands (2040–2120, 2240–2430, and 3200–3340 Å) were usually observed, the exact positions being dependent on solvent and substituents. Hartley[775] has analyzed the spectrum of $[C_3H_5PdCl]_2$ into five bands, three weak, which he assigned to d–d transitions, and two strong, one assigned to a $Cl_{p\pi}$ to Pd transition and the other to a Pd to allyl π^* transition. No correlations for substituted complexes were made.

Gubin et al.[773, 776] have determined the half-wave reduction potentials for some complexes polarographically. Two waves were observed, the first, corresponding to

$$C_3H_5PdX + e^- \rightarrow C_3H_5Pd \cdot + X^-$$

was very dependent on X, whereas the second, corresponding to

$$C_3H_5Pd \cdot + e^- + H^+ \rightarrow CH_3CH\!=\!CH_2 + Pd$$

was independent of X.

The mass spectra of some allylic rhodium and palladium complexes have been determined by Lupin and Cais.[777] Metal-containing fragments did not often occur.

D. BONDING IN π-ALLYL COMPLEXES

The allyl group has three molecular π orbitals, one, ψ_1, bonding; one, ψ_2, nonbonding; and one, ψ_3, antibonding (Fig. V-2).

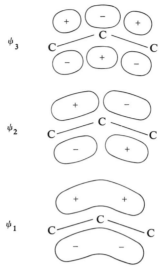

FIG. V-2. The molecular orbitals of C_3H_5.

The problem of bonding to the metal has been discussed by Kettle and Mason.[778] They suggest that two extremes of bonding, one in which the three carbons of the allyl and the metal are coplanar (Fig. V-3a) and the other in which the allylic C_3 plane is perpendicular (Fig. V-3b) to the coordination plane of the metal, can be postulated. In the first case, (Fig. V-3a), effective overlap can occur between ψ_1 and d_{xz} (or p_x) and between ψ_2 and d_{xy}, while in the second case, (Fig. V-3b), effective overlap can occur between ψ_1 and d_{z^2},

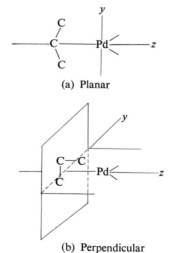

FIG. V-3. The bonding of the allyl group to Pd(II).

s, and p_z and between ψ_2 and p_y and d_{yz}. Overlap integrals involving the ψ_3 orbital were shown to be small, and hence back-bonding from the metal to the ligand is smaller than in metal–olefin complexes.

Although at first sight only the bonding scheme in Fig. V-3b might appear to be relevant, the observation that the dihedral angle between the plane of the π-allyl group and the coordination plane of the metal is never exactly 90°, suggests that bonding as in Fig. V-3a may play a role too. Kettle and Mason have calculated that the energy of the Pd–allyl bond is maximized for a dihedral angle between 102° and 114°, depending on the relative importance of the interactions between the metal orbitals and ψ_1 and ψ_2. This is in surprisingly good agreement with the results of the X-ray determination on allylpalladium chloride (111.5°)[208] and also for other π-allylic complexes.

The observation that the methyl substituents in the 2-methyl-, 1,3-dimethyl-, and 1,1,3,3-tetramethylallyl palladium chloride dimers are not coplanar with the allylic carbons has been interpreted by Mason and Wheeler to imply that the allylic carbons are not truly trigonal, but have a hybridization in-between trigonal and tetrahedral.[216, 217]

An *ab initio* MO calculation on bis(allyl)nickel and a "self-consistent charge with electron interaction" calculation on bis(allyl)palladium have been reported.[779, 780]

E. ASYMMETRIC π-ALLYLIC COMPLEXES OF THE TYPE (all)PdXY

Allylic complexes of palladium normally show a proton NMR spectra of the AM_2X_2 type, as discussed in Section C. This was first established by Dehm and Chien in 1960[781]; these authors soon afterward found that this type of spectrum was not always obtained and that particularly striking changes occurred when the spectrum was measured in a coordinating solvent such as DMSO.

In the case of allylpalladium chloride itself, the spectrum in DMSO was of the type AX_4, the central proton on the allyl group was seen as a quintet, coupled to four now equivalent terminal protons, which themselves appeared as a doublet. In other words, the molecule was undergoing a fast exchange process of some kind under these conditions which effectively averaged out the syn and anti protons and made them equivalent on the NMR time scale. This problem has been the subject of very intensive research since 1965, but the process by which this occurs, though understood in outline, still has many baffling details which are not yet resolved.

This phenomenon is not by any means restricted to the allyls of palladium; it was observed earlier with allyl Grignards by Nordlander and Roberts,[782] and has since been seen for a wide variety of π-bonded organotransition metal complexes. Molecules exhibiting such behavior have been termed "fluxional" or "sterically nonrigid". Many inorganic complexes, especially 5-coordinated ones, and a number of organic compounds such as bullvalene also show this behavior.

The reaction of allylic palladium complexes with bases has been investigated by Shaw and Vrieze and their co-workers as well as by a number of other groups.

Powell, Robinson, and Shaw[783] originally observed that if one mole of triphenylphosphine (per palladium) was added to a solution of an allylic palladium halide, the spectrum changed to one which indicated that the allylic ligand was now asymmetrically π-bonded to the metal (in other words, the two syn and the two anti protons were no longer equivalent). Addition of more phosphine gave the species which showed the AX_4 spectrum.

The asymmetric π-allylic complexes were isolated from these solutions and several X-ray structure determinations have been carried out on them. The first, by Mason and Russell,[211] on the triphenylphosphine adduct of 2-methylallylpalladium chloride was shown to have the structure indicated in Fig. V-4a.

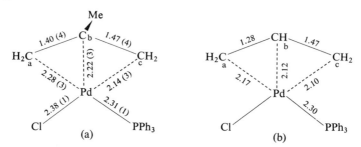

FIG. V-4. The structures of (a) chloro(triphenylphosphine)-2-methylallylpalladium, and (b) chloro(triphenylphosphine)allylpalladium. (Bond lengths in Å.)

A more recent abstract by Smith on chloro(triphenylphosphine)allylpalladium summarizing the results of measurements at −150° is in complete agreement with this.[224] Better accuracy was claimed for this work, but no e.s.d.'s were given (Fig. V-4b). In each case carbon atom C_a, trans to the phosphine, was significantly further from the metal than C_c, and there appeared to have been a real change in the allylic C–C bond lengths in that C_aC_b was now shorter than C_bC_c. In addition, Smith noted that the temperature factors for C_a and C_b were larger than for C_c, again indicating that C_a and C_b were more loosely bound to the metal and hence could vibrate more rapidly. Smith also claimed that, in agreement with Shaw's suggestion, C_aC_b was now a localized double bond, whereas C_bC_c had much more single bond character.

An adequate representation of this is difficult in a formula; from the NMR spectra, it is evident that a σ bond allowing free rotation is not present between the metal and C_c and that there is still some interaction between the metal and C_a and C_b. Cotton et al.[784] have pointed out that a σ,π structure, (V-34), is not feasible, and Powell and Shaw[327] have proposed the formulation (V-35) to meet this objection.

(V-34)   (V-35)

It has been suggested that the asymmetry in the π-allyl–metal bonding is caused by the presence of two different ligands on the other side of the metal, but this is probably not the only reason, though it may be a contributory cause. The greater trans influence of the phosphine by contrast to that of Cl is presumably important in helping to labilize $C_a–C_b$.

This is shown by the NMR spectrum (which is of the AM_2X_2 type) and crystal structure of the complex (V-36), both of which indicate that the allyl group is still symmetrically π-bonded to the metal despite the absence of a plane of symmetry in the complex.[218, 224] However, the NMR spectrum of (V-36) at low temperatures has not been reported; it may also show asymmetric bonding not detectable from the X-ray results.

(V-36)   (V-37)

On the other hand, the crystal structure of the monothiodibenzoylmethanato complex (V-37) again indicates an asymmetrical π-allylic ligand, the Pd–C bond trans to the S [2.21(1) Å] being significantly longer than that trans to the O [2.08(1) Å].[225] It therefore appears that a situation where one ligand atom is a good potential σ donor to Pd, while the other is not, results in a very asymmetric π-allyl. In this context, Powell and Shaw suggest that strongly electronegative elements such as Cl (or O) are poor σ donors, whereas P, C, S, and Sn are good σ donors.[327]

NMR investigations of the reactions of $(allPdX)_2$ with various Lewis bases have been carried out. Since the results obtained depend to some degree on the nature of L, the reactions with different types of L will be discussed separately.

F. REACTIONS OF (allPdX)₂

1. Dynamic Behavior in (allPdX)₂

a. *In the Presence of Tertiary Phosphines*

Although investigated by a number of groups, the most detailed accounts are by Vrieze *et al.*[785-787] and Powell and Shaw.[327] A number of different allyls have been examined, particularly by the latter workers, but most attention has been given to the complexes of the 2-methylallyl ligand since the spectra are more easily interpretable.†

Proton NMR spectra of a large number of allPdXL complexes have been reported[327, 787]; that of chloro(triphenylphosphine)-2-methylallylpalladium at −40° is typical[787]: H_1, 5.50 dd $(J_{1, p} 6.5, J_{1, 4} 2.9\text{ Hz})$; H_2, 6.38 d $(J_{2, p} 9.8\text{ Hz})$; H_3, 7.20 s; H_4, 7.11 d $(J_{4, 1} 2.9\text{ Hz})$; Me, 1.96 s. All workers appear to agree on

these assignments of chemical shifts, which are arrived at from arguments of proximity of protons to the metal (p. 199) and to the general observation that in complexes ³¹P only couples appreciably with nuclei trans to itself. Only small differences are seen when other phosphines replace PPh₃.

For allPdXL where the allyl is asymmetrically substituted [i.e., at $C_a(1)$], the NMR spectra are best interpreted in terms of the phosphine being trans to the more heavily substituted carbon of the allyl; this may arise from electronic or steric factors.[749]

In the absence of excess dimer, (allPdX)₂, or phosphine, Statton and Ramey[790] and others[327, 784] found that as the temperature was raised, the four H resonances first coalesced to two broad singlets at the average positions of $H_1 + H_4$ and $H_2 + H_3$, respectively. Finally, in *o*-dichlorobenzene at 140°,

---

† It should be pointed out, however, that the allyl- and 2-methylallyl groups will not necessarily behave in exactly the same way. Shaw and Singleton[788, 789] have shown that in some Ir(III) complexes, the allyl tended to form σ-bonded, and the 2-methylallyl, π-bonded structures. The same is true for allylic Pt(II) complexes.[762a, 762b]

one very broad singlet, characteristic of the "dynamic σ-allyl"† was observed at the average of the chemical shifts of H$_1$, H$_2$, H$_3$, and H$_4$. Since these same effects are obtained more easily at lower temperatures, simply by addition of either dimer, (allPdX)$_2$, or phosphine, and probably arise from the same causes, their investigation under these conditions was continued.

Two systems were investigated: (*i*) that where [R$_3$P]: ["Pd"] > 1, i.e., no (allPdX)$_2$ (dimer) was present, and (*ii*) where [R$_3$P]: ["Pd"] < 1 and the solution contained only dimer and allPd(R$_3$P)X.

i. [*R*$_3$*P*]: [*"Pd"*] > 1. For the 2-methylallyl complexes at −25° with [Ph$_3$P]: [Pd] between 1.01 and 1.1 four resonances, due to H$_1$, H$_2$, H$_3$, and H$_4$, are seen. On raising the temperature to 23° these coalesce to two resonances, due to H$_1$ + H$_4$ and H$_2$ + H$_3$. Coupling of H$_1$ and H$_2$ to ^{31}P also disappears, therefore exchange of coordinated with free Ph$_3$P must be occurring.[785, 786] These observations are accounted for by a bimolecular reaction.

(V-38a)   (V-39)   (V-38b)

This process has been explained by Vrieze *et al.* in terms of the observed geometry of the 2-MeC$_3$H$_4$PdClPPh$_3$ complex (Fig. V-4a). As described by Mason and Russell, there is a dihedral angle of 116° between the plane of the allyl and the coordination plane of the metal. Vrieze *et al.*[785] visualize this as an incomplete half-sandwich complex, similar to benzenechromium tricarbonyl, except that three carbon atoms of the organic ligand and one ligand trans to them are missing. The positions of these are indicated by dashed lines. A justification for this is that in C$_6$H$_6$Cr(CO)$_3$, the dihedral angle between the benzene plane and the plane containing the chromium and two carbonyls is 125°, close to that observed for 2-MeC$_3$H$_4$PdClPPh$_3$.

† This term was first introduced by Wilke[54] to describe the system responsible for the spectra discussed.

Thus 2-MeC$_3$H$_4$PdClPPh$_3$ (**V-38a**) can be thought of as having a vacant coordination site, shown by the dashed line; when this is filled, as in the 5-coordinate intermediate 2-MeC$_3$H$_4$PdCl(PPh$_3$)$_2$, (**V-39**), the resulting molecule has pseudotrigonal symmetry and the π-allyl group can rotate in its own plane around the sixfold axis. If the added PPh$_3$ group now comes off, the original complex, (**V-38a**) is regenerated. On the other hand, if the original PPh$_3$ group comes off and is accompanied by a 120° rotation of the π-allyl, then (**V-38b**) is produced. This has the effect of making H$_1$ and H$_4$, as well as H$_2$ and H$_3$, equivalent on the NMR time scale. The 5-coordinate intermediate is postulated to have a very short lifetime and is not observed; the only observable is the equivalence of both the syn and both the anti protons, since the steps leading to and from (**V-39**) are very fast.

Powell and Shaw[327] have pointed out that the refinement that the π-allyl group rotates in its own plane, is not necessary to explain this result, as it is explicable solely in terms of rapid phosphine exchange accompanied by some small reorganization at the metal but not involving the allyl group. That very rapid phosphine exchange occurs is shown by the disappearance (at 34°) of $J_{1, P}$ and $J_{2, P}$ on the addition of as little as 0.0014 mole of PPh$_3$ per mole of 2-MeC$_3$H$_4$PdClPPh$_3$ (see also Tibbetts and Brown[791]).

Although 5-coordinate π-allylic complexes of palladium of the type (**V-39**) are not known, Walter and Wilke[792] have isolated C$_3$H$_5$NiBr(PR$_3$)$_2$ (R = Ph or PhO), and Churchill and O'Brien[179] have carried out a crystal structure determination on

which has an approximately square pyramidal coordination about the metal.

The kinetics of the reaction of allPdXPR$_3$ with PR$_3$ showed an anomalous dependence on the phosphine concentration. This has been shown by Vrieze et al. to be due to a further series of reactions in which the ionic complex (**V-40**) was formed [very slowly, by comparison with reactions leading to (**V-38a** or **b**)] from 2-MeC$_3$H$_4$Pd(PR$_3$)$_2$Cl. Complexes of this type have been isolated and characterized by Powell and Shaw.[51] They were formed most readily from (allPdX)$_2$ and the phosphine in aqueous acetone and have also been prepared from (Ph$_3$P)$_2$PdCl$_2$ and the allyl Grignard. The NMR spectrum of complex (V-40) (BPh$_4^-$ in place of Cl$^-$) and its analogs, in the absence of added phosphine, showed a symmetrically π-bonded methylallyl group to be present. From conductivity measurements of the reaction

$$\text{(allPdX)}_2 + 4L \rightarrow 2\text{(allPdL}_2\text{)}^+X^-$$

the relative rates of formation of the cations were estimated to decrease in the order, L = PMe$_2$Ph $\sim$ PEt$_2$Ph $\sim$ PEt$_3$ > PPh$_3$ > AsPh$_3$ > SbPh$_3$ > py; all = 2-MeC$_3$H$_4$ > C$_3$H$_5$ > 1-MeC$_3$H$_4$ > 1,3-Me$_2$C$_3$H$_3$ > 1,1-Me$_2$C$_3$H$_3$; and X = Cl > Br > 1. Similar cationic complexes with chelating amines (bipy, en) were already reported in 1965 by Paiaro and Musco.†[793]

(V-40)

Evidence for the presence of (V-40) at [R$_3$P]:[Pd] $\geqslant$ 1 was obtained from the NMR spectra and conductivities of these solutions.[787] Even at $-50°$, for R$_3$P = Ph$_3$P, or Ph$_2$MeP, in the presence of an excess of the phosphine the allylic hydrogens of (V-40) appeared as one singlet. At low temperatures spectra of complexes of the more basic phosphines (PhMe$_2$P, n-Bu$_3$P) showed the allylic protons as two resonances, the lower field due to H$_1$ and H$_4$, and the higher to H$_2$ and H$_3$.‡ At higher temperatures (30°) these coalesced to a singlet again. The positions of the two resonances at low temperature and the one at higher temperature were different to the positions of H$_1$–H$_4$ in 2-MeC$_3$H$_4$PdClPR$_3$, another reason why it was believed that that complex was itself not involved in these rate processes.

The low-temperature spectra of the 2-MeC$_3$H$_4$PdClPR$_3$ complexes with an excess of the more basic phosphine imply the presence of a symmetrically bonded π-allylic group; at higher temperatures, or even at low temperatures with a less basic and bulkier phosphine such as Ph$_3$P, the spectrum observed (a singlet for the allylic hydrogens) is characteristic of a reaction in which a dynamic σ-allyl system is present. The extent to which a species such as (V-41)

---

† Although platinum shows a great reluctance to form π-allylic complexes, a number of the type (π-allPtL$_2$)$^+$ have been described.[51, 120, 137, 138] The complexes prepared by Baird and Wilkinson[120] turn out to be covalent and 5-coordinate in chloroform and ionic (and

$$\text{(Ph}_3\text{P)}_3\text{Pt} + \text{CH}{=}\text{CHCH}_2\text{X} \rightarrow \text{(Ph}_3\text{P)}_2\text{PtC}_3\text{H}_5\text{X}   \text{(X = NCS or Cl)}$$

4-coordinate) in nitromethane.

‡ This is seen as a triplet, owing to coupling to two equivalent ^{31}P nuclei.

is actually present is a matter for discussion. Powell and Shaw[327] pointed out that if the exchange processes are very fast, the stationary concentration of (V-41) can be negligible. Cotton et al.[784] reported that the infrared spectra of complexes in solution which contain more than two moles of Ph_3P, Me_2S, or DMSO (per Pd) showed weak or very weak absorptions in the region 1611–1650 cm^{-1}, where an uncoordinated double bond would be expected. They interpreted this as implying that, although the main species in such solutions were π-bonded allyls, substantial amounts of σ-allyls such as (V-41) were also present. Powell and Shaw[327] have also shown from osmometric molecular weight studies on the reaction of $[2\text{-}MeC_3H_4PdCl]_2$ with excess PMe_2Ph that a third molecule of ligand (per Pd) was weakly coordinated with the metal. This is compatible with the existence of an ion-paired species such as $[\sigma\text{-}2\text{-}MeC_3H_4Pd(PMe_2Ph)_3]^+Cl^-$ in solution.

(V-40a)   (V-41)   (V-40b)

(L = solvent, Cl$^-$, or PR$_3$)

In the intermediate (V-41), the σ-allylic group can be twisted about C_bC_c; a small movement of the metal with respect to C_c accompanied by a 180° twist leads to the π-allylic complex (V-40b) in which H_3 and H_4 have exchanged positions by comparison with (V-40a). H_1 and H_2 can also exchange by an exactly analogous process in which a σ bond is formed from the metal to C_a. For more recent evidence in favor of σ-bonded intermediates in such reactions see van Leeuwen et al.[793a] and Alexander et al.[793b]

The system allylpalladium chloride–R_3P appears to parallel very closely the one just described. However, some di- and trimethylallyl complexes show instructive differences in behavior. The symmetrical syn-1,2,syn-3-trimethylallylpalladium chloride complex, obtained as a single isomer, when treated with PPh_3 (or $AsPh_3$) at $-80°$ gave the asymmetric π-allyl complex (V-42).[787]

(V-42)

At 10° in the presence of excess Ph_3E the resonances of the protons of the methyl groups A and D and of the protons H_2 and H_3 collapsed to their respective weighted means. This process was reversed on lowering the temperature and was clearly due to ligand exchange of the type already discussed for the 2-methylallyl system with a small excess of phosphine. At higher temperatures or higher ligand-to-metal ratios, isomerization of syn and anti groups occurred.† This again implied that a dynamic σ-allyl intermediate was present.

The asymmetric 1,1,2-trimethylallylpalladium chloride (**V-43a**) gave the complex (**V-44a**) with Ph$_3$P or Ph$_3$As. On raising the temperature or increasing the amount of ligand present, coalescence of the H$_3$ and H$_4$ resonances occurred, but no perceptible broadening of the methyl resonances A or B was observed.[787]

(**V-43a**) (R = Me)   (**V-44a**) (R = Me)
(**V-43b**) (R = H)   (**V-44b**) (R = H)

Powell and Shaw[327] have investigated the analogous 1,1-dimethylallyl complex (**V-43b**), which with Ph$_3$P gave (**V-44b**). On addition of a small amount of phosphine to the latter, a similar effect to that noted for the trimethylallyl complex (**V-43a**) by Vrieze *et al.* was also seen. H$_3$ and H$_4$ collapsed to a doublet [coupled to R(H)], but Me$_A$ and Me$_B$ remained unaltered and still coupled to the phosphorus. Neither rotation of the allylic ligand nor a simple phosphine exchange mechanism via a 5-coordinate intermediate can explain this result, since the methyls still remain trans to the phosphine (as shown by their coupling constants to ^{31}P). Powell and Shaw have suggested that for steric reasons a second phosphine (Ph$_3$P*) cannot add easily to give a 5-coordinate complex, and that in the presence of excess PPh$_3$ a 4-coordinate σ-allyl complex (**V-45**) is formed. In this complex rotation in the σ-allyl group followed by loss of Ph$_3$P* again could account for the observed equivalence of H$_3$ and H$_4$, while Me$_A$ and Me$_B$ remain distinct and still coupled to the PPh$_3$. This would imply that Ph$_3$P* was, under these circumstances, less strongly bonded than the PPh$_3$.

With larger amounts of PPh$_3$ collapse of the Me–P couplings was observed and complete exchange occurred (involving loss of Ph$_3$P as well). At a ratio of Ph$_3$P:Pd of 2:1, the NMR spectrum showed the presence of two σ-allylic complexes, not of the dynamic type. Powell and Shaw suggested that they

---

† A similar process is probably responsible for the observation of syn and anti isomers in the PPh$_3$ and AsPh$_3$ complexes of 1-acetyl-2-methylallylpalladium chloride by Fong and Kitching.[794] See also Section F,1,c.

were $CMe_2{=}CH \cdot CH_2PdCl(PPh_3)_2$ and $cis\text{-}CH_2{=}CH \cdot CMe_2PdCl(PPh_3)_2$ in a 5:1 ratio. Increasing substitution at the allylic carbons decreased the tendency for these exchange and reorganization reactions to occur. For example, the allylic complexes, (V-29a) (halide in place of acetylacetonate), showed little tendency to undergo these reactions.

(V-45)

A simple molecular orbital description of the process which occurs when $[allPdX]_2$ is cleaved by L to $[allPdClL]$ has been given by Vrieze et al.[785] The main effect, these authors suggest, is to increase the negative charge density at the metal. This causes an increase in the energy of the metal orbitals, which, in turn, decreases the interaction between the metal s and d_{z^2} orbitals and ψ_1 while increasing the interaction between d_{xz} and ψ_3 (Figs. V-2 and V-3a). This implies a decrease in the π bonding and an increase in the population of the antibonding π^* allylic orbital, which makes the allylic group more negative, and therefore more likely to σ-bond.

Solutions containing $[allPdCl]_2 + 4Ph_3P$ disproportionated rapidly to give $(Ph_3P)_2PdCl_2$ plus a solution, which from its NMR spectrum, still contained a dynamic σ-allyl species, possibly $(Ph_3P)_2Pd(\sigma\text{-all})_2$, which has, however, not been characterized.[327]

Powell and Shaw[51] have shown that $[allPdCl]_2$ with an excess of PPh_3 in aqueous acetone gave $(Ph_3P)_4Pd$ and an allylic phosphonium salt. This reaction has also been studied by Hüttel and König[795] who obtained a yellow solid from mixing $[C_3H_5PdCl]_2$ and PPh_3 in benzene at 20°; when this was decomposed with water, allyltriphenylphosphonium chloride was obtained. If the solid was first heated in benzene, the isomeric propenyltriphenylphosphonium chloride was isolated. 1-Phenylallylpalladium chloride reacted with triphenylphosphine to give a 1:1 mixture of the isomeric compounds **(V-46a)** and **(V-46b)**.

$$\left[\langle\!\!\!\!-\text{PdCl}\atop\text{Ph}\right]_2 + \text{Ph}_3\text{P} \longrightarrow \text{PhCH}_2\text{CH}=\text{CHPPh}_3{}^+\text{Cl}^- + \underset{\underset{\text{Ph}}{|}}{\text{MeCH}=\text{CPPh}_3{}^+\text{Cl}^-}$$

$$\qquad\qquad\qquad\qquad\qquad\qquad\qquad \textbf{(V-46a)} \qquad\qquad \textbf{(V-46b)}$$

ii. $[R_3P]:["Pd"] < 1$. The second type of reaction which has been investigated in some detail is the behavior of allPdXPR$_3$ in the presence of [allPdX]$_2$.[785-787] At $-40°$ the spectrum of 2-MeC$_3$H$_4$PdClPR$_3$ ("monomer") in the presence of [2-MeC$_3$H$_4$PdCl]$_2$ ("dimer") is that expected of such a mixture and the separate resonances of H$_1$–H$_4$ in the monomer as well as the singlets due to the syn (τ 6.15) and anti (τ 7.14) protons in the dimer are clearly seen. Between $-40°$ and $-20°$, while the dimer resonances and H$_1$ and H$_2$ of the monomer remain unchanged, the H$_3$ and H$_4$ resonances coalesce. Between $-20°$ and $25°$ the monomer signals of H$_1$ and H$_2$ also broaden and finally coalesce with H$_3$ and H$_4$ to form a broad band. Neither the dimer resonances nor the methyl resonances of the monomer change in line width. Therefore up to $25°$ there is no *exchange* between monomer and dimer; however, above $25°$ the resonances due to the dimer and that due to the methyl of the monomer all broaden and hence exchange between monomer and dimer is now occurring.

The rate of exchange of H$_3$ and H$_4$ between $-40°$ and $0°$ was measured by Vrieze et al. for a variety of phosphines other than PPh$_3$, and found to be proportional to the dimer concentration (at constant monomer concentration) and independent of the concentration of monomer at constant dimer concentration.[787]

The model proposed to explain these observations involves an association of dimer and monomer to a short-lived "association complex" which must then regenerate the initial monomer and dimer since no exchange of phosphine was observed to occur. Activation energies and frequency factors for this process were measured, but no more concrete reaction path has been proposed to account for the exchange of H$_3$ and H$_4$ which presumably again occurs via some type of dynamic σ-allyl system. It would appear plausible that a chlorine of the dimer is acting as a ligand to the monomer in the association complex, giving rise to a situation somewhat analogous to that proposed by Powell and Shaw for the reaction of 1,1-Me$_2$C$_3$H$_3$PdClPPh$_3$ with a small excess of phosphine (see below). In that case, other chloro complexes should be able to give rise to similar effects. van Leeuwen et al.[796,797] have recently shown that [1,5-CODRhCl]$_2$ does catalyze the movement of the allylic group in 2-MeC$_3$H$_4$PdClPPh$_3$.

The reactions which occur between $-20°$ and $20°$ are qualitatively very similar to those below $-20°$ except that all four allylic protons of the monomer are now involved in the exchange reaction. Neither the methyl group of the monomer nor any of the resonances of the dimer are broadened. Vrieze et al.[786] showed that the rate of this reaction is also dependent on the concentration of dimer and independent of that of monomer (for $PR_3 = PPh_3$). The rate is given by $k \approx 10^{15}(\exp -17{,}500/RT)$, where the frequency factor is higher by ca. 10^5 than in the reactions below $-20°$. To explain this, these authors postulate a preequilibrium involving the dimer to give a new species which may be either the dimer with one chlorine bridge broken (V-47) or (less probably) two $[2\text{-}MeC_3H_4PdCl]_1$ units in a solvent cage. Again, attack by a coordinated Cl (either a bridging Cl or a terminal Cl of the dimer or activated dimer) at the vacant coordination site of $2\text{-}MeC_3H_4PdClPR_3$, (V-38), followed by the same types of rearrangement postulated below may explain these results. The authors point out, though, that the temperatures at which these reactions occur are very much lower than those involving additional phosphine or arsine ligands.

(V-47)   (V-48)

Above $25°$ all the signals (both of monomer and dimer) broaden and chemical exchange between the two species occurs. The observed kinetics are consistent with a small amount of dissociation of the dimer to (V-48) which is now the active species.

Powell and Shaw[327] have also studied the reactions of the dimer at $[R_3P]:[Pd] < 1$, but at only one temperature ($34°$). Although their results are therefore oversimplified, it appears that their model, with modification to allow for the participation of species such as $[2\text{-}MeC_3H_4PdCl]_2$ and (V-47), as well as (V-48) as they suggest, can account for the observations. This is shown in the scheme, where Z–Cl* can be either the "dimer," (V-47), or (V-48),

(V-49)

depending on the temperature and the amount of dimer present (i.e., ratio of PR$_3$:Pd). The reactions on the left-hand side which involve only exchange of H$_3$ and H$_4$ are important at low temperatures, while those on the right-hand side become significant at -20 to $20°$. Exchange involving the methylallyl group in Z becomes important above $20°$ and may also proceed through the common intermediate (V-49). More kinetic data are needed before these pathways can be delineated more precisely.

b. *In the Presence of Tertiary Arsines*

These reactions have again been carefully studied by Vrieze *et al.* The chief points of difference to the triphenylphosphine reactions arise from the As–Pd bond being weaker than the P–Pd bond.[786, 787] Hence the formation of ionic species [allPdL$_2$]$^+$ is much less important for L = AsR$_3$ than for PR$_3$. Furthermore, in reactions in which the dimer, [allPdX]$_2$, is present the concentration of free R$_3$P is negligible, but that of free R$_3$As is more substantial. Therefore, 2-MeC$_3$H$_4$PdCl, formed by dissociation of 2-MeC$_3$H$_4$PdClAsPh$_3$, becomes a kinetically significant entity here. The 2-methylallyl complex was again the one more carefully investigated.[785,798]

i. [*Ph$_3$As*]:[*"Pd"*] $\geqslant$ *1*. The NMR spectrum of 2-MeC$_3$H$_4$PdClAsPh$_3$ again showed an asymmetrically π-bonded methylallyl group at $-73°$, H$_1$, τ 5.46 (s); H$_2$, 6.43 (s); H$_3$, 7.12 (s); and H$_4$, 6.74 (s).

$$\text{Me}-\left\langle\begin{matrix} \overset{\displaystyle \text{H}_4}{\overset{|}{}}-\text{H}_3 \\ \\ \underset{\displaystyle \text{H}_1}{\underset{|}{}}-\text{H}_2 \end{matrix}\right. \quad \text{Pd}\underset{\displaystyle \text{Cl}}{\overset{\displaystyle \text{AsPh}_3}{\diagup\negmedspace\diagdown}}$$

Between $-70°$ and $-40°$, in the presence of a small amount of Ph$_3$As, slow exchange between H$_1$ and H$_4$ and between H$_2$ and H$_3$ occurred; this was complete at around $0°$ and two sharp singlets (H$_1$ + H$_4$) and (H$_2$ + H$_3$) were observed. This reaction was first-order in Ph$_3$As and in 2-MeC$_3$H$_4$PdClAsPh$_3$, and was due to

$$[2\text{-MeC}_3\text{H}_4\text{PdClAsPh}_3] + \text{As*Ph}_3 \;\rightleftharpoons\; [2\text{-MeC}_3\text{H}_4\text{PdClAs*Ph}_3] + \text{AsPh}_3$$

The rate constant for this reaction was found to be $k = 5.5 \times 10^{11}(\exp{-6400}/RT)$ liter mole^{-1} sec^{-1}. Again, a similar mechanism for this process to the one illustrated for the phosphine reaction has been proposed, involving the intermediate 2-MeC$_3$H$_4$PdCl(AsPh$_3$)$_2$. Vrieze *et al.* estimated that K for

$$2\text{-MeC}_3\text{H}_4\text{PdClL} + \text{L} \;\rightleftharpoons\; [2\text{-MeC}_3\text{H}_4\text{PdL}_2]^+\text{Cl}^-$$

was about 2.0 to 0.5 between $-30°$ and $20°$ (K decreased with rise in temperature) for $L = PPh_3$ and less than 0.05 for $L = AsPh_3$; hence the ionic species was negligible here, particularly at higher temperature.

In the presence of a small excess of the arsine, H_1, H_2, H_3, and H_4 all became equivalent between $20°$ and $80°$, and the reaction leading to this result was also first-order in both Ph_3As and $2\text{-}MeC_3H_4PdClAsPh_3$. The activation energy of the process, again involving a dynamic σ-allyl intermediate, was similar to that of the lower temperature reactions (6–7 kcal/mole), but the frequency factor was now only about 10^7. Vrieze *et al.* suggest that in this reaction attack by the incoming ligand to form the dynamic σ-allyl intermediate may therefore be along a sterically unfavorable path. No detailed reaction path was suggested, but one involving ionic intermediates, such as (V-40) or (V-41) which were considered for the phosphine reactions, is unlikely here.

    ii. $[Ph_3As]:[\text{"}Pd\text{"}] < 1$. Below $-50°$ in $CDCl_3$ resonances arising both from the monomer, $2\text{-}MeC_3H_4PdCl(AsPh_3)$, and the dimer, $[2\text{-}MeC_3H_4PdCl]_2$, were observed. The allylic proton resonances of both monomer and dimer broadened at temperatures up to $0°$. Above this, coalescence of resonances due to H_1 and H_4 of both the monomer and the dimer and of H_2 and H_3 of the dimer to two signals was observed. The rate of coalescence of the monomer peaks was proportional to $[2\text{-}MeC_3H_4PdCl(AsPh_3)]^{1/2}$, and that of the coalescence of the dimer signals was proportional to $[2\text{-}MeC_3H_4PdCl(AsPh_3)]$.[1] Both appeared to be essentially independent of dimer concentration. These results were interpreted in terms of the following reactions.

$$[2\text{-}MeC_3H_4PdCl]_2 + 2Ph_3As \overset{k_1}{\rightleftharpoons} 2[2\text{-}MeC_3H_4PdClAsPh_3]$$

$$2\text{-}MeC_3H_4PdClAsPh_3 \overset{k_2}{\rightleftharpoons} [2\text{-}MeC_3H_4PdCl]_1 + Ph_3As$$

$$2\text{-}MeC_3H_4PdClAsPh_3 + As^*Ph_3 \overset{k_3}{\rightleftharpoons} 2\text{-}MeC_3H_4PdClAs^*Ph_3 + AsPh_3$$

The *equilibrium* constant for the second reaction, K_2, was estimated as $120(\exp -12{,}600/RT)$ mole·liter^{-1}, and the *rate* constant for the first, k_1, as $10^{11}(\exp -200/RT)$ liter^2·mole^{-2}·sec^{-1}. The first reaction, that of the dimer with the arsine, was presumed to go in two stages, via the formation of a short-lived intermediate such as

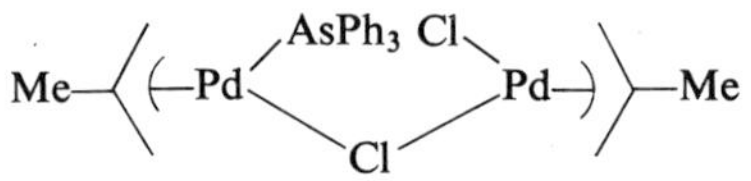

$$Me\text{—}\left(\text{—Pd}\overset{AsPh_3}{\underset{Cl}{\diagdown}}\overset{Cl}{\diagup}Pd\text{—}\right)\text{—Me}$$

which then reacted with more arsine to give $2\text{-}MeC_3H_4PdCl(AsPh_3)$. However, to account for the second-order dependence (effectively) on $[Ph_3As]$, the second step must be slower than the first.

An indication that an intermediate of this type is not unrealistic is provided by the isolation of a complex from cyclohexanone oxime and allylpalladium chloride dimer by Imamura et al.[223] This was shown to have the structure

by an X-ray determination.

The third reaction, involving exchange of ligand and (possible) rotation of the π-allyl group in its plane has already been discussed in Section F,1,a and is, of course, responsible for the equivalence of H$_1$ and H$_4$, and of H$_2$ and H$_3$ in the monomer.

A dynamic σ-allyl intermediate is also observed under these conditions but at a much higher temperature than in the presence of excess arsine. No rates have been measured for this process.

c. With Other Nucleophiles

Phenomena related to those described for phosphines and triphenylarsine have been observed for other ligands. Chien and Dehm[799] and later Ramey and Statton[764] showed that similar reactions occurred with DMSO. At low temperatures, where the allPdCl(EPh$_3$) was frozen into the asymmetric π-allylic configuration, allPdCl(DMSO) still showed the equivalence of H$_1$ and H$_4$, H$_2$ and H$_3$. This indicated that exchange of ligand occurred much more easily here than for PPh$_3$ or AsPh$_3$. At higher temperatures, spectra consistent with the existence of a dynamic σ-allyl intermediate were observed here too.

An exchange mechanism which averages H$_1$ and H$_4$, and H$_2$ and H$_3$, in allPdCl(amine) has been shown by Musco and co-workers[800, 801] and by Faller et al.[802, 803] to involve loss of amine and the formation of some dimer, which was observed at low temperatures.

However, DeCandia et al.[800] and Corradini et al.[804] have reported the isolation, at low temperatures, of enantiomorphs of the asymmetric syn-1-acetyl-2-methylallylpalladium chloride using (S)-α-phenylethylamine. The complex epimerized rapidly in acetone between $-30°$ and $-5°$ in a first-order reaction. The free energy of activation, ΔG^+ (18.6 ± 1.3 kcal/mole), and the enthalpy of the reaction, ΔH^+ (19.9 ± 1.3 kcal/mole), were measured polarimetrically. The exchange reaction just mentioned cannot account for the observed epimerization, even if the π-allyl group rotates around the sixfold axis in its plane.

Faller et al.[805] have examined the behavior of a complex crystallized from 1-acetyl-2-methylallylpalladium chloride and (S)-α-phenylethylamine in carbon tetrachloride at low temperatures. From the NMR spectrum and coupling constants observed, they concluded that at $-50°$ in $CDCl_3$ and in the solid the isomer obtained was one epimer of anti-1-acetyl-2-methylallyl[(S)-α-phenylethylamine]palladium chloride. On raising the temperature to $-20°$ a new series of resonances appeared, arising from the other epimer of the anti isomer and at higher temperatures two new sets of resonances, due to both epimers of the complex with a syn-1-acetyl group, appeared. Cis–trans isomerism of the amine with respect to the acetyl group was not observed and at least 90% of the complex was present as one isomer, probably with the amine trans. The rate of epimerization for the anti isomer ($\Delta G^+ \sim 18.5$ kcal/mole) was similar to that observed by DeCandia, and was close to that for the syn isomer ($\Delta G^+ \sim 18.8$ kcal/mole). ΔG^+ for the isomerization anti to syn was ca. 21.2 kcal/mole. From spin saturation experiments, Faller et al. concluded that epimerization occurred via a σ-bonded intermediate,

and that syn and anti isomerization occurred with inversion of configuration, via the other σ-bonded allylic intermediate,

These workers concluded from these results and others that the ease of formation of the σ-allylic intermediate in these isomerization reactions depended on the size of the substituent at the carbon which became σ-bonded. An increase in ΔG^+ of 1 to 3 kcal mole^{-1} per substituent was estimated. Furthermore, a bulky substituent at the 2- position on the allyl favored the anti configuration for the 1-substituent since non-bonded interactions were minimized.[794, 805] These results appear to be of general applicability and open up the possibility of carrying out stereospecific syntheses using π-allylic complexes. (However, see also Ref. 724.)

Tibbetts and Brown[767] have examined the NMR spectra of 2-methylallyl-palladium thiocyanate. At −61° the spectrum revealed that the syn and anti protons exist as pairs of lines, suggesting that an asymmetric π-allyl complex was again present.

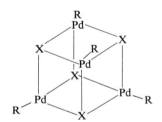

At about −15° coalescence of H$_1$ and H$_4$, and of H$_2$ and H$_3$ was observed, owing to a reaction which was *second*-order in the dimer (E_{act} 6.8 kcal/mole: ΔS^+ −19 e.u.).

A similar observation was made regarding the mixed halide complex, 2-MeC$_3$H$_4$PdICIPdC$_3$H$_4$-2-Me, and its reaction with [2-MeC$_3$H$_4$PdI]$_2$ and [2-MeC$_3$H$_4$PdCl$_2$]$_2$. Tibbetts and Brown suggested that breaking the bridge, to give (2-MeC$_3$H$_4$PdX) (X = SCN, NCS, I, or Cl), did not actually occur, but that the intermediate in question was a dimer of the dimer.

(R = 2-methylallyl; X = SCN, I, or Cl)

Since this tetramer can dissociate to give dimers in any of three different ways, this would be an effective method for transfer of bridging groups between allyl–Pd moieties. It remains to be seen whether mechanisms of this type play any part in the reactions of (allPdLX) and (allPdX)$_2$ discussed above.

Powell[806] has recently examined π-allylic palladium acetate complexes by NMR at different temperatures and has obtained evidence for a process involving inversion of acetate bridges, e.g.,

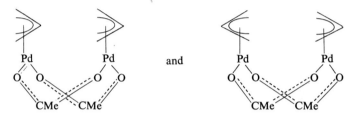

In addition, an interchange between

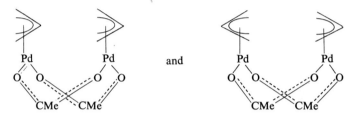

and

was observed, which may proceed via a bimolecular reaction involving a tetramer, (allPdOAc)$_4$. See also van Leeuwen.[806a]

d. Dynamic Behavior in Allylic Complexes of Other Metals

The behavior described here is by no means confined to palladium, and similar NMR spectra have been observed for allylic complexes of Pt(II),[120, 137, 138, 807] Rh(III),[137, 648, 807–810] Mo(II),[811] and Ni(II).[53, 792]

π-Allylic complexes of Pt(II) analogous to (allPdX)$_2$ are known,[762b] and a number of cationic complexes [allPtL$_2$]$^+$ have been prepared. The NMR spectrum of [C$_3$H$_5$Pt(PPh$_3$)$_2$]$^+$Cl$^-$ showed the allyl group to be π-bonded below −10°; above this temperature, the syn and anti proton signals coalesced, indicating that a dynamic σ-allyl intermediate was present. Phosphine exchange was also indicated.[138, 807] σ-, π-Bridged allyl complexes are also well known for Pt(II).[762a, 762b]

The NMR spectra of a series of Rh(III) complexes

were determined by Vrieze and Volger.[807] At low temperatures, spectra characteristic of a slightly asymmetrically π-bonded 2-methylallyl group were obtained; between −50° and −20° coalescence to give the typical singlet (for H$_1$–H$_4$) of the dynamic allyl occurred. Kinetic parameters were calculated for these reactions for a variety of ligands, L, and it was found that the activation energies decreased in the order Ph$_3$Sb > Ph$_3$As > (p-Me$_2$NC$_6$H$_5$)$_3$As > Ph$_3$P. It appears that increase in electron donor capacity of L is reflected in a decrease in E_{act}.

A closely related observation was made by Walter and Wilke[792] for the reaction of allylnickel bromide with phosphines. The complex

$$\left\langle\!\!\left(\!-NiBr(PR_3)_2\right.\right.$$

shows a π-allyl spectrum when R = Ph or PhO and a dynamic σ-allyl spectrum for the more basic phosphine, R = Et. Wilke et $al.$[54] have also mentioned, but with complete absence of detail, that in the presence of a third molecule of Et$_3$P and AlBr$_3$, a σ-allyl complex [CH$_2$=CHCH$_2$Ni(PEt$_3$)$_3$]$^+$AlBr$_4^-$ was obtained.

This suggests that the ease with which a σ-allyl intermediate is formed increases from Ni to Pd and is so great for Pt that relatively few symmetric π-allyl complexes are known. This is in agreement with other qualitative data on the stability of M–C σ bonds in the nickel triad.

2. Further Reactions with Tertiary Phospines and Similar Ligands

As mentioned on p. 216, Powell and Shaw[327] found that a solution of allPdCl(PPh$_3$) and one mole of PPh$_3$ in chloroform disproportionated to give (Ph$_3$P)$_2$PdCl$_2$ and a solution which may well contain (σ-all)$_2$Pd(PPh$_3$)$_2$. However, these reactions depend very greatly upon conditions since Hüttel and König[795] and Powell and Shaw[51] showed that in more polar solvents allyl- and propenyltriphenylphosphonium salts could be isolated. In addition, Becconsall et $al.$[53] found that bis(π-allyl)palladium gave (Ph$_3$P)$_4$Pd when treated with 5 moles of triphenylphosphine in pentane at 20°. Wilke et $al.$[54] had previously observed the same reaction for bis(π-allyl)nickel; the other product in this case was characterized as 1,5-hexadiene.

$$\left\langle\!\!\left(\!-Ni\!-\!\right)\!\right\rangle + 4PPh_3 \longrightarrow (Ph_3P)_4Ni + \left|\!\!\left[\overbrace{\qquad}\right.\!\!\right|$$

However, bis(π-allyl)platinum gave a bis σ-allyl complex, (Ph$_3$P)$_2$Pt (CH$_2$CH=CH$_2$)$_2$.[53]

The reaction of bis-allylic nickel complexes to form an organic (coupled) product and a Ni(0) complex is fairly general; Wilke et al.[54] showed that at −40° bis(1-methylallyl)nickel reacted with CO to give nickel carbonyl and 98% trans,trans-2,6-octadiene (V-50) together with 2% 3-methyl-1,5-heptadiene (V-51). At higher temperatures the coupling was no longer stereospecific and appreciable amounts of (V-51) and (V-52) were formed.

An instructive and amusing example of this reaction is the ring closure,[52, 90]

π-Allylic nickel halide dimers and similar complexes also undergo disproportionation quite readily. The halides react with ammonia,[91]

$$[C_3H_5NiX]_2 + 6NH_3 \rightarrow (C_3H_5)_2Ni + Ni(NH_3)_6X_2$$

and Corey et al.[812] have shown this to be a general reaction for a number of nitrogen bases (DMF, N-methylpyrrolidone, triethylenetetramine); yields of better than 90% bis-π-allyl complex were obtained by using a Zn/Cu couple as a reducer in DMF or hexamethylphosphoramide. These workers also indicated that the coupling of allylic halides with π-allylic nickel halide complexes in ethereal solvents (to give the biallyl and NiX_2) did not proceed via a bis(π-allyl)nickel intermediate.[813]

The reactions of π-allylic nickel halide complexes with phosphines, at least, is not as simple as might be anticipated from the above.

Heimbach[17] and later Porri et al.[18] showed that $(PPh_3)_3NiX$, a complex of Ni(I), is formed in the reaction.

$$[C_3H_5NiX]_2 + PPh_3 \rightarrow (Ph_3P)_3NiX$$

This reaction has, so far, no parallel in palladium chemistry.

Ketley and Braatz[588] and Lupin et al.[736] have shown that a number of chloro-substituted π-allylic palladium halide complexes lose halogen from the

π-allylic group (to give a diene) in the presence of phosphines or even DMSO. For example,

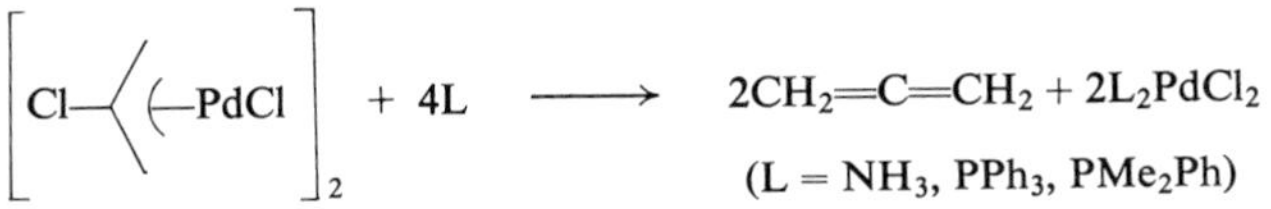

$$\left[Cl\!-\!\!\left\langle\!\!\left(\!-PdCl\right.\right.\right]_2 + 4L \longrightarrow 2CH_2\!=\!C\!=\!CH_2 + 2L_2PdCl_2$$

$$(L = NH_3, PPh_3, PMe_2Ph)$$

(V-53)

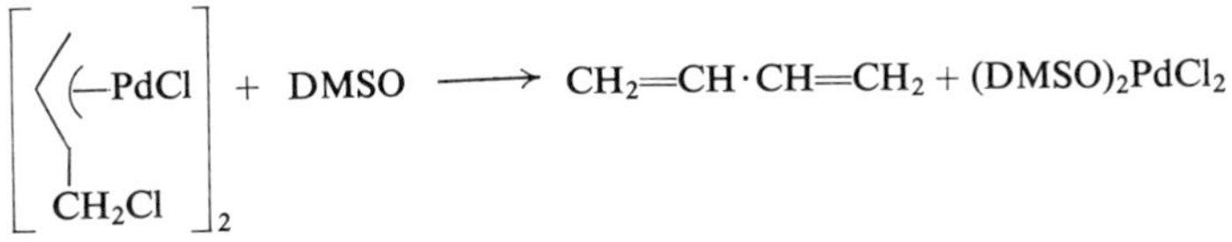

$$\left[\left\langle\!\!\left(\!-PdCl \atop CH_2Cl\right.\right.\right]_2 + DMSO \longrightarrow CH_2\!=\!CH\!\cdot\!CH\!=\!CH_2 + (DMSO)_2PdCl_2$$

The decomposition of the 2-chloroallylpalladium chloride complex, **(V-53)**, with phosphines was studied in some detail by Lupin *et al.* They were able to isolate the asymmetric π-allylic complex 2-ClC$_3$H$_4$PdClL for L = PPh$_3$ and obtained NMR evidence for its presence for L = PMe$_2$Ph. In the former case addition of one further mole of PPh$_3$ gave a spectrum corresponding to a σ-2-chloroallyl complex, which on the addition of further PPh$_3$ rapidly gave allene and (Ph$_3$P)$_2$PdCl$_2$. The σ-2-chloroallyl intermediate could not be observed when PMe$_2$Ph was used, instead only allene and a mixture of *cis*- and *trans*-(PMe$_2$Ph)$_2$PdCl$_2$ were obtained. The asymmetric π-allylic complex, 2-ClC$_3$H$_4$PdClPPh$_3$, could also be decomposed thermally at 200° to allene and [Ph$_3$PPdCl$_2$]$_2$. However, the complex **(V-53)** did not break down to give allene in DMSO solution; a spectrum characteristic of a dynamic σ-allyl complex was obtained instead.[736, 799]

The above reactions are very important for understanding both the chemistry of π-allylic complexes and of a number of catalytic reactions and will repay closer investigation.

3. Reactions in Which the Allylic Group Remains π-Bonded to the Metal

Apart from the reactions described in Section F,1, a number of other reactions are known, in which the other ligands are exchanged or a substitution occurs at an allylic carbon atom, which do not result in decomposition of the complex.

a. *Exchange of Halide in* (allPdX)$_2$

The halide, usually chloride, in (allPdhalide)$_2$ complexes is readily exchanged for other halides (I displaces Br, which displaces Cl) or thiocyanate simply by reaction with an alkali metal halide (or thiocyanate) in a suitable solvent.[634, 639, 650, 653, 713, 767] The isolation of complexes such as $[C_3H_5PdCl_2]^-$ in the presence of a large cation[814] suggests that similar species may be intermediate in these replacement reactions. An unusual example of this type of reaction involves the treatment of a steroid π-allylic palladium chloride complex with methyl iodide. The iodo complex, rather than a coupled product was obtained.[715]

The π-cyclopentadienyl and the acetylacetonate groups are other favorites since they give rise to monomeric products, normally of greater solubility. These groups have been introduced using sodium cyclopentadienide,[649, 672, 728, 815, 816] or acetylacetone in the presence of base. A useful alternative is to use the thallium(I) cyclopentadienide[650, 736] or acetylacetonate.[639, 650, 736] The only other product then is the thallium(I) halide which is quite insoluble.

Both of these ligands are readily cleaved off, for example, by HCl, halogen, N-bromosuccinimide, metal halides ($AlCl_3$, $FeCl_3$, $HgCl_2$), etc , to regenerate the allylic palladium halide complex.[773, 817]

Robinson and Shaw[765] have also prepared various allylic palladium carboxylate complexes, from the halo complex and the silver carboxylate.

The acetate reacted with trifluoroacetic acid to give the trifluoroacetato complex, with acetylacetone to give the acac complex, and with HCl to give the chloro complex.

$SnCl_2$ has been inserted into a Pd–Cl linkage to give $C_3H_5Pd(SnCl_3)$ (PPh_3).[218, 818]

As extensively discussed earlier, a wide variety of ligands will break the halogen bridges in $(allPdX)_2$. Phosphines and arsines will usually give stable 1:1 adducts; amines sometimes do too, especially pyridine, quinoline, isoquinoline,[639] and some primary amines.[650, 800, 804]

A number of authors have prepared π-allylic palladium complexes containing an extra mole of PdCl$_2$ per dimer. Moiseev et al.[757] suggested a structure

$$(\pi\text{-all})Pd\underset{X}{\overset{X}{<}}>Pd\underset{X}{\overset{X}{<}}>Pd(\pi\text{-all})$$

in which the extra PdX$_2$ bridges the two metal atoms via halide bridges.

Donati and Conti[708] have criticized it on the basis of their far-infrared measurements. For example the π-octenyl complex, which molecular weight measurements and analyses indicated to be $(\pi\text{-C}_8\text{H}_{15})_2\text{Pd}_3\text{Cl}_4$, showed bands at 251 and 327 cm^{-1}, ascribed to ν_{PdCl}. While the former was in the range for a bridging PdCl vibration,[770] the latter was regarded as arising from a terminal Pd–Cl vibration. If this assignment is right, then structures such as Moiseev's are obviously excluded, and Donati and Conti prefer a structure (unspecified) in which the palladiums are in a cluster linked by weak metal–metal bonds. The nature of this is difficult to envisage and it may be that in this case the far-infrared spectra are misleading. An X-ray determination of such a complex would be very useful.

Donati and Conti also reported that the extra PdCl$_2$ was easily removed by various ligands, such as 1,5-COD, DMF, or KCl, to give the normal π-allylic complexes. Similar observations have been made in the tetraphenylcyclo-butadienepalladium halide complexes[625,652] (Chapter IV, Section D,3, this volume).

b. *Substitution at Carbon*

Nucleophilic substitution at an allylic carbon usually leads to decomposition of the complex. However, nucleophilic displacement occurs rather easily at carbons *attached* to C-1 or C-3 of the allylic group, and may well also be the case for carbons attached to C-2. Examples of such reactions have been described by Shaw[649,650] and others.

(V-54)   (V-55)

The facility of this reaction is explained by the stability of the carbonium ion intermediate (**V-54**), particularly when R = Me. This system is analogous to that of ferrocenylcarbonium ions ($C_5H_5FeC_5H_4CH_2^+$), where similar effects have been noted. It is also conceivable, in view of the ease with which these systems undergo syn–anti isomerization at C-1, that structures of the type (**V-55**) may play a part too. A similar suggestion has been made by Lukas and Kramer who have isolated cations of type (**V-55**) (R = H) in the presence of SbF_5 at low temperatures, but not for R = Me.

A very interesting observation by Smutny *et al.* is that the ion (**V-54**) can also carry out an electrophilic displacement at carbon in sufficiently reactive aromatic molecules such as phenol. Substitution largely appears to be para.[93]

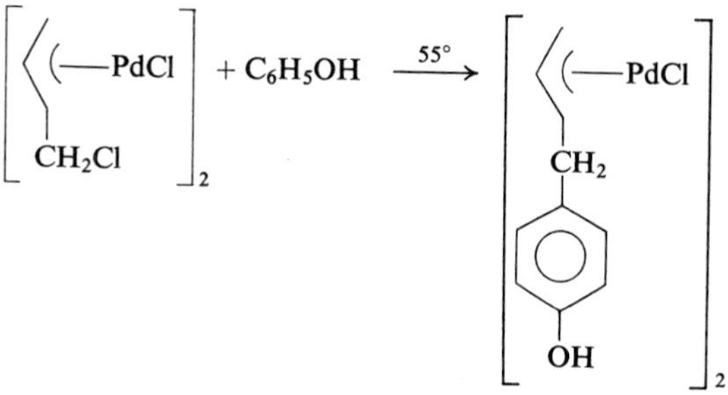

$$\left[\left\langle\!\!\left(\!-PdCl \atop CH_2Cl \right)\right]_2 + C_6H_5OH \xrightarrow{55°} \left[\left\langle\!\!\left(\!-PdCl \atop CH_2 \right)\right]_2 \right.$$

Yet another related example, reported by Keim,[819] involves nucleophilic displacement at a carbon conjugated to the π-allylic system:

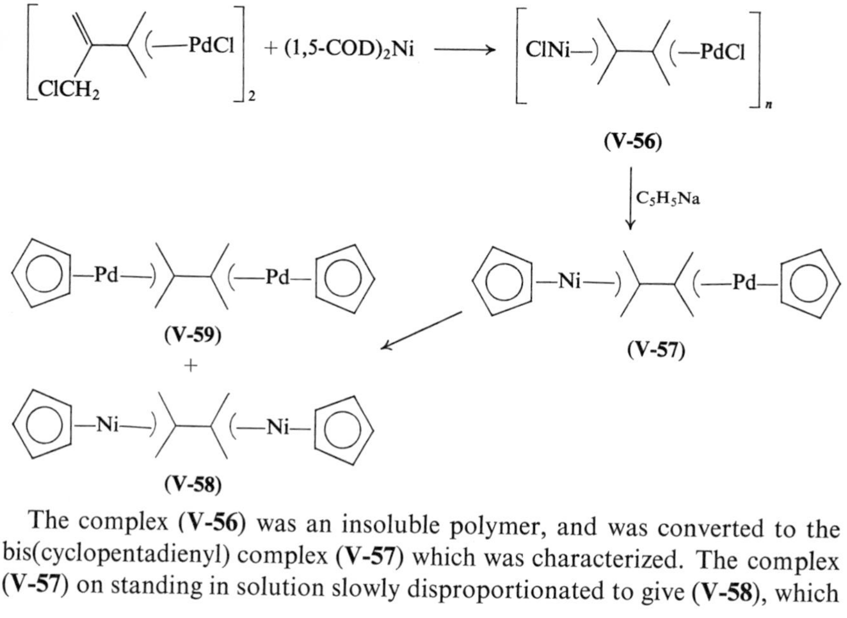

$$\left[{ClCH_2 \atop } \!\!\right\rangle\!\!\left\langle\!\!\left(\!-PdCl \right) \right]_2 + (1,5\text{-}COD)_2Ni \longrightarrow \left[ClNi\!\!-\!\!\right\rangle\!\!\left\langle\!\!\left(\!-PdCl \right) \right]_n$$

(**V-56**)

$$\Big\downarrow C_5H_5Na$$

(**V-59**)

\+

(**V-58**)

(**V-57**)

The complex (**V-56**) was an insoluble polymer, and was converted to the bis(cyclopentadienyl) complex (**V-57**) which was characterized. The complex (**V-57**) on standing in solution slowly disproportionated to give (**V-58**), which was isolated, and presumably, (**V-59**). This latter complex has been prepared directly by Hughes and Powell.[737]

4. Reactions in Which the Allylic Group Is Transferred to Another Metal

In view of the great lability which the allylic group shows in the palladium complexes, it is perhaps not surprising that transfer of the allylic group from palladium onto other metals proceeds relatively easily.

Mercury metal reacts with a number of π-allylic complexes to give either mono- or diallylmercury complexes, e.g.,[820, 821]

$$\left[\left\langle\!\!\!\!\text{(}\!\!\!-\text{PdX}\right]_2 + \text{Hg} \longrightarrow \text{CH}_2\!\!=\!\!\text{CHCH}_2\text{HgX} + \text{Pd}$$

$$\left\langle\!\!\!\!\text{(}\!\!\!-\text{Pd-Y} + \text{Hg} \longrightarrow (\text{CH}_2\!\!=\!\!\text{CHCH}_2)_2\text{Hg} + \text{Pd}$$

$$(\text{X} = \text{halide, OAc; Y} = \pi\text{-C}_3\text{H}_5, \text{acac})$$

Neither, $\text{C}_3\text{H}_5\text{PdClPPh}_3$, $\text{C}_3\text{H}_5\text{PdC}_5\text{H}_5$, nor allylnickel complexes reacted, but $(\text{C}_3\text{H}_5)_2\text{Pt}$ did. The allylic palladium complexes were reported not to react with magnesium, copper, or gallium. The suggested mechanism involved insertion of mercury into, for example, a Pd–Cl bond, followed by an intramolecular transfer of the allyl group.

$$[\text{C}_3\text{H}_5\text{PdCl}]_2 + \text{Hg} \longrightarrow \left\langle\!\!\!\!\text{(}\!\!\!-\text{Pd}\!\!\!\begin{smallmatrix}\text{HgCl}\\\text{Cl}^-\end{smallmatrix} \longrightarrow \left\langle\!\!\!\!\text{(}\!\!\cdots\!\text{Pd}\!\!\!\begin{smallmatrix}\text{HgCl}\end{smallmatrix}\right.$$

$$\text{Pd} + \text{CH}_2\!:\!\text{CHCH}_2\text{HgCl} \longleftarrow$$

Heck showed that a number of allylic palladium halide complexes reacted with tetracarbonylcobaltate to give the allylic cobalt tricarbonyl (usually isolated and identified as the more stable dicarbonyltriphenylphosphine complexes).[730]

$$\left[\left\langle\!\!\!\!\text{(}\!\!\!-\text{PdCl}\right]_2 + \text{NaCo(CO)}_4 \longrightarrow \left\langle\!\!\!\!\text{(}\!\!\!-\text{Co(CO)}_3 \xrightarrow{\text{PPh}_3} \left\langle\!\!\!\!\text{(}\!\!\!-\text{Co(CO)}_2\text{PPh}_3$$

However, the 1,1,2-trimethylallylpalladium chloride complex gave a product, containing both cobalt and palladium, for which structure (**V-60**) was suggested. See also footnote, p. 202.

Maitlis and Grindrod[822] and Nesmeyanov et al.[823] showed that the allylic group could, in some cases, also be transferred to iron by reaction with

$$\left[\text{Me}\overset{\displaystyle \diagdown}{\underset{\displaystyle \text{Me}}{\diagup}}\Big(\text{-----}\overset{\displaystyle \text{--Me}}{}\text{PdCl} \right]_2 + \text{NaCo(CO)}_4 \xrightarrow{\text{PPh}_3} \text{Me}\overset{\displaystyle \diagdown}{\underset{\displaystyle \text{Me}}{\diagup}}\Big(\text{-----}\overset{\displaystyle \text{--Me}}{}\text{Pd}\overset{\displaystyle \nearrow\text{PPh}_3}{\searrow\text{Co(CO)}_4}$$

(V-60)

$Fe_2(CO)_9$ giving **(V-62)**, allyl(chloro)tricarbonyliron. The yields were low. Nesmeyanov *et al.*[824] subsequently undertook an investigation of this reaction. They found that only half or less of the allylpalladium chloride reacted, even in the presence of excess iron carbonyl. However, the iron carbonyl was very efficiently decomposed (catalytically), to metal and CO, during the reaction. By a combination of Mössbauer, infrared, and room temperature NMR studies Nesmeyanov *et al.* were able to detect the formation of an intermediate, which they believed to be **(V-61)**. The NMR of this showed the CH_2 protons as a doublet, and suggested that a dynamic allyl species was present. The mechanism outlined was proposed.

$$Fe_2(CO)_9 \rightleftharpoons Fe(CO)_4 + Fe(CO)_5$$

$$[C_3H_5PdCl]_2 + 2Fe(CO)_4 \longrightarrow \pi\text{-}C_3H_5Pd\overset{\displaystyle \nearrow\text{Fe(CO)}_4}{\searrow\text{Cl}}$$

(V-61)

$$Fe(CO)_4Cl + Pd \longleftarrow$$

$$\Big|\,-\text{CO}$$

$$\overset{\displaystyle \diagdown}{\diagup}\Big(\text{-----Fe(CO)}_3\text{Cl}$$

(V-62)

An intermediate somewhat similar to **(V-61)** may also occur in the reaction with tetracarbonyl(diphenylphosphine)iron described by Benson *et al.* H-transfer must then take place to give the observed product, **(V-63)**, and propylene.[825]

$$\left[\overset{\displaystyle \diagdown}{\diagup}\Big(\text{---PdCl} \right]_2 + 2(\text{HPh}_2\text{P})\text{Fe(CO)}_4 \longrightarrow \underset{(OC)_4Fe}{\overset{Ph_2P}{\diagdown}}\text{Pd}\underset{Cl}{\overset{Cl}{\diagdown}}\text{Pd}\underset{PPh_2}{\overset{Fe(CO)_4}{\diagup}}$$

(V-63)

5. Thermal Decomposition

A number of workers have described the thermal decomposition, usually without solvent, of allylic palladium halide complexes. Zaitsev et al.[826] reported a thermogravimetric study of the decomposition and noted that an endothermic decomposition took place at 160°–200° for some chloro complexes. Palladium metal was obtained, but the organic products were not identified. Hüttel et al.[498, 599] showed that 2-methylallylpalladium chloride gave metal, 2-methylallylchloride (96%), and isobutene (1.7%) at 250°. Similarly the 2-phenylallyl complex gave 1-chloro-2-phenyl-1-propene (57%) and α-methylstyrene (12%),

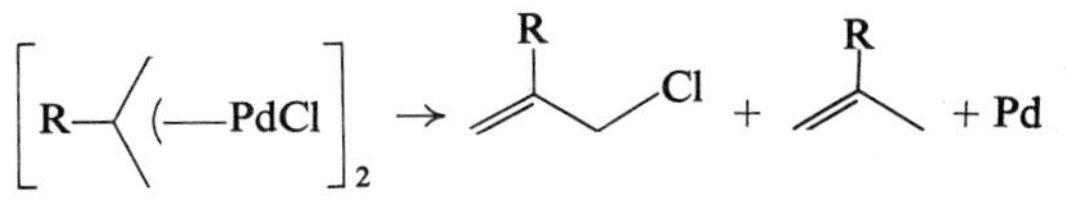

(R = Ph, Me)

Allylpalladium chloride gave allyl chloride,[498] while the acetate gave allylacetate, propylene, and unidentified products.[765]

Donati and Conti,[827] however, reported that π-allylic complexes derived from straight-chain olefins, C_nH_{2n}, where $n \geqslant 4$ gave largely conjugated dienes, together with HCl and metal, on pyrolysis at 160° and 10^{-2} mm, e.g.,

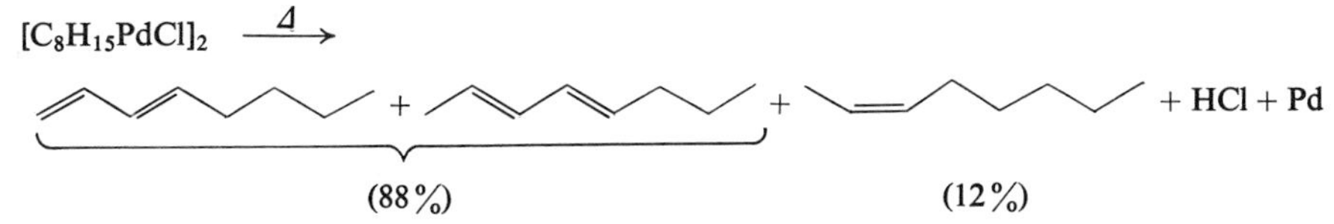

(88%)   (12%)

A similar result was obtained by Howsam and McQuillin[715] on heating a steroidal π-allylic complex in hexane (Chapter III, Volume II, Section E).

Blomquist and Maitlis[612] reported that decomposition in vacuo (or in benzene at 80°) of the cyclobutenyl complex (V-28a) gave a 1:2 mixture of the triphenylbenzofulvene and the ethoxytriphenylbenzofulvene.

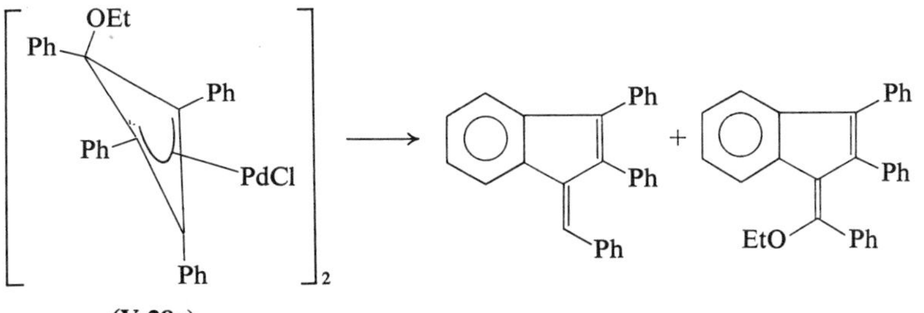

(V-28a)

6. Nucleophilic Attack at Carbon

Numerous examples of reactions which, explicitly or implicitly, involve nucleophilic attack at an allylic carbon are known. Tsuji *et al.*[828, 829] reported on the reactions of some common nucleophiles with allylpalladium chloride; the anions of diethylmalonate and ethyl acetoacetate gave mixtures of mono- and diallyl compounds.

$$[C_3H_5PdCl]_2 + (RCOCHCOOEt)^- \xrightarrow{\text{DMSO}}$$

$$CH_2{=}CHCH_2CH(COR)COOEt + (CH_2{=}CHCH_2)_2C(COR)COOEt$$

$$(R = Me, EtO)$$

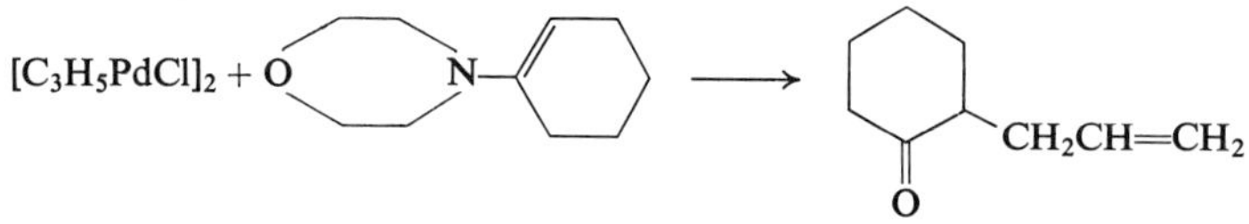

$$[C_3H_5PdCl]_2 + OR^- \rightarrow CH_2{=}CHCH_2OR$$

$$(R = Ac, Ph)$$

Enamines also reacted, but acetate and phenoxide only gave low yields of the expected allylic products. Saegusa *et al.*[586] have reported the methoxycarbonyl-ation using ClHgCOOMe. Tsuji *et al.* suggested that the carbanions attacked

$$[C_3H_5PdCl]_2 + ClHgCOOMe \rightarrow CH_2{=}CHCH_2COOMe$$

at the allylic carbons, whereas the other reagents attacked preferentially at the metal with a subsequent rearrangement.

An interesting but mechanistically not very enlightening reaction is that of carbon monoxide with allylpalladium acetylacetonate. The product, allyl-

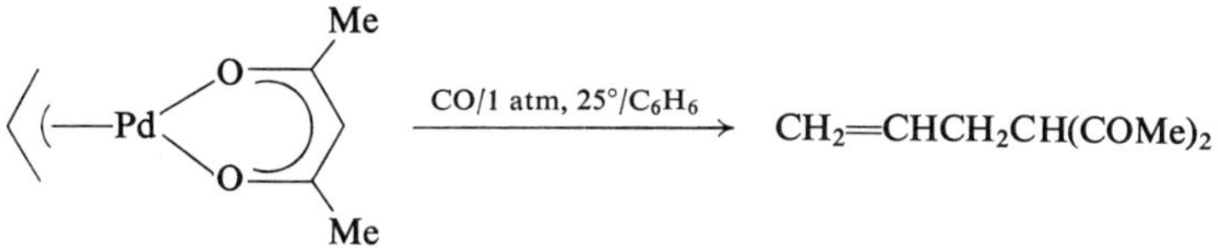

$$CH_2{=}CHCH_2CH(COMe)_2$$

acetylacetone, obviously arises by a CO-induced coupling and Takahashi *et al.*[830] suggested that the reactions described by Tsuji *et al.* involved similar intermediates. However, this does not necessarily follow as the conditions, especially solvent, were quite different in the two reactions. It is also of interest that other good π-acceptor ligands such as butadiene do not give the same product with $C_3H_5Pdacac$ that carbon monoxide does (see p. 245).

Hüttel *et al.* have shown that π-allylic palladium halides were relatively cleanly degraded to olefins on treatment with methanolic potassium hydroxide.[599,707,831,832] They also showed that the source of the extra hydrogen was *not* the methyl group of CH_3OH by deuterium labeling studies.[832] Isomers were frequently obtained and they postulated that the isomerization arose in a further step which was neither base- nor palladium-catalyzed. For example, 1,2,3-trimethylallylpalladium chloride gave a mixture of *cis*- and *trans*-2-methyl-2-pentenes, 3-methyl-1-pentene, and 2-ethyl-1-butene. The first step was suggested to be attack at the metal by the nucleophile; this releases the allylic anion which can either be protonated directly or undergo hydride shifts followed by protonation, for example,[832]

(7%)   (63%)

There are a number of serious objections to this mechanism. A recent study by Schenach and Caserio[833] showed that π-allylpalladium chloride dissolved in aqueous base to give a new π-allylic complex, which then decomposed, slowly at 20° but rapidly on heating. When the reaction was carried out with NaOD in D_2O, about 60% of the olefin was obtained, but this was only 7% monodeuterated for R = H and 13% for R = Me. The preponderance of

(R = H, Me)

undeuterated olefin in the product rules out the formation of an allylic anion followed by protonation. The authors suggested that an intermolecular H

$$2RC_3H_4 \rightarrow RC_3H_3 + MeCR{=}CH_2$$

transfer from one allylic group to another occurred and the RC_3H_3 fragment then added oxygen or H_2O to give oxidized products. In fact, alcohols and aldehydes were detected (see also Chapter III, Volume II, Section D). This intriguing reaction will certainly repay further study.

Some cyclic π-allylic complexes have also been examined, e.g.,[707, 831]

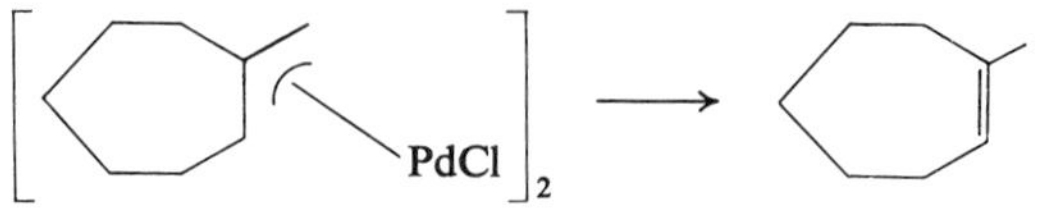

(76%)

In contrast, Tsuji and Imamura,[710] reported the isolation of the diene, *trans*, *trans*-diethyl muconate from the complex (**V-64**) on treatment with sodium dimsyl.

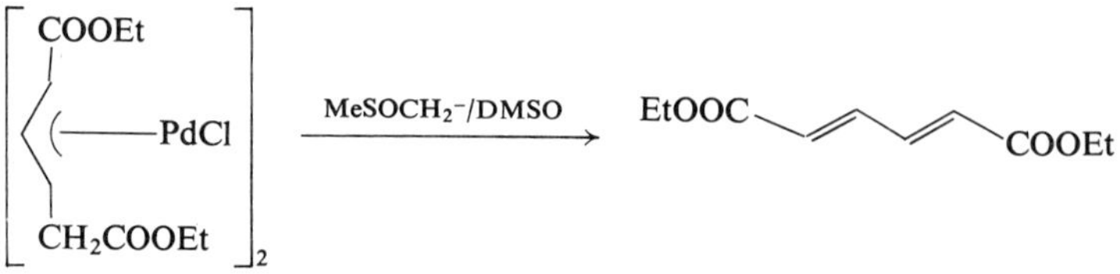

(**V-64**)

The reaction may take this course because of the active methylenic hydrogens in (**V-64**); this reaction may be compared to one described in Chapter II, Section C,3,d,iii, this volume.

7. Oxidation of π-Allylic Complexes to Aldehydes and Ketones

Three different kinds of reactions may be distinguished. The first, considered in more detail in Chapter II, Volume II, is the formation of aldehydes and ketones in the reaction of olefins with $PdCl_2$ under conditions in which π-allylic complexes are also formed, for example, in aqueous acetic acid. In some cases these are the same as are produced in the second type, when a π-allylic complex is itself oxidized, for example, by palladium chloride. The third is the so-called hydrolysis reaction, in which some π-allylic complexes are decomposed by water. It is impossible at present to discuss these reactions very rationally since the effects of numerous variables, such as the pH of the reaction, the role which atmospheric oxygen, and the presence of two potential ligands, acetate and chloride, have not been sufficiently investigated.

Hüttel and his collaborators[599] have described a few such reactions, and Hüttel and Christ[705] later noted that oxidation of allylic complexes usually gave unsaturated aldehydes (or ketones), whereas oxidation of the parent

$$CH_2=CHCH_2Cl + PdCl_2 + HOAc/H_2O \xrightarrow{60°} \left[\left\langle\!\!\left(\!\!-PdCl \right. \right]_2$$

$$\xrightarrow{100°} MeCOCHO$$

$$[C_3H_5PdCl]_2 + PdCl_2 + HOAc/H_2O \xrightarrow{100°} MeCOCHO + CH_2=CHCHO \text{ (trace)}$$

$$[C_3H_5PdCl]_2 + H_2O \xrightarrow{100°} CH_2=CHCHO + CH_2=CHMe$$

olefins usually gave the saturated carbonyls. Oxidation of π-allylic complexes at different pH values sometimes gave different products, for example,

The reactions were explained in terms of attack by OH⁻ at an allylic carbon, followed by loss of H⁻ to the uncomplexed Pd^{II}. The position of attack of the OH⁻ was assumed to be determined by the degree of hindrance at the various sites and hydride loss was assumed to be easier at a methyl-substituted allylic carbon.

However, this is probably too naive a picture. Other oxidizing agents (MnO₂, CrO₇²⁻) gave similar results, but as these were more powerful oxidants, they were able to oxidize the more heavily substituted π-allylic complexes which did not react with palladium chloride,[499, 705] e.g.,

(38% with CrO_7^{2-}, 29% with MnO_2)

Vallarino and Santarella[653] reported that both the *exo*- and *endo*-cyclo-butenyl complexes (**V-28**), were oxidized by either nitric acid or hydrogen peroxide in acetic acid to *cis*-dibenzoylstilbene.

(**V-28a**) (R = HO, R′ = Ph)
(**V-28b**) (R = Ph, R′ = EtO)

8. Halogenation

Few reports of the halogenation of π-allylic complexes have appeared. Belov *et al.*[720] have shown that allylpalladium chloride, when reacted with bromine in methanol containing sodium bromide gave allyl bromide. More bromine led to 1,2,3-tribromopropane.

$$[C_3H_5PdCl]_2 + Br_2 \xrightarrow{\text{MeOH/NaBr}} CH_2{=}CHCH_2Br + PdBr_4^{2-}$$

Blomquist and Maitlis[612] obtained tetraphenylfuran from the bromination of a cyclobutenyl complex.

9. Hydrogenation and Reduction

Hydrogenation of π-allylic complexes (usually, but not always, in the absence of a catalyst) leads either to the monoolefin or to the completely saturated organic compound. For example,

$$\left[\text{ClCH}_2\text{CH}_2\text{—}\!\!\left<\!\!\left(\text{—PdCl}\right)\right]_2 \xrightarrow{\text{H}_2/60°} \begin{array}{c} \text{CH}_2 \\ \diagdown \\ \text{Me} \end{array}\!\!\text{C·CH}_2\text{CH}_2\text{Cl} + \text{Pd} \;^{740}$$

$$\left<\!\!\left(\text{—Pd}\begin{array}{c}\text{OAc}\\ \diagdown \\ \diagup \\ \text{OAc}\end{array}\text{Pd—}\right)\!\!\right> \xrightarrow{\text{H}_2} \text{CH}_3(\text{CH}_2)_{10}\text{CH}_3 + \text{HOAc} + \text{Pd} \;^{326,\,833}$$

Reduction of a steroidal π-allylic complex to a saturated steroid by hydrogen has been reported by Howsam and McQuillin.[715] They also noted that in the presence of SnCl_2 the hydrogenation was inhibited and an olefinic steroid resulted.

Reduction with LiAlH_4 or NaBH_4 leads to similar products to those obtained on hydrogenation.[495]

$$\xrightarrow[\text{H}_2/\text{catalyst}]{\text{LiAlH}_4 \text{ or NaBH}_4 \text{ or}} \text{RCH}_2\text{CHPh·CH}_2\text{R}'$$

(R = R' = Ph; R = Bu^t; R' = Me)

However, in some cases (RO endo) ring-opening[653]

$$\xrightarrow[\text{H}_2/\text{Pd}]{\text{NaBH}_4 \text{ or}}$$

(R = Me, Et)

or (RO exo) complex rearrangements can occur.[653]

10. Reactions with Acids

The exact products from π-allylic complexes and proton acids are the subject of some dispute that is still unresolved. Hüttel *et al.* originally reported[599] that 2-methylallylpalladium chloride was converted to propylenepalladium chloride quantitatively by HCl in ether.

This result has been widely accepted and used in the interpretation of a number of reactions in terms of an equilibrium between the π-allyl and π-olefin complexes[291, 706]

$$[C_nH_{2n-1}PdX]_2 \underset{-HX}{\overset{+HX}{\rightleftharpoons}} [C_nH_{2n}PdX_2]_2$$

However, Ketley and Braatz[600] have reported their inability to repeat the work of Hüttel *et al.*; they only obtained nonstoichiometric products with uninterpretable NMR spectra. It seems likely that the results of Hüttel *et al.* are basically correct, but since π-olefin complexes are very labile and reactive, the conditions under which they can be isolated are extremely critical. A very slight excess of acid or a slight change in the reaction conditions will probably decompose the product.

Equilibria of this type have been used by Lukas *et al.*[724] to explain why 2,4,4-trimethyl-2-pentene gives 2-neopentylallylpalladium chloride (**V-65**) in the presence of sodium carbonate in chloroform, while in acetic acid in the presence of sodium acetate the major products are the *syn*- or *anti*-1-*t*-butyl-2-methylallyl complexes.

The lower effective basicity of the $Na_2CO_3/CHCl_3$ system allows the establishment of equilibria, as described in Section B,1, leading to the thermodynamically favored product.

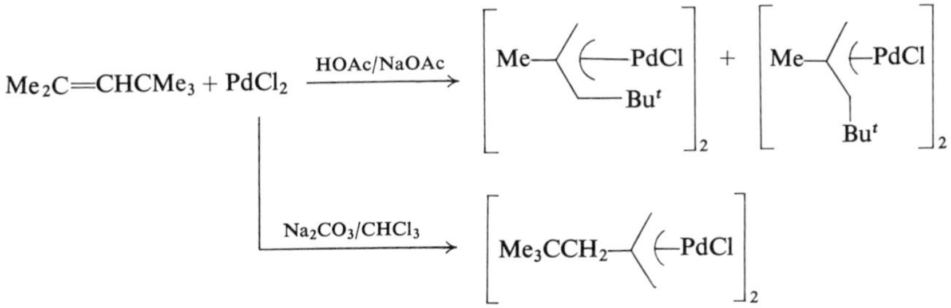

$$Me_2C=CHCMe_3 + PdCl_2 \xrightarrow{HOAc/NaOAc} \left[Me-\!\!\left<\!\!\underset{Bu^t}{\overset{PdCl}{}}\right]_2 + \left[Me-\!\!\left<\!\!\underset{Bu^t}{\overset{PdCl}{}}\right]_2 \right.$$

$$\xrightarrow{Na_2CO_3/CHCl_3} \left[Me_3CCH_2-\!\!\left<\!\!-PdCl\right]_2 \right.$$

(V-65)

Medema et al.[326, 834] have made the interesting observation in this connection that methanol is a strong enough proton acid to decompose bis(π-allyl)-palladium.

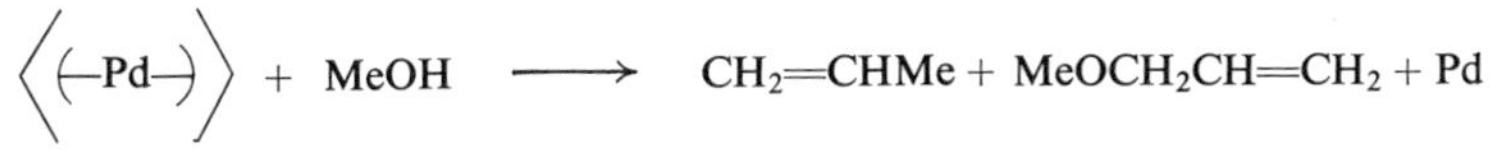

$$\left<\!\!-Pd-\!\!\right> + MeOH \longrightarrow CH_2=CHMe + MeOCH_2CH=CH_2 + Pd$$

Belov et al.[834a] have recently reported that allylpalladium chloride dimer reacted with aqueous HCl to give allyl chloride in the presence of benzoquinone; palladium chloride was also formed.

The endo- and exo-alkoxycyclobutenyl complexes (V-28), investigated by Malatesta[611, 653] and Maitlis[612, 625, 672, 731, 732] and their co-workers, all react with acid to give the cyclobutadiene complexes (V-66). The cyclobutadiene complexes are insensitive to acid and can therefore easily be isolated. The reaction presumably involves protonation of the alkoxy groups. See also Chapter IV, Section D,3 and F,2 and Section B,2,a in this chapter.

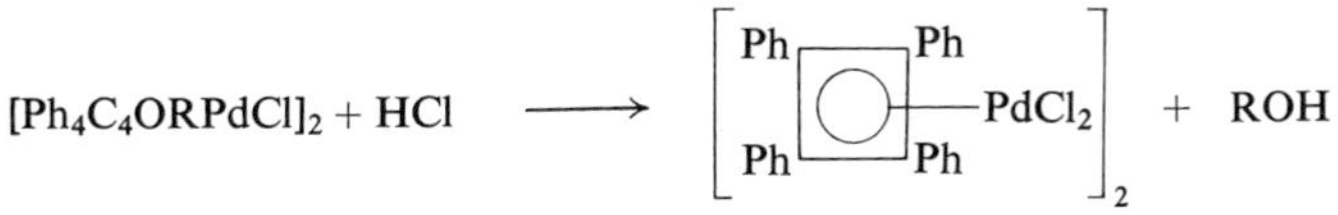

$$[Ph_4C_4ORPdCl]_2 + HCl \longrightarrow \left[\underset{Ph}{\overset{Ph}{\square}}\!\!-PdCl_2 \right]_2 + ROH$$

(V-28)　　　　　　　　　　　(V-66)

Similar reactions for an exo-hydroxy-di-t-butyldiphenylcyclobutenyl complex have been described by Hosokawa and Moritani.[605]

Johnson et al.[463] have briefly reported the action of some Lewis acids, especially the triphenylmethylcarbonium ion, with various complexes, including some allylic palladium acetylacetonates,

11. Reactions with Carbon Monoxide

Powell and Shaw[327] showed that CO acted on π-allylic complexes anal-ogously to other Lewis bases, such as phosphines, to give a species which showed a dynamic σ-allyl NMR spectrum. The ν_{CO} of the complex derived from 2-methylallylpalladium chloride in solution was very high, at 2121 cm^{-1}, and indicated that the CO was weakly coordinated. No solid could be isolated as the CO was again removed on evaporation of the solvent. (See also Volger et al.[834b]).

Reaction with CO, however, proceeds further and has been studied by Tsuji and co-workers,[709, 835-839] Long and Whitfield,[840] and Medema et al.[326] The initial product is usually the butenoyl chloride which is isolated as the ester in the presence of an alcohol, or as the acid in the presence of water.

These reactions are reminiscent of those of π-allylic complexes with excess tri-phenylphosphine in polar solvents to give the allyltriphenylphosphonium salts, as discussed in section F,1,a,i. However, here attack appears invariably to be at the *least* substituted allylic carbon.

Tsuji *et al.* and others have also studied the reactions of some chloroallylic complexes, in particular (V-67), with CO. In this case, the main reaction path at low temperatures involves the formation of 1,4-dichloro-2-butene, (V-68); the expected product, (V-69), predominates at higher temperatures.[839]

$$\left[\begin{array}{c} \diagdown\!\!\!\diagup\!\!-PdCl \\ | \\ CH_2Cl \end{array} \right]_2 + CO \xrightarrow{C_6H_6} ClCH_2CH\!\!=\!\!CHCH_2Cl + ClCH_2CH\!\!=\!\!CHCH_2COCl$$

$\qquad$ (V-67) $\qquad\qquad\qquad\qquad\qquad$ (V-68) $\qquad\qquad\qquad$ (V-69)

The formation of products in which the ligands have coupled and released the metal is even more noticeable when oxy ligands are on the metal.[326, 830] No carbonylated products are obtained.

$$C_3H_5Pdacac \xrightarrow{CO/C_6H_6/25°} CH_2\!\!=\!\!CHCH_2CH(CMeO)_2 + Pd$$

$$\left[\begin{array}{c} \diagdown\!\!\!\diagup\!\!-PdCl \\ \diagup\!\!-R \\ R' \end{array} \right]_2 + AgOAc + CO \longrightarrow RR'CH\!\!=\!\!CHCH_2OAc + RR'\underset{\underset{OAc}{|}}{C}CH\!\!=\!\!CH_2$$

The sensitivity of these reactions to substituents and ligands is further illustrated by the reaction of (V-70), derived from isoprene. In this case, the major product, (V-71) at 20°, was derived from carbonylation at the least hindered allylic carbon while that at 100° was (V-72), where attack by CO has also occurred at the other end of the carbon chain. However, the species present at

$$\left[Me\diagdown\!\!\!\diagup\!\!\begin{array}{c} -PdCl \\ | \\ CH_2OEt \end{array} \right]_2 \xrightarrow{CO/C_6H_6/20°, \ EtOH} EtOCH_2CH\!\!=\!\!\underset{\underset{Me}{|}}{C}CH_2COOEt$$

$\qquad$ (V-70) $\qquad\qquad\qquad\qquad\qquad\qquad$ (V-71)

$$\xrightarrow{CO/C_6H_6/100°, \ EtOH} EtOOC\cdot CH_2CH\!\!=\!\!\underset{\underset{Me}{|}}{C}CH_2COOEt$$

$\qquad\qquad\qquad\qquad\qquad\qquad\qquad\qquad$ (V-72)

100° are probably quite different from those at 20° and π-allylic complexes may not even be involved. Minor amounts of other carbonylated products were also obtained and identified.†[838] An amusing synthetic sequence has been described by Tsuji and Imamura,[710]

† These include compounds in which isomerization of the double bond, perhaps Pd-catalyzed, has occurred. Tsuji and Suzuki[837] have also described the reactions of π-allylic complexes derived from allene with CO.

$$[C_3H_5PdCl]_2 \xrightarrow{CO/EtOH} CH_2{=}CH{\cdot}CH_2COOEt \xrightarrow{PdCl_2} \left[\left\langle\!\!\!\left\langle {\raise2pt\hbox{$\overset{\displaystyle\ }{\underset{\displaystyle COOEt}{}}$}} {-}PdCl \right.\right.\right]_2 \xrightarrow{CO/EtOH}$$

$$EtOOCCH_2CH{=}CH{\cdot}COOEt \xrightarrow{PdCl_2} \left[{\raise2pt\hbox{$COOEt$}} \left\langle\!\!\!\left\langle {-}PdCl \right.\right. {\raise-2pt\hbox{$COOEt$}} \right]_2$$

The mechanism of the carbonylation reaction has not been investigated in detail but it presumably involves the intermediacy of a σ-allylic or similar species, which undergoes CO insertion to give the acyl complex and which then couples with the halogen and extrudes the metal; but see Ref. 834b.

$$\left[\left\langle\!\!\!\left\langle \underset{R}{-PdCl} \right.\right.\right]_2 \xrightleftharpoons{CO} \left\langle\!\!\!\left\langle \underset{R}{-Pd}\overset{CO}{\underset{Cl}{\big\langle}} \right.\right. \xrightleftharpoons{CO} \left\langle\!\!\!\left\langle \underset{R}{-Pd(CO)_2Cl} \right.\right. \rightleftharpoons$$

$$RCH{=}CH{\cdot}CH_2{-}\overset{\overset{O}{\overset{\|}{C}}}{\underset{\searrow\overset{}{\underset{O}{C}}}{Pd}}{-}Cl \xrightarrow{CO} \underset{\underset{RCH=CHCH_2}{CO}}{\overset{\overset{O}{\overset{\|}{C}}}{OC{-}Pd{-}Cl}} \rightarrow 2CO + Pd + RCH{=}CHCH_2COCl$$

This reaction can also be carried out catalytically, as discussed in Chapter I, Volume II, Section B,2.[326, 841] The catalysts are deactivated by the formation of an insoluble complex, $[\pi{-}allPdCl_2PdCO]_2$, which shows a low frequency ν_{CO} at ca. 1950 cm^{-1}.[326, 328, 838] The formation of $[Pd_2Cl(CO)_2]$ has also been reported.[326]

O'Brien[763] has reported that SO$_2$ will insert into a Pd–allyl bond.

$$(C_3H_5)_2Pd + SO_2 \longrightarrow \left\langle\!\!\!\left\langle {-}Pd\overset{SO_2}{\diagdown} CH_2CH{=}CH_2 \right.\right. \rightleftharpoons \left\langle\!\!\!\left\langle {-}Pd\overset{SO_2\diagdown CH_2}{\diagdown\underset{CH_2}{\diagup} CH} \right.\right.$$

At lower temperatures the double bond is coordinated to the metal.

Isonitriles also insert into Pd–allyl bonds and again complexes are isolated,[341b]

$$[C_3H_5PdCl]_2 + C_6H_{11}NC \longrightarrow \left[CH_2{=}CHCH_2\underset{\underset{NC_6H_{11}}{\|}}{\overset{C_6H_{11}NC\diagdown}{C}}{>}PdCl \right]_2$$

12. Reactions with Dienes

The reactions of π-allylic complexes with both 1,2-dienes (allenes) and 1,3-dienes are also of considerable interest. The first report of these reactions in the open literature came from Takahashi et al.,[830, 842] who showed that allyl-palladium acetylacetonate reacted with butadiene to give a butenyl-π-allylic complex. This reaction has been studied in great detail by Medema et al.,[326, 834] Bright et al.,[749] Takahashi et al.,[842] and in part by Smutny et al.[93] The basic reaction with 1,3-dienes is

$$\left[R'-\left\langle \underset{R}{\overset{}{-}PdX} \right. \right]_2 + CH_2{=}CHCR''{=}CH_2 \xrightarrow{C_6H_6/20°/1\ atm} \left[\left\langle \begin{array}{l} {-}PdX \\ {-}R'' \\ CH_2CHRCR'{=}CH_2 \end{array} \right. \right]_2$$

as shown by Bright et al.,[749] using 1,1,4,4-d_4-butadiene and isoprene with allyl-palladium acetate.

In contrast to the reaction with CO, attack here occurs at the *most* heavily substituted π-allylic carbon.[326, 749, 834] The rate of the reaction is first-order with respect to diene and (allPdX)$_2$, and depends strongly on the nature of X (CF$_3$COO > AcO > OBz > NO$_3$ > Cl > Br > NCS > I). Acetylacetonate is also a strongly activating ligand for this reaction, but Takahashi et al.[842] have noted some anomalies. For example, isoprene did react (at 70°) with both allyl- and 2-methylallylpalladium chloride, but only reacted (at 20°) with the unsubstituted allylpalladium acetylacetonate. Similarly 1-methylallylpalladium acetylacetonate did not react with butadiene, whereas the chloride did.

Electron-withdrawing substituents on the π-allylic ligand increased the rate considerably, for example the 1- (or 2-) chloroallyl complex reacted ca.10^5 as fast as the unsubstituted complex. Increasing substitution (by methyl) on both the allyl group and the diene slowed the reaction in the order C_3H_5 > 2-MeC_3H_4 > 1,1,2-$Me_2C_3H_2$; and CH_2:CH·CH:CH_2 > CH_2:CMe·CH:CH_2 > CH_2:CCl·CH:CH_2 > CH_2:CMe·CMe:CH_2. The addition of one mole of triphenylphosphine to the reaction mixture deactivated it and all$PdXPPh_3$ was inactive.

These results suggest that the rate-determining step in the reaction is the coordination of the diene to the π-allylic complex. Medema et al.[326, 834] visualize the intermediate here to be (V-73), where the diene has coordinated

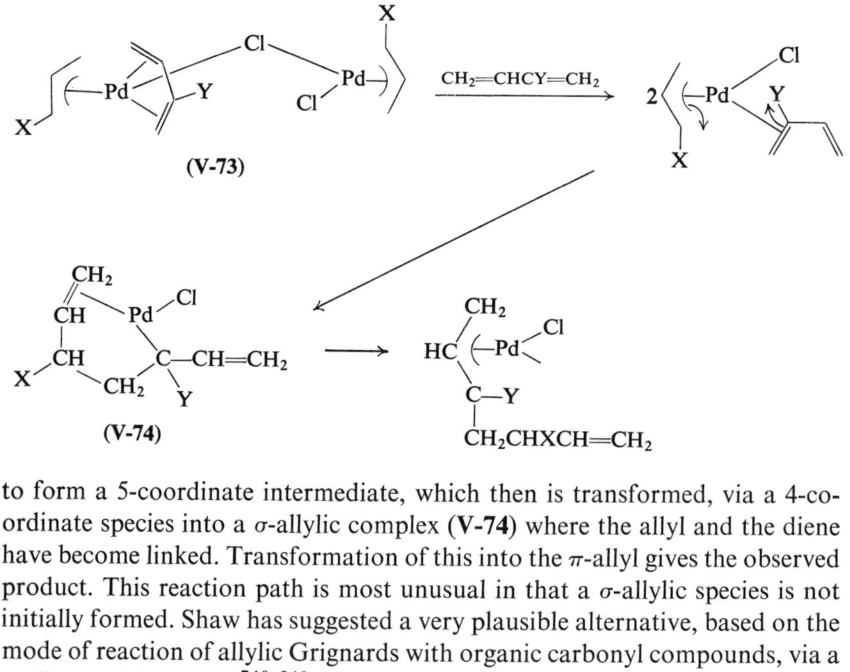

(V-73)

(V-74)

to form a 5-coordinate intermediate, which then is transformed, via a 4-co-ordinate species into a σ-allylic complex (V-74) where the allyl and the diene have become linked. Transformation of this into the π-allyl gives the observed product. This reaction path is most unusual in that a σ-allylic species is not initially formed. Shaw has suggested a very plausible alternative, based on the mode of reaction of allylic Grignards with organic carbonyl compounds, via a cyclic transition state,[749, 843]

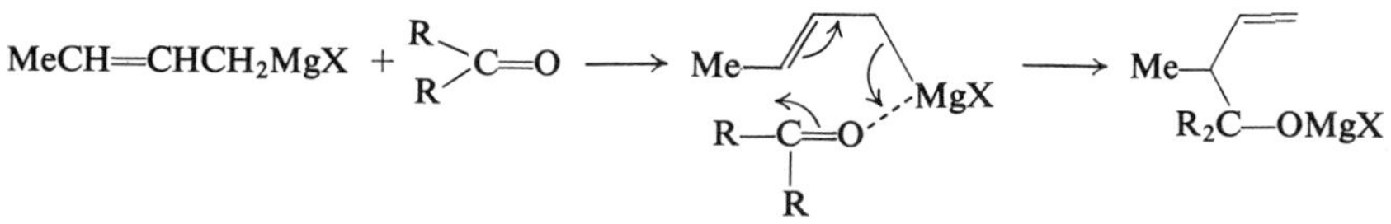

In this case a σ-allylic complex is again formed, the less heavily substituted carbon being σ-bonded to the metal. A similar suggestion was made by Takahashi et al.[842]

The reaction can also proceed further, and a second molecule of diene is added more slowly. These complexes are converted to hydrocarbons by reduction or to acids (or esters) on carbonylation.

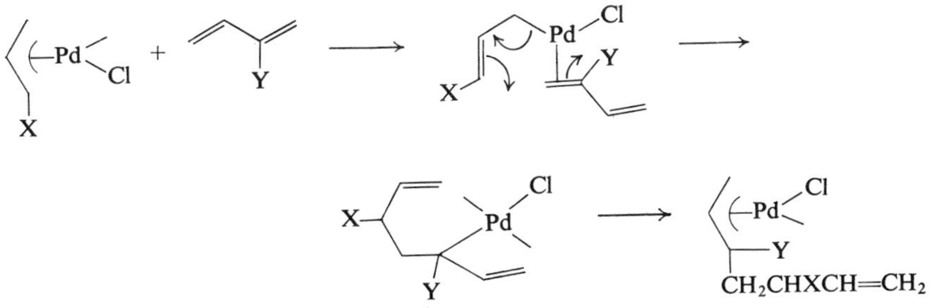

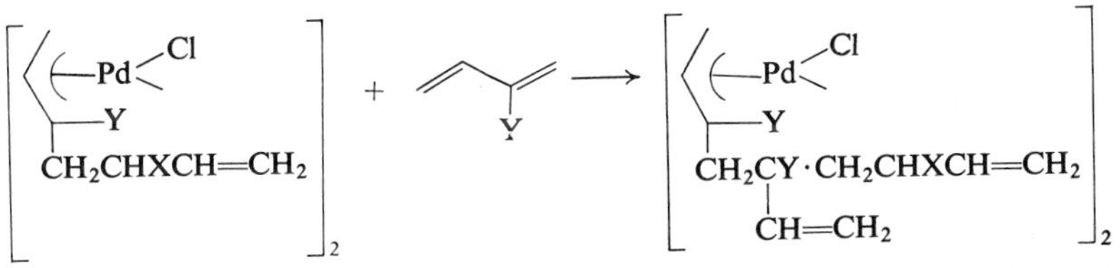

With acetate (or nitrate or trifluoroacetate), the above reaction takes place as far as the first stage. However, if the reaction is not stopped at this stage, a different further reaction, involving displacement of the C$_7$ ligand (as a 1:1 mixture of 1,3,6-heptatriene and 1,5-heptadiene) and the formation of a new complex, (V-75), occurs. If this reaction is carried out at 50° and higher butadiene concentration, 1,3,6,10-dodecatetraene is obtained catalytically by trimerization of the butadiene.[834] A similar type of reaction has been described by Smutny et al.,[93] in which butadiene is dimerized in acetic acid (or phenols) to give 1-acetoxy-2,7-octadiene or 1,3,7-octatriene.

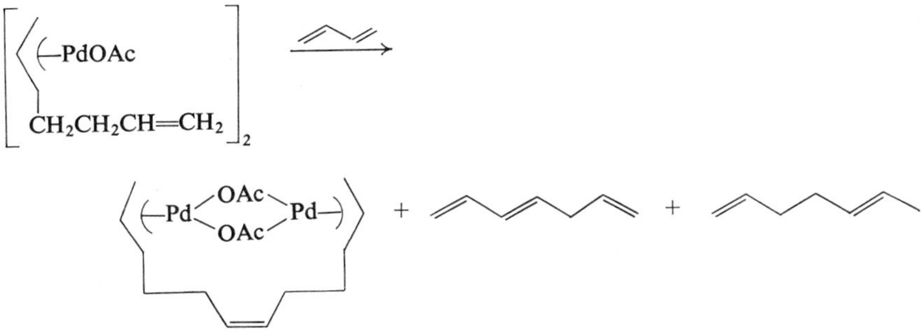

(V-75)

In methanol allylpalladium acetate reacted with butadiene to give a complex for which structure (V-76) was proposed.[845]

It has been proposed that (V-76) and analogous complexes are intermediates in the formation of (V-75) and the heptadiene and heptatriene and in the catalytic formation of alkoxyoctadienes (from butadiene and an alcohol)[93, 834] (see Chapter I, Volume II, Section C,3,c).

$$\left[\left(\!\!\left\langle\!-PdOAc\right.\right]_2 + CH_2=CHCH=CH_2 \xrightarrow{MeOH} \left\langle\!-Pd\!\!\begin{array}{c}OAc\\O\end{array}\!\!Pd-\right.$$

$$\underset{H}{}\quad \underset{Me}{O}$$

(V-76)

$$\textbf{(V-76)} + HCl \longrightarrow \left[\left(\!\!\left\langle\!-PdCl\right.\right]_2 + \left[\left\langle\!-PdCl\right.\right]_2 + HOAc + MeOH$$

$$\underset{Me}{}$$

(V-76) + CH$_2$=CMeCH$_2$Cl $\longrightarrow$

$$CH_2=CHCH=CH_2 + \left[\left(\!\!\left\langle\!-PdCl\right.\right]_2 + \left[Me\left\langle\!-PdCl\right.\right]_2$$

A very similar type of reaction was also described by Wilke *et al.*,[54] who used bis(π-allyl)palladium to catalytically trimerize butadiene to n-dodecatetraene (probably the 1,3,6,10-isomer). In the presence of 1 mole of PCl$_3$, only dimerization of the butadiene to vinylcyclohexene occurred. Wilke *et al.* suggested that the latter arose via a σ-allyl-π-allyl intermediate.

Rather analogous reactions are possible using nickel catalysts; these have been extensively reviewed by Wilke *et al.*[54,90,91] See also Chapter I, Volume II, Section C,3,c.

An interesting variation is the reaction with allene, described by Shier[844] and Medema *et al.*[326,834] The form of this reaction is

$$\left[R'\!\!\left\langle\!-PdCl\right.\right]_2 + CH_2=C=CH_2 \longrightarrow \left[RCH=CR'CH_2\!\!\left\langle\!-PdCl\right.\right]_2$$

In contrast to the reaction of butadiene and other 1,3-dienes, attack is by C-2 of the allene at the *least* hindered site of the π-allylic group. In the latter respect, this is similar to the reactions with carbon monoxide. Substitution at R or R′ on the allylic ligand has the same effect on the rate as for the reactions with 1,3-dienes; hence it is believed that the rate-determining step here too is the coordination of the diene to the allylic complex. (See also Section B,1,b.)

The mechanism can therefore be written as

$$\left[R'-\!\!\!\left\langle \!\!\!-PdX \atop R \right. \right]_2 + CH_2\!=\!C\!=\!CH_2 \longrightarrow R'-\!\!\!\left\langle \!\!\!-Pd \atop R \right. \begin{matrix} CH_2 \\ \| \\ C \\ \diagdown CH_2 \\ X \end{matrix}$$

$$\downarrow L$$

$$CHR\!=\!CR'CH_2\!-\!C\diagup^{CH_2}_{\diagdown CH_2-PdXL_n} \longleftarrow CHR\!=\!CR'\!-\!CH_2\text{------}C\diagup^{CH_2}_{\|} \atop L\diagup Pd\text{---}CH_2 \atop X$$

$$\downarrow -L_n$$

$$\left[RCH\!=\!CR'CH_2\!-\!\!\!\left\langle \!\!\!-PdX \right. \right]_2$$

(L = solvent or ligand)

π-Allene complexes are well known for other metals, e.g., Pt(II)[845] and Rh(I).[845–847]

This is supported by the observation of Shier[844] that a complex such as 2-phenylallylpalladium acetate (V-77) is stable by itself in acetic acid at 50°. In the presence of allene, however, decomposition products, (V-78) and (V-79), together with the normal products of the catalytic reaction of allene with palladium acetate are obtained (Chapter I, Volume II, Section C,3,b).

$$\left[Ph\!\!\left\langle \!\!\!-PdOAc \right. \right]_2 + CH_2\!=\!C\!=\!CH_2 \xrightarrow{HOAc/50°} \begin{matrix} Ph \\ \diagdown \\ Me \end{matrix}\!\!C\!=\!CH_2 + \begin{matrix} Ph \\ \diagdown \\ AcOCH_2 \end{matrix}\!\!C\!=\!CH_2 + \cdots$$

(V-77)   (V-78)   (V-79)

                                                     (88%)   (7%)

The complex (V-77) reacted with 1,5-cyclooctadiene to give the complex (V-80), isolated as the hexafluorophosphate salt.[844] The cation decomposed in acetic acid to (V-78), (V-79), and cycloocten-4-one.

$$[2\text{-PhC}_3\text{H}_4\text{PdOAc}]_2 + 1,5\text{-COD} \longrightarrow \left[Ph\!\!\left\langle \!\!\!-Pd\!\!\left\langle \right. \right. \right]^+$$

(V-77)   (V-80)

These reactions are notable examples of the very much higher reactivity of palladium complexes with weakly coordinating ligands (OCOR, NO$_3$, SO$_4$, F, etc.) compared to that of the palladium halide complexes.

G. π-ALLYLIC COMPLEXES IN CATALYTIC REACTIONS

π-Allylic palladium complexes have been clearly implicated as reaction intermediates in a number of catalytic reactions, and can frequently be used as efficient catalysts themselves. These reactions are discussed further in the appropriate sections and will merely be summarized here.

(1) Carbonylation of allylic halides to derivatives of the vinylacetic acid.[326, 841, 848-850] Variations on this include

$$CH_2{=}CH \cdot CH_2Cl + CH_2{=}CH \cdot CH{=}CH_2 + CO \; \rightarrow \; CH_2{=}CH(CH_2)_2CH{=}CH \cdot CH_2COCl \;^{834}$$

$$CH_2{=}C{=}CH_2 + CO + EtOH \; \rightarrow \; EtOOC \cdot CH_2C({=}CH_2)COOEt \;^{837}$$

$$CH_2{=}CH \cdot CH_2OH + CO + EtOH \; \rightarrow \; CH_2{=}CH \cdot CH_2COOEt \;^{835}$$

(2) The linear di- and trimerization of butadiene.[54, 93, 326, 834, 851]

(3) The acetylation of olefins and allene (accompanied by coupling).[291, 844]

A remarkable example of allylpalladium chloride acting as a catalyst is in the decomposition of ethyl diazoacetate.[852] Copper and copper salts act similarly, but higher temperatures (50° or above) are normally required than with allylpalladium chloride (0°–10°). In the absence of a reactive substrate diethyl fumarate is formed (Cu catalysts give diethyl maleate), whereas in the presence of an olefin or acetylene, addition of the carbene occurs. The allyl complex is recovered. A mechanism has been suggested involving splitting of the halide bridge by $N_2CHCOOEt$, followed either by direct attack on the substrate accompanied by loss of N_2 or loss of N_2 followed by attack on the substrate by the carbene.

These reactions are probably quite closely related to the recently discovered insertions of carbenes into Pd–Cl bonds when the diazo compounds are decomposed in the presence of $PdCl_2$ (p. 81). In that case the insertion product is favored over release of carbene. A somewhat similar reaction has been described for $(Ph_3P)_2IrCOCl$.[853] Nickelocene also catalyzes the decomposition of diazo compounds.[854]

H. π-ALLYLIC COMPLEXES OF OTHER METALS

Many of the methods of synthesis of allylic palladium complexes are also applicable, with modification, to syntheses of other allylic metal complexes. In addition, some other routes are also important, e.g.,

$$CH_2=CHCH_2Cl + NaMn(CO)_5 \longrightarrow$$
$$CH_2=CHCH_2Mn(CO)_5 \xrightarrow[-CO]{h\nu} \left\langle\!\!\left(\!\!-Mn(CO)_4 \right.\right. ^{855}$$

$$trans\text{-}(Et_3P)_2PtH(NO_3) + CH_2=C=CH_2 \longrightarrow \left[\left\langle\!\!\left(\!\!-Pt(PEt_3)_2 \right.\right.\right]^{+ \ 442a}$$

$$[(EtO)_3P]_4NiH^+ + CH_2=CHCH=CH_2 \longrightarrow$$
$$\left[\left\langle\!\!\left(\!\!-Ni[P(OEt)_3]_3 \right.\right.\right]^+ + \left[\left\langle\!\!\left(\!\!-Ni[P(OEt)_3]_3 \right.\right.\right]^{+ \ 482a}$$
$$\quad\quad Me \quad\quad\quad\quad\quad\quad\quad\quad\quad\quad\quad -Me$$

Allyl complexes without other ligands can be prepared by reaction of allyl Grignards with the metal halides at low temperatures.[54, 91, 762b]

The existence of symmetrical π-, unsymmetrical π-, σ-allylic, and dynamic σ-allylic forms have also been observed for the other elements. The differences between Ni, Pd, and Pt allyls in this respect have already been commented on.

Allylic complexes, particularly of nickel, have also found wide application in syntheses[812, 813, 856] and catalytic reactions.[54, 91, 857, 858]

I. CYCLOPROPENYL COMPLEXES

Moiseev et al.[757] have reported the preparation of a triphenylcyclopropenyl π complex, $Ph_3C_3Pd_3Cl_4Ph_3C_3$, (V-22) (p. 196), but no structural studies to prove the formulation have been undertaken.

Gowling and Kettle[859] reported the formation of a triphenylcyclopropenyl-nickelcarbonyl halide complex.

$$Ph\underset{Ph}{\overset{Ph}{\diamond}}X^- + Ni(CO)_4 \xrightarrow{MeOH} [Ph_3C_3NiCOX]_n$$

The pyridine derivative of this complex (X = Cl) was shown by an X-ray study[860] to contain a 3-membered ring symmetrically π-bonded to the metal.

$$Ph\underset{Ph}{\overset{Ph}{\diamond}}-Ni\underset{py}{\overset{Cl}{\underset{}{-py}}}$$

By analogy with the other C_3 ligands, this is most conveniently considered as a 5-coordinated Ni(II) complex of the (unknown) triphenylcyclopropenyl anion, $Ph_3C_3^-$, a 4π electron ligand.

Ofele[861] has recently described the diphenylcyclopropenyl complex (pre-

pared from dichlorodiphenylcyclopropene and palladium) in which the diphenylcyclopropenyl is acting as a carbenoid ligand. (See also Chapter II, Section B,3, this volume.)

Chapter VI
Cyclopentadienyl
and Benzene Complexes

A. DICYCLOPENTADIENYL COMPLEXES

A characteristic of the transition metals is the formation of cyclopentadienyl complexes of the type Cp_2M ($Cp = C_5H_5$). These are also called metallocenes. The best known is, of course, ferrocene (dicyclopentadienyliron), but these complexes are also known for all the 3d elements (bar copper) and many of the 4d and 5d elements. The tendency for them to be formed appears to decrease down any given triad, if one may judge from reports of successful syntheses.

A characteristic of the 3d metallocenes is that they are all, except ferrocene and the cobalticinium cation, (Cp_2Co^+), paramagnetic. Nickelocene (Cp_2Ni) is a good example; it has a magnetic moment of 2.86 B.M. corresponding to the presence of two unpaired electrons. Unlike cobaltocene (Cp_2Co) which readily loses one electron to give the exceedingly stable Cp_2Co^+ ion, nickelocene does not lose two electrons to give the (hypothetical) diamagnetic Cp_2Ni^{2+} ion. Instead, nickelocene has a high tendency to (effectively) lose one cyclopentadienyl ring. This can occur in a number of ways.

In fact, with some ligands both cyclopentadienyl rings are removed[57]; for example,

$$Cp_2Ni + 4KCN \rightarrow 2K^+Cp^- + K_2Ni(CN)_4$$

$$Cp_2Ni + 6NH_3 \rightarrow Ni(NH_3)_6^{2+} + 2Cp^-$$

$$Cp_2Ni + 4PPh_3 \rightarrow (Ph_3P)_4Ni$$

The palladium and the platinum analogs of nickelocene remain unknown. Wilkinson[865] prepared, but did not fully characterize, a red complex, $[(C_5H_5)_2Pd]_n$, from sodium cyclopentadienide and palladium acetylacetonate. It was thermally very unstable. Fischer and Schuster-Woldan[866] obtained $(C_5H_5)_4Pt_2$, a more stable green complex, in low yield from sodium cyclopentadienide and $PtCl_2$ in hexane. The NMR spectrum showed a singlet, with satellites arising from the coupling of all the Cp protons to both ^{195}Pt nuclei; this indicated the presence of a Pt–Pt bond. From the infrared and ultraviolet spectra, however, these authors concluded that the molecule had σ- and π-bonded C_5H_5 groups, and was presumably a fluxional molecule in solution.

Bonding in ferrocene and its analogs has been extensively discussed and a good summary has been given by Rosenblum.[867] In simple terms each cyclopentadienyl ligand acts as a 6-electron donor [to M(II)] and can be thought of as occupying three coordination positions about the metal. Hence, the metallocenes with two rings parallel to each other and with the metal atom sandwiched between them can be regarded as pseudooctahedral complexes. For ferrocene, the effective atomic number formalism is obeyed. Nickelocene is usually regarded as isostructural with ferrocene, though suggestions have been made (to account for some of the reactions described above) that the two un-paired electrons may be localized to some extent on two carbons of one ring.

The apparent dislike which Pd(II) and Pt(II) have to forming dicyclopenta-dienyl complexes may be understood in terms of the much lower tendency for these metals to adopt octahedral (or pseudooctahedral) stereochemistry, in contrast to Ni(II) for which this is quite common. It should be pointed out, however, that although the dicyclopentadienyls are known for Rh and Ir, and have been studied, they are exceedingly unstable thermally and dimerize (via C–C bonds) even at very low temperatures,[868, 869]

(M = Rh, Ir)

Fritz and Schwartzhans[870] have reported the cis and trans isomers of $(Me_2S)_2Pt(C_5H_5)_2$. The C_5H_5 rings are probably σ-bonded here, and these complexes should be regarded as simple derivatives of square planar Pt(II). The NMR spectrum, however, showed the Cp proton resonances as a singlet and indicated this molecule to be fluxional. X-Ray investigations and low temperature NMR studies are necessary to completely characterize this complex and $(C_5H_5)_4Pt_2$.

B. MONOCYCLOPENTADIENYL COMPLEXES

Quite a number of these are known for all these elements in the (II) oxida-tion state and also for Pt(IV). Complexes of the type CpNiXL (L = R_3P, CO, etc; X = alkyl, aryl, halide, etc.) are well known.[367–370, 370a] Recently Cross and Wardle[370b] have reported the synthesis of the analogous palladium and platinum complexes, e.g.,

$$(Ph_3P)_2Pd_2Br_4 + CpTl \longrightarrow CpPdBr(PPh_3) \xrightarrow{\text{PhMgBr}} CpPdPh(PPh_3)$$

These complexes are all relatively stable.

Fischer and his co-workers[871, 872] have described the preparation of the cyclopentadienylpalladium and cyclopentadienylplatinum nitrosyls.

$$M(NO)Cl + CpNa \xrightarrow{\text{pentane}} \text{Cp}-M-NO \qquad (M = Pd, Pt)$$

The nickel analog has been known for a long time[873] and has been shown to be a symmetric top molecule, with a linear Ni–N–O group, in the gas phase.[874] Fischer and Schuster-Woldan concluded from the infrared spectra of the three complexes that they were isostructural and that the M–NO bonding increased from Ni to Pt.[872]

Both the nickel and platinum cyclopentadienyl carbonyl dimers are known,[863, 875] but their structures are quite different. The nickel complex has bridging carbonyls (**VI-1**), whereas the platinum one has not (**VI-2**). Similar trends are observed in many di- and polynuclear carbonyl complexes, for example, [CpM(CO)$_2$]$_2$, for M = Fe, Ru, and Os.[876]

(VI-1)   **(VI-2)**

The analogous palladium complex has not yet been prepared.

Smidt and Jira in 1959 described a brown complex, from cyclopentadiene and aqueous palladium chloride, which they proposed to be C$_5$H$_5$PdCl.[877] However, the nature of this compound is not clear, and it may well be a non-stoichiometric polymer. Robinson and Shaw[634] reported the reaction between cyclopentadiene and methanolic sodium chloropalladate to give a yellow precipitate which darkened rapidly. No pure product could be isolated. In view of the preparation of pentamethylcyclopenta*diene*palladium chloride (**VI-3a**) (which is stable) by Balakrishnan and Maitlis,[616] the yellow solid prepared by Robinson and Shaw may be cyclopenta*diene*palladium chloride (**VI-3b**) which then reacts further to give unknown products.

(VI-3a) (R = Me)   **(VI-3b)** (R = H)

By far the most common cyclopentadienyl complexes of palladium are those with π-allylic, π-cyclobutadiene, π-1,5-cyclooctadiene, or π-enyl ligands. Some examples include

$$\left[\langle\!\langle\text{—PdCl} \right]_2 + \text{C}_5\text{H}_5\text{Na} \longrightarrow 2 \langle\!\langle\text{—Pd—}\bigcirc\!\!\!\!\!\bigcirc \quad ^{728,\,815}$$

$$\left[\begin{array}{c} \text{Ph} \quad \text{Ph} \\ \square \text{—MBr}_2 \\ \text{Ph} \quad \text{Ph} \end{array} \right]_2 + \text{C}_5\text{H}_5\text{Fe(CO)}_2\text{Br} \longrightarrow \left[\begin{array}{c} \text{Ph} \quad \text{Ph} \\ \square \text{—M—}\bigcirc \\ \text{Ph} \quad \text{Ph} \end{array} \right]^+ \text{FeBr}_4^- \;\; ^{672}$$

(M = Ni, Pd)

$$\left[\begin{array}{c} \text{OMe} \\ \searrow\text{MCl} \end{array} \right]_2 + 2\text{C}_5\text{H}_5\text{Tl} \longrightarrow \begin{array}{c} \text{OMe} \\ \searrow\text{M—}\bigcirc \end{array} \quad ^{469}$$

(M = Pd, Pt)

Apart from sodium cyclopentadienide (in ethereal solvents) cyclopentadienylthallium and cyclopentadienyliron dicarbonyl bromide have been used as sources of the Cp group. The latter reagents offer some advantages in that they can be used under milder conditions than $Na^+C_5H_5^-$. For example, tetramethylcyclobutadienenickel chloride undergoes attack at carbon as well as at nickel with $Na^+C_5H_5^-$ [673-675] (this volume, Chapter IV, Section F,2).

The structure of allylcyclopentadienylpalladium has been determined by Minasyants and Struchkov.[214] They showed that both groups were π-bonded to the metal, and that all the Cp carbons and all the allylic carbons were at the same distances from the metal (2.25 and 2.05 Å, respectively), but that the two ligands were not parallel.

NMR studies on a variety of cyclopentadienylpalladium complexes of this type showed that the Cp group was symmetrically bonded. There was no evidence for fluxional behavior. The ultraviolet spectrum and the polarographic behavior of C_3H_5PdCp have been reported.[773, 776]

The cyclopentadienyl group is also easily cleaved from palladium. For example, Maitlis et al.[672] showed that whereas the alkoxycyclobutenylnickel complex (VI-4a) reacted with HBr to give the (π-cyclobutadiene)(π-cyclopentadienyl)nickel cation (VI-5), the palladium complex (VI-4b), even under the mildest conditions, lost the cyclopentadienyl ring to give (VI-5).

This work was extended by Gubin et al.,[773] who showed that the cyclopentadienyl group in C_3H_5PdCp was easily displaced by a variety of reagents to give allylpalladium complexes. The authors say that this cleavage occurs with both electrophilic and nucleophilic reagents under acidic, neutral, or basic conditions.

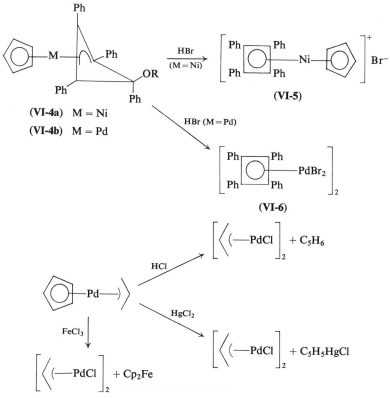

(VI-4a) M = Ni

(VI-4b) M = Pd

(VI-5)

(VI-6)

Some transformations which do not involve cleavage of the Cp group are possible under basic or neutral conditions.[672]

Maitlis *et al.*[672] were also able to transfer the cyclopentadienyl and the cyclobutadiene group onto cobalt by reaction of [Ph$_4$C$_4$PdCp]Br with Co$_2$(CO)$_8$.

Robinson and Shaw in 1963 reported the preparation of (π-cyclopentadienyl)trimethylplatinum(IV) from iodotrimethylplatinum tetramer.[878]

$$[Me_3PtI]_4 + CpNa \longrightarrow$$

The complex was shown to be monomeric, and on the basis of its NMR spectrum (two singlets, with satellites due to coupling to ^{195}Pt), Robinson and Shaw suggested the Cp ring was π-bonded. This was disputed by Fritz and Schwartzhans[870] on the basis of the chemical shifts of the resonances, an unreliable criterion at best. Semion et al.[879] have briefly reported an X-ray crystal structure determination of CpPtMe$_3$ and have shown that the Cp ring was indeed π-bonded and parallel to the plane of the three methyls. Approximate distances were Pt–C (cyclopentadienyl) 2.2 Å, Pt–C (methyl) 2.05 Å.

No palladium analogs of this complex are known.

C. BENZENE COMPLEXES

While mono- and bisarene complexes are well established among elements in the early and middle groups of the transition metals, they become progressively more rare in group VIII. Here, in fact, only the very electron-rich hexamethylbenzene (HMB) will act as a ligand to give isolable complexes. Bis-HMB complexes of Fe(0), Fe(I), Co(0), Co(I), Co(II), Rh(I), and Rh(II) have been described by Fisher and Lindner.[880] These authors have also recently described the synthesis of the first such nickel complex, $(HMB)_2Ni^{2+}$, and have obtained evidence for the formation of a very unstable (and not isolated) $[(HMB_2Ni]^+$ ion.[881]. These complexes are paramagnetic [μ_{eff} for the Ni(II) complex is 3.0 B.M.], since they have more electrons than are required by the effective atomic number formalism, and are very sensitive to hydrolysis. No details of the structures of these complexes are known.

The nickel(II) complex was prepared by heating HMB with NiBr$_2$ in the presence of AlBr$_3$ and was characterized as the hexachloroplatinate(IV), $[(HMB)_2Ni]^{2+}(PtCl_6)^{2-}$.[881]

It is perhaps, therefore, a little surprising that benzene complexes of palladium, formally in the (I) oxidation state, are known. In fact, these complexes probably play a very important role in the coupling and oxidation reactions of arenes with Pd(II) salts.

Allegra et al. have reported the formation of two complexes of this type by the classical Fischer synthesis of arene–metal complexes.[157, 158] When the melt obtained by heating $PdCl_2$, aluminum, aluminum trichloride, and benzene at 80° was allowed to cool, large brown crystals of diamagnetic complexes with formula $(C_6H_6PdAlCl_4)_2$, or $(C_6H_6PdAl_2Cl_7)_2$, were obtained. The latter complex was isolated when a large excess of $AlCl_3$ was used.

The complexes were only stable in the solid; they were insoluble in hydrocarbons and disproportionated in tetrahydrofuran to give equal amounts of palladium and $PdCl_2$.

$$(C_6H_6PdAlCl_4)_2 \xrightarrow{THF} 2C_6H_6 + Pd + PdCl_2 + 2AlCl_3$$

X-Ray structure determinations on both complexes have been carried out (Figs. VI-1 and VI-2).[157, 158] The basic features are similar. In each case two Pd atoms, 2.57(1) Å apart, are sandwiched between two benzene rings. The metal–metal distances observed are much shorter than in palladium metal, 2.75 Å.

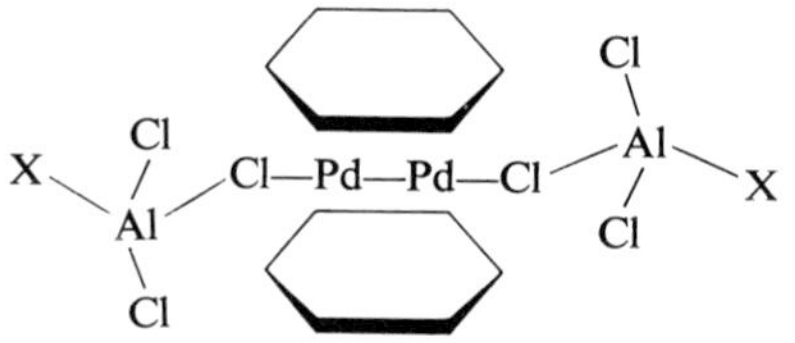

FIG. VI-1. The structures of $(C_6H_6PdAlCl_4)_2$ (X = Cl) and $(C_6H_6PdAl_2Cl_7)_2$ (X = $AlCl_4$). After Allegra *et al.*[157, 158]

(a)   (b)

FIG. VI-2. (a) The arrangement of the metal atoms with respect to the benzene rings in $(C_6H_6PdAl_2Cl_7)_2$. (b) A projection perpendicular to the average ring plane of the complex $(C_6H_6PdAlCl_4)_2$.[158]

Each metal atom is also bonded to one Cl atom [at 2.45(1) Å; the array Cl–Pd–Pd–Cl is very nearly linear] which is also attached to an $AlCl_3$ or Al_2Cl_7 group.

The bonding of the metal atoms to the benzene rings is most unusual. In $(C_6H_6PdAl_2Cl_7)_2$, the benzene rings have at least two different orientations with respect to the Pd–Pd axis, one of which is shown in Fig. VI-2a. The other is the mirror image of this. The metal atoms are not symmetrically arranged with respect to the benzenes, but each metal atom appears to be within a reasonable bonding distance (2.2–2.5 Å) of two carbon atoms and out of the bonding range (2.6–3.5 Å) of the remainder in each ring. However, the e.s.d.'s of the Pd–C distances are large, 0.03Å.

In $(C_6H_6PdAlCl_4)_2$, the structure of which is better resolved, the $(C_6H_6)_2Pd_2$ core is of D_{2h} symmetry. Again (Fig. VI-2b) the metal atoms are not arranged symmetrically with respect to each ring. In this complex each metal is within reasonable bonding range of two carbon atoms of each ring [2.26(2), 2.37(2) Å] and out of range of the remainder (2.77, 3.03, 3.38, 3.50 Å). The e.s.d.'s on the C–C bond lengths are too large to allow any conclusions to be drawn about the benzene ring, but it does appear to be distorted slightly away from planar, two of the carbons being bent 7° out of the plane of the other four *toward* the metal atoms.

These complexes have no direct parallel elsewhere. The bonding of each metal atom to the benzene is more reminiscent of the edge-bonded copper and silver complexes $C_6H_6CuAlCl_4$ and $C_6H_6AgAlCl_4$ (**VI-7**)[882] than of the symmetrically π-bonded complexes such as dibenzenechromium (**VI-8**) or benzenechromium tricarbonyl.[883,884] The metal–carbon bond lengths in the silver

Cl
Ag—Cl
Cl   Cl—Al—

(**VI-7**)

Cr

(**VI-8**)

complex (**VI-7**) are much longer [2.47(6), 2.57(6) Å] than in either of the benzenepalladium complexes, and the orientation of the metal with respect to the ring is also different. There appears to be a wide variety of behavior possible for benzene rings complexed to a metal.[885,886]

Davidson and Triggs[159] have studied the oxidative coupling of benzene to biphenyl, "catalyzed" by palladium acetate. When the reaction was carried out in acetic acid in the presence of 0.5 M perchloric acid below 60°, no acetoxylation or precipitation of metal occurred, but the color of the solution turned magenta. From this solution the authors isolated, on addition of acetic anhydride, the very explosive Pd(I) complex, $(C_6H_6Pd \cdot H_2O \cdot ClO_4)_n$, together with a 50% yield of biphenyl.

The complex decomposed above 80° in acetic acid, but benzene was not oxidized, nor were acetoxy radicals formed. On addition of chloride ion, the complex disproportionated with the formation of metal and $PdCl_4{}^{2-}$. Oxidizing agents such as oxygen, bromine, or permanganate oxidized the metal to Pd(II), but Cr(VI) or Pb(IV) did not, except in the presence of halide ion. The complex appeared to be paramagnetic since no benzene resonances were seen until after halide was added and disproportionation to Pd(0) and Pd(II) had occurred. This suggested the absence of metal–metal bonding between Pd(I) atoms in the complex. The ultraviolet–visible spectrum was also in agreement with the presence of a d^9 ion.

The complex could also be obtained from palladium acetate and 1-hexene or $PhB(OH)_2$ (phenylboronic acid) provided benzene and perchloric acid were present. This particular complex does not therefore appear to be an intermediate in the reaction $2C_6H_6 \rightarrow C_6H_5C_6H_5$, but is formed simultaneously, and probably plays a vital role in the overall process,

$$C_6H_6 + Pd(II) \longrightarrow PhPd(II) + H^+$$

$$2PhPd(II) \xrightarrow{C_6H_6} Ph_2 + 2Pd(I)$$

$$2Pd(I) \xrightarrow{Cl^-} Pd^0 + PdCl_4{}^{2-}$$

The decomposition of two PhPd(II) species to biphenyl appears to be best regarded as a concerted reaction since no phenyl radicals are produced. The benzene appears to be necessary in order to stabilize Pd(I).

However, benzene π complexes are probably intermediates in the electrophilic palladation reactions. These and the above reactions are more fully discussed in Chapters II and III, Volume II, Section D.

Glossary

Ac	Acetyl	CH$_3$CO—
acac	Acetylacetonate	
bipy	Bipyridyl, 2,2′-dipyridyl	
B.M.	Bohr magneton (magnetic moment)	
Bu	Butyl	C$_4$H$_9$—
Cp	Cyclopentadienyl	
COD	Cyclooctadiene	C$_8$H$_{12}$
COT	Cyclooctatetraene	
D	Debye (dipole moment)	
diars	o-Phenylenebis(dimethylarsine)	
dien	Diethylenetriamine	NH$_2$CH$_2$CH$_2$NHCH$_2$CH$_2$NH$_2$
diphos	1,2-Bis(diphenylphosphino)ethane	Ph$_2$PCH$_2$CH$_2$PPh$_2$
DMAC	N,N′-Dimethylacetamide	Me$_2$NCOMe
DMF	N,N′-Dimethylformamide	Me$_2$NCHO
DMSO	Dimethyl sulfoxide	Me$_2$SO
DTA	Differential thermal analysis	
E	Usually phosphorus, arsenic, or antimony	
e.s.d.	Estimated standard deviation †	
ESR	Electron spin resonance (or electron paramagnetic resonance)	

† E.s.d's are expressed in parentheses after the bond length. A Pd–C bond length of 2.132(7)Å (or 2.132 ± 0.007Å) implies that the true value is about 70% certain to be within this range (2.125–2.139Å) and 99% certain to be within *three* times the e.s.d. (2.111–2.153Å)

Et	Ethyl	CH_3CH_2-
Et$_4$dien	*N,N,N,N*-Tetraethyldiethylenetriamine	$Et_2NCH_2CH_2NHCH_2CH_2NEt_2$
Me	Methyl	CH_3-
NMR	Nuclear magnetic resonance, usually of protons, ^{1}H	
NQR	Nuclear quadrupole resonance	

Ph   Phenyl

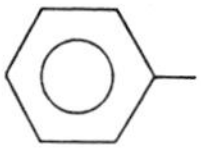

o-phen   *o*-Phenanthroline

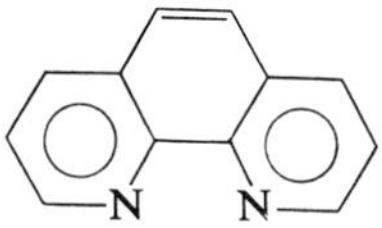

| PMR | Proton magnetic resonance | |
| Pr | Propyl | C_3H_7- |

py   Pyridine

R	Organic radical	
R$_f$	Perfluoroalkyl or -aryl	
TAS	2,6,10-Trimethyl-2,6,10-triarsaundecane	$Me_2As(CH_2)_3\overset{\displaystyle Me}{\overset{\displaystyle |}{As}}(CH_2)_3AsMe_2$
X	Usually halide	

Bibliography†

1. W. H. Wollaston, *Phil. Trans. Roy. Soc. London* **94**, 419 (1804); **95**, 316 (1805).
2. U.S. Dept. of the Interior, Bureau of Mines, *Miner. Yearb.* Vol. I/II, p. 341 (1966).
3. R. Gilchrist, *Chem. Rev.* **33**, 277 (1943).
4. E. M. Wise, "Palladium: Recovery, Properties and Applications." Academic Press, New York, 1968.
5. "Interatomic Distances," *Chem. Soc., Spec. Publ.* **11** (1958); **18** (1965).
6. C. Duval, "Nouveau traité de chimie minérale," Vol. 19. Masson, Paris 1958.
7. H. Moissan, *Ann. Chim. Phys.* [2] **24**, 249 (1891).
8. N. Bartlett and M. A. Hepworth, *Chem. Ind. (London)*, p. 1425 (1956).
9. O. Ruff and E. Ascher, *Z. Anorg. Allg. Chem.* **183**, 207 (1929).
10. N. Bartlett and J. W. Quail, *J. Chem. Soc.* p. 3728 (1961).
11. C. C. Addison and B. G. Ward, *Chem. Commun.* p. 155 (1966).
12. R. S. Nyholm and M. L. Tobe, *Advan. Inorg. Chem. Radiochem.* **5**, 1 (1963).
13. L. F. Warren and M. F. Hawthorne, *J. Amer. Chem. Soc.* **89**, 470 (1967).
14. K. A. Jensen, *Z. Anorg. Allg. Chem.* **229**, 265 (1936); K. A. Jensen, B. Nygaard, and C. T. Pedersen, *Acta Chem. Scand.* **17**, 1126 (1963).
15. P. Kreisman, R. Marsh, J. R. Preer, and H. B. Gray, *J. Amer. Chem. Soc.* **90**, 1067 (1968).
16. D. W. Meek, E. C. Alyea, J. K. Stalick, and J. A. Ibers, *J. Amer. Chem. Soc.* **91**, 4920 (1969).
17. P. Heimbach, *Angew. Chem., Int. Ed. Engl.* **3**, 648 (1964).
18. L. Porri, M. C. Gallazzi, and G. Vitulli, *Chem. Commun.* p. 228 (1967).
19. F. Calderazzo, R. Ercoli, and G. Natta, *in* "Organic Syntheses via Metal Carbonyls" (I. Wender and P. Pino, eds.), p. 1. Wiley (Interscience), New York, 1968.
20. G. Booth, J. Chatt, and P. Chini, *Chem. Commun.* p. 639 (1965).
21. K. I. Matveev, L. N. Rachkovskaya, and N. K. Eremenko, U.S.S.R. Pat. 210,838 (1968); *Chem. Abstr.* **69**, 4065x (1968).
22. T. Krück and K. Baur, *Z. Anorg. Allg. Chem.* **364**, 192 (1969).
23. T. Kashiwagi, N. Yasuoka, T. Ueki, N. Kasai, M. Kakudo, S. Takahashi, and N. Hagihara, *Bull. Chem. Soc. Jap.* **41**, 296 (1968).
24. V. Albano, P. L. Bellon, and V. Scatturin, *Abstr., Inorg. Chim. Acta Symp. 1968* p. B6 (1968).
25. V. G. Albano, G. M. Basso Ricci, and P. L. Bellon, *Inorg. Chem.* **8**, 2109 (1969).
26. V. G. Albano, P. L. Bellon, and M. Sansoni, *Chem. Commun.* p. 899 (1969).
27. J. Chatt, F. A. Hart, and H. R. Watson, *J. Chem. Soc.* p. 2537 (1962).
28. J. F. Nixon and M. D. Sexton, *Inorg. Nucl. Chem. Lett.* **4**, 275 (1968).

† A number of references are given to Russian journals in their English versions. Russ. Chem. Rev. is the translation of Uspekhi Khimii (Chemical Society, London); Russ. J. Inorg. Chem. is the translation of Zh. Neorg. Khim. (Chemical Society, London); Dokl. Akad. Nauk SSSR (English translation) refers to Proceedings of the Academy of Sciences of the USSR, translated by Consultants Bureau, New York; Izv. Akad. Nauk SSSR (English translation) refers to Bulletin of the Adademy of Sciences of the USSR, Division of Chemical Sciences, translated by Consultants Bureau. J. Struct. Chem. USSR is the translation of Zh. Strukt. Khim. (Consultants Bureau), and "Kinetics and Catalysis" of Kinetika i Kataliz (Consultants Bureau). In all these cases the page referred to is that of the English translation.

29. C. A. Tolman, *J. Amer. Chem. Soc.* **92**, 2953 and 2956 (1970).
30. M. Meier, F. Basolo, and R. G. Pearson, *Inorg. Chem.* **8**, 795 (1969).
31. T. Krück, K. Baur, and W. Lang, *Chem. Ber.* **101**, 138 (1968).
32. J. Chatt, F. A. Hart, and D. T. Rosevear, *J. Chem. Soc.* p. 5504 (1961).
33. T. Krück and K. Baur, *Angew. Chem., Int. Ed. Engl.* **4**, 521 (1965).
34. L. Malatesta and M. Angoletta, *J. Chem. Soc.* p. 1186 (1957).
35. J. Chatt and B. L. Shaw, *J. Chem. Soc.* p. 5075 (1962).
36. J. C. Bailar and H. Itatani, *Inorg. Chem.* **4**, 1618 (1965).
37. A. F. Clemmit and F. Glockling, *J. Chem. Soc., A* p. 2163 (1969).
38. L. Malatesta and C. Cariello, *J. Chem. Soc.* p. 2323 (1958).
39. R. J. Cross, *Organometal. Chem. Rev.* **2**, 97 (1967).
40. R. Ugo, *Coord. Chem. Rev.* **3**, 319 (1968).
41. G. C. Dobinson, R. Mason, G. B. Robertson, R. Ugo, F. Conti, D. Morelli, S. Cenini, and F. Bonati, *Chem. Commun.* p. 739 (1967).
42. A. Sacco, R. Ugo, and A. Moles, *J. Chem. Soc., A* p. 1670 (1966).
43. F. Basolo and R. G. Pearson, "Mechanisms of Inorganic Reactions," 2nd ed. Wiley, New York, 1967.
44. G. Wilke, E. W. Müller, M. Kröner, P. Heimbach, and H. Breil (Studiengesellschaft Kohle m.b.H.), French Pat,, 1,320,729 (1963); *Chem. Abstr.* **59**, 14026c (1963).
45. E. O. Greaves, C. J. L. Lock, and P. M. Maitlis, *Can. J. Chem.* **46**, 3879 (1968).
46. E. O. Fischer and H. Werner, *Chem. Ber.* **95**, 703 (1962).
47. E. O. Fischer and H. Werner, Ger. Pat., 1,181,708 (1964); *Chem. Abstr.* **62**, 4947b (1965).
48. K. Bittler, N. V. Kutepow, D. Neubauer, and H. Reiss, *Angew. Chem., Int. Ed. Engl.* **7**, 329 (1968).
49. G. Calvin and G. E. Coates, *J. Chem. Soc.* p. 2008 (1960).
50. B. F. G. Johnson, T. Keating, J. Lewis, M. S. Subramanian, and D. A. White, *J. Chem. Soc., A* p. 1793 (1969).
51. J. Powell and B. L. Shaw, *J. Chem. Soc., A* p. 774 (1968).
52. G. Wilke and B. Bogdanovic, *Angew. Chem.* **73**, 755 and 756 (1961).
53. J. K. Becconsall, B. E. Job, and S. O'Brien, *J. Chem. Soc., A* p. 423 (1967).
54. G. Wilke, B. Bogdanovic, P. Hardt, P. Heimbach, W. Keim, M. Kröner, W. Oberkirch, K. Tanaka, G. Steinrücke, D. Walter, and H. Zimmerman, *Angew. Chem., Int. Ed. Engl.* **5**, 151 (1966).
55. J. R. Olechowski, C. G. McAlister, and R. F. Clark, *Inorg. Chem.* **4**, 246 (1965).
56. J. R. Olechowski, C. G. McAlister, and R. F. Clark, *Inorg. Syn.* **9**, 181 (1967).
57. H. Behrens and K. Meyer, *Z. Naturforsch. B* **21**, 489 (1966).
58. J. F. Nixon, *J. Chem. Soc., A* p. 1136 (1967).
59. T. A. Manuel, *Advan. Organometal. Chem.* **3**, 181 (1965).
60. L. Malatesta and M. Angoletta, *Atti Accad. Naz. Lincei, Cl. Sci. Fis., Mat. Natur., Rend.* [8] **19**, 43 (1955).
61. L. Malatesta, *J. Chem. Soc.* p. 3924 (1955).
62. S. Takahashi and N. Hagihara, *Nippon Kagaku Zasshi* **88**, 1306 (1967); *Chem. Abstr.* **69**, 27514g (1968).
63. S. Takahashi, K. Sonogashira, and N. Hagihara, *Nippon Kagaku Zasshi* **87**, 610 (1966).
64. T. Krück, *Angew. Chem.* **79**, 27 (1967).
65. T. Krück, K. Baur, K. Glinka, and M. Stadler, *Z. Naturforsch. B* **23**, 1147 (1968).
66. T. Krück and M. Höfler, *Angew. Chem., Int. Ed. Engl.* **6**, 563 (1967).
67. A. J. Mukhedkar, M. Green, and F. G. A. Stone, *J. Chem. Soc., A* p. 3023 (1969).
68. J. W. Eastess and W. M. Burgess, *J. Amer. Chem. Soc.* **64**, 1187 (1942).

69. J. J. Burbage and W. C. Fernelius, *J. Amer. Chem. Soc.* **65**, 1484 (1943).
70. R. Nast and H. D. Moerler, *Chem. Ber.* **102**, 2050 (1969).
71. G. W. Watt, J. L. Hall, G. R. Choppin, and P. S. Gentile, *J. Amer. Chem. Soc.* **76**, 373 (1954).
72. R. Nast and K. Vester, *Z. Anorg. Allg. Chem.* **279**, 146 (1955).
73. R. Nast and W. Horl, *Chem. Ber.* **95**, 1470 (1962).
74. R. Nast and W. D. Heinz, *Chem. Ber.* **95**, 1478 (1962).
75. R. Nast, *Angew. Chem., Int. Ed. Engl.* **4**, 366 (1965).
76. L. Malatesta, R. Ugo, and S. Cenini, *Advan. Chem. Ser.* **62**, 318 (1967).
77. S. Otsuka, A. Nakamura, and Y. Tatsuno, *J. Amer. Chem. Soc.* **91**, 6994 (1969).
78. S. Otsuka, A. Nakamura, and T. Yoshida, *J. Amer. Chem. Soc.* **91**, 7196 (1969).
79. V. Albano, P. L. Bellon, and V. Scatturin, *Chem. Commun.* p. 507 (1966).
80. L. A. Woodward and J. R. Hall, *Spectrochim. Acta* **16**, 654 (1960).
81. V. G. Myers, F. Basolo, and K. Nakamoto, *Inorg. Chem.* **8**, 1204 (1969).
81a. J. C. Marriott, J. A. Salthouse, M. J. Ware, and J. M. Freeman, *Chem. Commun.* p. 595 (1970).
82. J. P. Day, F. Basolo, and R. G. Pearson, *J. Amer. Chem. Soc.* **90**, 6927 and 6933 (1968).
83. F. Basolo, *Trans. N.Y. Acad. Sci.* [2] **31**, 676 (1969).
84. R. Ugo, F. Cariati, and G. LaMonica, *Chem. Commun.* p. 868 (1966).
85. R. D. Gillard, R. Ugo, F. Cariati, S. Cenini, and F. Bonati, *Chem. Commun.* p. 869 (1966).
86. R. Ugo, F. Cariati, and G. LaMonica, *Inorg. Syn.* **11**, 105 (1968).
87. R. Ugo and S. Cenini, *Abstr., Inorg. Chim. Acta Symp.*, *1968* p. D9 (1968).
87a. R. Ugo, G. LaMonica, F. Cariati, S. Cenini, and F. Conti, *Inorg. Chim. Acta* **4**, 390 (1970).
88. P. Fitton and J. E. McKeon, *Chem. Commun.* p. 4 (1968).
89. J. A. Osborn, *Chem. Commun.*, p. 1231 (1968).
90. G. Wilke, *Angew. Chem., Int. Ed. Engl.* **2**, 105 (1963).
91. U. Birkenstock, H. Boennemann, B. Bogdanovic, D. Walter, and G. Wilke, *Advan. Chem. Ser.* **70**, 250 (1968).
92. A. Misono, Y. Uchida, M. Hidai, and K. Kudo, *J. Organometal. Chem.* **20**, P7 (1969).
93. E. J. Smutny, H. Chung, K. C. Dewhirst, W. Keim, T. M. Shryne, and H. E. Thyret, *Amer. Chem. Soc., Div. Petrol. Chem., Prepr.* **14**, B100, B112 (1969).
94. A. D. Allen and C. D. Cook, *Can. J. Chem.* **42**, 1063 (1964).
95. J. P. Birk, J. Halpern, and A. L. Pickard, *J. Amer. Chem. Soc.* **90**, 4491 (1968); J. Halpern and A. L. Pickard, *Inorg. Chem.* **9**, 2798 (1970).
96. J. P. Day, F. Basolo, R. G. Pearson, L. F. Kangas, and P. M. Henry, *J. Amer. Chem. Soc.* **90**, 1925 (1968).
97. J. P. Collman, *Accounts Chem. Res.* **1**, 136 (1968).
98. J. P. Collman and W. R. Roper, *Advan. Organometal. Chem.* **7**, 53 (1968).
99. S. Carra and R. Ugo, *Inorg. Chim. Acta Rev.* **1**, 49 (1967).
100. L. Vaska and J. W. DiLuzio, *J. Amer. Chem. Soc.* **84**, 679 (1962).
101. L. Vaska, *J. Amer. Chem. Soc.* **88**, 5325 (1966).
102. A. Sacco, M. Rossi, and C. F. Nobile, *Chem. Commun.* p. 589 (1966).
103. J. Chatt and S. A. Butter, *Chem. Commun.* p. 501 (1967).
104. M. A. Bennett, R. J. H. Clark, and D. L. Milner, *Inorg. Chem.* **6**, 1647 (1967).
105. J. P. Collman and C. T. Sears, *Inorg. Chem.* **7**, 27 (1968).
106. P. B. Chock and J. Halpern, *J. Amer. Chem. Soc.* **88**, 3511 (1966).
107. J. A. Osborn, F. H. Jardine, J. F. Young, and G. Wilkinson, *J. Chem. Soc., A* p. 1711 (1966).

108. D. R. Eaton and S. R. Suart, *J. Amer. Chem. Soc.* **90**, 4170 (1968); R. L. Augustine and J. F. van Peppen, *Chem. Commun.* p. 497 (1970); D. D. Lehman, D. F. Shriver, and I. Wharf, *ibid.* p. 1486 (1970).
109. M. A. Bennett and D. L. Milner, *Chem. Commun.* p. 581 (1967).
110. M. A. Bennett and D. L. Milner, *J. Amer. Chem. Soc.* **91**, 6983 (1969).
111. G. W. Parshall, W. H. Knoth, and R. A. Schunn, *J. Amer. Chem. Soc.* **91**, 4990 (1969); G. W. Parshall, *ibid.* **90**, 1669 (1969).
112. J. Chatt and J. M. Davidson, *J. Chem. Soc.* p. 843 (1965).
112a. D. M. Blake and C. J. Nyman, *Chem. Commun.* p. 483 (1969); *J. Amer. Chem. Soc.* **92**, 5359 (1970).
113. F. Cariati, R. Ugo, and F. Bonati, *Inorg. Chem.* **5**, 1128 (1966).
114. K. Kudo, M. Hidai, T. Murayama, and Y. Uchida, *Chem. Commun.* p. 1701 (1970).
115. W. C. Drinkard, D. R. Eaton, J. P. Jesson, and R. V. Lindsey, *Inorg. Chem.* **9**, 392 (1970); C. A. Tolman, *J. Amer. Chem. Soc.*, **92**, 4217 (1970).
116. R. A. Schunn, *Inorg. Chem.* **9**, 394 (1970).
117. P. Fitton, M. P. Johnson, and J. E. McKeon, *Chem. Commun.* p. 6 (1968).
118. D. T. Rosevear and F. G. A. Stone, *J. Chem. Soc.*, *A* p. 164 (1968).
119. C. D. Cook and G. S. Jauhal, *Can. J. Chem.* **45**, 301 (1967).
120. M. C. Baird and G. Wilkinson, *J. Chem. Soc.*, *A* p. 865 (1967).
121. W. J. Bland and R. D. W. Kemmitt, *J. Chem. Soc.*, *A* p. 1278 (1968).
122. G. Schmid, W. Petz, W. Arloth, and H. Nöth, *Angew. Chem.*, *Int. Ed. Engl.* **6**, 696 (1967).
123. A. J. Layton, R. S. Nyholm, G. A. Pneumaticakis, and M. L. Tobe, *Chem. Ind.* (*London*) p. 465 (1967).
124. A. D. Allen and C. D. Cook, *Proc. Chem. Soc.*, *London* p. 218 (1962).
125. J. Chatt, C. Eaborn, and P. N. Kapoor, *J. Organometal. Chem.* **13**, P21 (1968).
126. D. Morelli, A. Segre, R. Ugo, G. LaMonica, S. Cenini, F. Conti, and F. Bonati, *Chem. Commun.* p. 524 (1967).
127. B. Clarke, M. Green, R. B. L. Osborn, and F. G. A. Stone, *J. Chem. Soc.*, *A* p. 167 (1968).
128. T. L. Gilchrist, F. J. Graveling, and C. W. Rees, *Chem. Commun.* p. 821 (1968).
129. J. J. Levison and S. D. Robinson, *Chem. Commun.* p. 198 (1967).
130. G. A. Pneumaticakis, *Chem. Commun.* p. 275 (1968).
131. W. J. Bland, R. D. W. Kemmitt, I. W. Nowell, and D. R. Russell, *Chem. Commun.* p. 1065 (1968).
132. P. J. Hayward, D. M. Blake, C. J. Nyman, and G. Wilkinson, *Chem. Commun.* p. 987 (1969).
133. M. Green, R. B. L. Osborn, A. J. Rest, and F. G. A. Stone, *Chem. Commun.* p. 502 (1966).
134. W. J. Bland, J. Burgess, and R. D. W. Kemmitt, *J. Organometal. Chem.* **14**, 201 (1968).
135. W. J. Bland, J. Burgess, and R. D. W. Kemmitt, *J. Organometal. Chem.* **15**, 217 (1968).
136. C. H. Bamford, G. C. Eastmond, and K. Hargreaves, *Trans. Faraday Soc.* **64**, 175 (1968).
137. H. C. Volger and K. Vrieze, *J. Organometal. Chem.* **6**, 297 (1966).
138. H. C. Volger and K. Vrieze, *J. Organometal. Chem.* **13**, 495 (1968).
139. C. D. Cook and G. S. Jauhal, *Inorg, Nucl. Chem. Lett.* **3**, 31 (1967).
140. T. Kashiwagi, N. Yasuoka, N. Kasai, M. Kakudo, S. Takahashi, and N. Hagihara, *Chem. Commun.* p. 743 (1969); C. D. Cook, P. T. Cheng, and S. C. Nyburg, *J. Amer. Chem. Soc.* **91**, 2123 (1969).

141. C. J. Nyman, C. E. Wymore, and G. Wilkinson, *J. Chem. Soc., A* p. 561 (1968); P. J. Hayward, D. M. Blake, G. Wilkinson, and C. J. Nyman, *J. Amer. Chem. Soc.* **92**, 5873 (1970).
142. C. D. Cook and G. S. Jauhal, *J. Amer. Chem. Soc.* **89**, 3066 (1967).
143. J. J. Levison and S. D. Robinson, *Inorg. Nucl. Chem. Lett.* **4**, 407 (1968).
144. R. Ugo, F. Conti, S. Cenini, R. Mason, and G. B. Robertson, *Chem. Commun.* p. 1498 (1968).
145. G. Wilke, H. Schott, and P. Heimbach, *Angew. Chem. Int. Ed. Engl.* **6**, 92 (1967).
146. S. Takahashi, K. Sonogashira, and N. Hagihara, *Mem. Inst. Sci. Ind. Res., Osaka Univ.* **23**, 69 (1966).
147. S. Otsuka, A. Nakamura, and Y. Tatsuno, *Chem. Commun.* p. 836 (1967).
148. I. Bellucci and R. Corelli, *Atti. Reale Accad. Lincei, Rend., Cl. Sci. Fis., Mat. Natur.* [5] **22** Part II, 485 (1913); *Z. Anorg. Allg. Chem.* **86**, 88 (1914).
149. J. R. Miller, *Advan. Inorg. Chem. Radiochem.* **4**, 133 (1962).
149a. O. Jarchow, H. Schulz, and R. Nast, *Angew. Chem. Int. Ed. Engl.* **9**, 71 (1970).
150. J. P. Martella and W. C. Kaska, *Tetrahedron Lett.* p. 4889 (1968).
151. B. Corain, M. Bressan, P. Rigo, and A. Turco, *Chem. Commun.* p. 509 (1968); *Inorg. Chem.* **7**, 1623 (1968).
152. L. Porri, G. Vitulli, and M. C. Gallazzi, *Angew. Chem., Int. Ed. Engl.* **6**, 452 (1967); *J. Polym. Sci., Part B* **5**, 629 (1967).
153. M. Dubini and F. Montino, *Chim. Ind. (Milan)* **49**, 1283 (1967).
154. F. Glockling and K. A. Hooton, *J. Chem. Soc., A* p. 913 (1968).
155. A. C. Skapski and P. G. H. Troughton, *Chem, Commun.* p. 170 (1969).
156. M. A. Oranskaya and N. A. Mikhailova, *Zh. Neorg. Khim.* **5**, 12 (1960).
157. G. Allegra, A. Immirzi, and L. Porri, *J. Amer. Chem. Soc.* **87**, 1394 (1965).
158. G. Allegra, G. T. Casagrande, A. Immirzi, L. Porri, and G. Vitulli, *J. Amer. Chem. Soc.* **92**, 289 (1970).
159. J. M. Davidson and C. Triggs, *J. Chem. Soc., A* p. 1324 (1968).
160. I. I. Moiseev and S. V. Pestrikov, *Izv. Akad. Nauk SSSR* (*English Transl.*) p. 1690 (1965).
161. I. I. Moiseev and S. V. Pestrikov, *Dokl. Akad. Nauk SSSR* **171**, 722 (1966).
162. I. I. Moiseev and A. A. Grigor'ev, *Dokl. Akad. Nauk SSSR* (*English Transl.*) **178**, 132 (1968).
163. F. Klanberg and E. L. Muetterties, *J. Amer. Chem. Soc.* **90**, 3296 (1968).
164. H. J. Keller and H. Wawersik, *Z. Naturforsch. B* **20**, 938 (1965).
165. F. A. Cotton and G. Wilkinson, "Advanced Inorganic Chemistry," 2nd ed. Wiley (Interscience), New York, 1966.
166. H. Basch and H. B. Gray, *Inorg. Chem.* **6**, 365 (1967).
167. W. R. Mason and H. B. Gray, *J. Amer. Chem. Soc.* **90**, 5721 (1968).
168. P. L. Orioli and L. Sacconi, *Chem. Commun.* p. 1310 (1968).
169. R. N. Goldberg and L. G. Hepler, *Chem. Rev.* **68**, 229 (1968).
169a. R. M. Izatt, D. J. Eatough, C. E. Morgan, and J. J. Christensen, *J. Chem. Soc., A* p. 2514 (1970).
170. M. Bonamico and G. Dessy, *Chem. Commun.* p. 483 (1968).
171. C. M. Harris, R. S. Nyholm, and N. C. Stephenson, *Nature* (*London*) **177**, 1127 (1956).
172. G. Thiele, K. Brodersen, E. Kruse, and B. Holle, *Naturwissenschaften* **54**, 615 (1967).
173. R. Ettorre, G. Dolcetti, and A. Peloso, *Gazz. Chim. Ital.* **97**, 1681 (1967).
174. G. A. Mair, H. M. Powell, and D. E. Henn, *Proc. Chem. Soc., London* p. 415 (1960).
175. J. W. Collier, F. G. Mann, D. G. Watson, and H. R. Watson, *J. Chem. Soc.* p. 1803 (1964).

176. N. A. Bailey and R. Mason, *J. Chem. Soc.*, *A* p. 2594 (1968).
177. E. B. Fleischer and S. W. Hawkinson, *Inorg. Chem.* 7, 2312 (1968).
178. B. Bosnich, R. S. Nyholm, P. J. Pauling, and M. L. Tobe, *J. Amer. Chem. Soc.* 90, 4741 (1968).
179. M. R. Churchill and T. A. O'Brien, *Chem. Commun.* p. 246 (1968); *J. Chem. Soc.*, *A* p. 206 (1970).
180. G. A. Mair, H. M. Powell, and L. M. Venanzi, *Proc. Chem. Soc.*, *London* p. 170 (1961).
181. R. Cramer, R. V. Lindsey, C. T. Prewitt, and U. G. Stolberg, *J. Amer. Chem. Soc.* 87, 658 (1965).
182. A. D. Westland, *J. Chem. Soc.* p. 3060 (1965).
183. W. Hieber and V. Frey, *Chem. Ber.* 99, 2614 (1966).
184. J. P. Collman, F. D. Vastine, and W. R. Roper, *J. Amer. Chem. Soc.* 90, 2282 (1968).
185. C. M. Harris and R. S. Nyholm, *J. Chem. Soc.* p. 4375 (1956).
186. R. Ettorre, A. Peloso, and G. Dolcetti, *Gazz. Chim. Ital.* 97, 968 (1967).
187. C. M. Harris, R. S. Nyholm, and D. J. Phillips, *J. Chem. Soc.* p. 4379 (1960).
188. C. M. Harris, S. E. Livingstone, and I. H. Reece, *J. Chem. Soc.* p. 1505 (1959).
189. G. D. Watt, D. J. Eatough, R. M. Izatt, and J. J. Christensen, *Utah Acad. Sci.*, *Arts Lett.*, *Proc.* 42, 298 (1965); *Chem. Abstr.* 66, 59526s (1967).
190. S. Ahrland, J. Chatt, and N. R. Davies, *Quart. Rev.*, *Chem. Soc.* 12, 265 (1958).
191. A. J. Poe and D. H. Vaughan, *Inorg. Chim. Acta* 1, 255 (1967).
192. J. S. Coe, M. D. Hussain, and A. A. Malik, *Inorg. Chim. Acta* 2, 65 (1968).
193. F. Basolo, J. Chatt, H. B. Gray, R. G. Pearson, and B. L. Shaw, *J. Chem. Soc.* p. 2207 (1961); see also R. A. Reinhardt and W. W. Mark, *Inorg. Chem.* 9, 2026 (1970).
194. B. B. Smith and D. T. Sawyer, *Chem. Commun.* p. 1454 (1968).
195. C. H. Langford and H. B. Gray, "Ligand Substitution Processes." Benjamin, New York, 1966.
196. M. A. Bennett, G. J. Erskine, J. Lewis, R. Mason, R. S. Nyholm, G. B. Robertson, and A. D. C. Towl, *Chem. Commun.* p. 395 (1966).
197. P. G. Owston, J. M. Partridge, and J. M. Rowe, *Acta Crystallogr.* 13, 246 (1960).
198. J. D. Bell, D. Hall, and T. N. Waters, *Acta Crystallogr.* 21, 440 (1966).
199. R. Mason, G. B. Robertson, P. O. Whimp, B. L. Shaw, and G. Shaw, *Chem. Commun.* p. 868 (1968); R. Mason, G. B. Robertson, and P. O. Whimp, *J. Chem. Soc.*, *A* p. 535 (1970).
200. W. A. Whitla, H. M. Powell, and L. M. Venanzi, *Chem. Commun.* p. 310 (1966).
201. R. W. Siekman and D. L. Weaver, *Chem. Commun.* p. 1021 (1968); D. L. Weaver, *Inorg. Chem.* 9, 2250 (1970).
201a. E. Forsellini, G. Bombieri, B. Crociani, and T. Boschi, *Chem. Commun.* p. 1203 (1970).
202. J. M. Rowe, *Proc. Chem. Soc.*, *London* p. 66 (1962).
203. V. F. Levdik and M. A. Porai-Koshits, *Zh. Strukt. Khim.* 3, 472 (1962).
204. M. R. Churchill and R. Mason, *Nature (London)* 204, 777 (1964).
205. W. E. Oberhansli and L. F. Dahl, *J. Organometal. Chem.* 3, 43 (1965).
206. L. F. Dahl and W. E. Oberhansli, *Inorg. Chem.* 4, 629 (1965).
207. M. R. Churchill, *Chem. Commun.* p. 625 (1965).
208. A. E. Smith, *Acta Crystallogr.* 18, 331 (1965).
209. I. A. Zakharova, G. A. Kukina, T. S. Kuli-Zade, I. I. Moiseev, G. Yu Pek, and M. A. Porai-Koshits, *Russ. J. Inorg. Chem.* 11, 1364 (1966).
210. M. R. Churchill, *Inorg. Chem.* 5, 1608 (1966).
211. R. Mason and D. R. Russell, *Chem. Commun.* p. 26 (1966).

212. M. Kh. Minasyan, S. P. Gubin, and Yu. T. Struchkov, *Zh. Strukt. Khim.* **8**, 1108 (1967).

213. G. R. Davies, R. H. B. Mais, S. O'Brien, and P. G. Owston, *Chem. Commun.* p. 1151 (1967).

214. M. Kh. Minasyants and Yu. T. Struchkov, *Zh. Strukt. Khim.* **9**, 481 (1968).

215. R. Mason and A. G. Wheeler, *Nature (London)* **217**, 1253 (1968).

216. R. Mason and A. G. Wheeler, *J. Chem. Soc., A* p. 2549 (1968).

217. R. Mason and A. G. Wheeler, *J. Chem. Soc., A* p. 2543 (1968).

218. R. Mason, G. B. Robertson, P. O. Whimp, and D. A. White, *Chem. Commun.* p. 1655 (1968); R. Mason and P. O. Whimp, *J. Chem. Soc., A* p. 2709 (1969).

219. B. T. Kilbourn, R. H. B. Mais, and P. G. Owston, *Chem. Commun.* p. 1438 (1968).

220. K. Oda, N. Yasuoka, T. Ueki, N. Kasai, M. Kakudo, Y. Tezuka, T. Ogura, and S. Kawaguchi, *Chem. Commun.* p. 989 (1968); K. Oda, N. Yasuoka, T. Ueki, N. Kasai, and M. Kakudo, *Bull. Chem. Soc. Jap.* **43**, 362 (1970).

221. J. F. Malone, W. S. McDonald, B. L. Shaw, and G. Shaw, *Chem. Commun.* p. 869 (1968); J. F. Malone and W. S. McDonald, *J. Chem. Soc., A* p. 3124 (1970).

222. T. S. Kuli-Zade, G. A. Kukina, and M. A. Porai-Koshits, *Zh. Strukt. Khim.* **10**, 149 (1969).

223. S. Imamura, T. Kajimoto, Y. Kitano, and J. Tsuji, *Bull. Chem. Soc. Jap.* **42**, 805 (1969).

224. A. E. Smith, *Acta Crystallogr. Sect. A* **25**, Suppl., S161 (1969).

225. S. J. Lippard and S. M. Morehouse, *J. Amer. Chem. Soc.* **91**, 2504 (1969).

226. N. C. Baenziger, G. F. Richards, and J. R. Doyle, *Acta Crystallogr.* **14**, 303 (1961); **18**, 924 (1965).

227. C. V. Goebel, *Diss. Abstr.* **28**, 625B (1967).

228. F. Glockling and E. H. Brooks, *Amer. Chem. Soc., Div. Petrol. Chem., Prepr.* **14**, B135 (1969) (quoting results of H. M. M. Shearer and M. Schneider).

229. O. Hassel and B. F. Pedersen, *Proc. Chem. Soc., London* p. 394 (1959).

230. H. C. Freeman, J. F. Geldard, F. Lions, and M. R. Snow, *Proc. Chem. Soc., London* p. 258 (1964).

231. C. K. Prout and A. G. Wheeler, *J. Chem. Soc., A* p. 1286 (1966).

232. J. R. Wiesner and E. C. Lingafelter, *Inorg. Chem.* **5**, 1770 (1966).

233. N. C. Stephenson, J. F. McConnell, and R. Warren, *Inorg. Nucl. Chem. Lett.* **3**, 553 (1967).

234. M. Tanimura, T. Mizushima, and Y. Kinoshita, *Bull. Chem. Soc. Jap.* **40**, 2777 (1967).

235. V. W. Day, M. D. Glick, and J. L. Hoard, *J. Amer. Chem. Soc.* **90**, 4803 (1968).

236. W. L. Duax, *Diss. Abstr.* **28**, 3239B (1968).

237. D. E. Williams, G. Wohlauer, and R. E. Rundle, *J. Amer. Chem. Soc.* **81**, 755 (1959).

238. M. Calleri, G. Ferraris, and D. Viterbo, *Inorg. Chim. Acta* **1**, 297 (1967).

239. B. T. Kilbourn and R. H. B. Mais, *Chem. Commun.* p. 1507 (1968).

240. A. F. Wells, *Proc. Roy. Soc., Ser. A* **167**, 169 (1938).

241. D. A. Langs, C. R. Hare, and R. G. Little, *Chem. Commun.* p. 1080 (1967).

242. M. J. Bennett, F. A. Cotton, D. L. Weaver, R. J. Williams, and W. H. Watson, *Acta Crystallogr.* **23**, 788 (1967).

243. D. L. Sales, J. Stokes, and P. Woodward, *J. Chem. Soc., A* p. 1852 (1968).

244. N. R. Kunchur, *Acta Crystallogr., Sect. B* **24**, 1623 (1968).

245. M. A. Hepworth, K. H. Jack, R. D. Peacock, and G. J. Westland, *Acta Crystallogr.* **10**, 63 (1957).

246. N. Bartlett and R. Maitland, *Acta Crystallogr.* **11**, 747 (1958).

247. J. R. Holden and N. C. Baenziger, *J. Amer. Chem. Soc.* **77**, 4987 (1955).

248. J. N. Dempsey and N. C. Baenziger, *J. Amer. Chem. Soc.* **77**, 4984 (1955).

249. D. J. Robinson and C. H. L. Kennard, *Chem. Commun.* p. 1236 (1967).

250. A. F. Wells, *Z. Kristallogr., Kristallgeometrie, Kristallphys., Kristallchem.* **100**, 189 (1938).

251. C. M. Harris, S. E. Livingstone, and N. C. Stephenson, *J. Chem. Soc.*, p. 3697 (1958).

252. K. Brodersen, G. Thiele, and H. Gaedcke, *Z. Anorg. Allg. Chem.* **348**, 162 (1966).

253. G. Thiele, K. Brodersen, E. Kruse, and B. Holle, *Chem. Ber.* **101**, 2771 (1968).

254. N. A. Bailey, J. M. Jenkins, R. Mason, and B. L. Shaw, *Chem. Commun.* pp. 237 and 296 (1965).

255. O. S. Mills and E. F. Paulus, *Chem. Commun.* p. 738 (1966).

256. J. Chatt, L. A. Duncanson, B. L. Shaw, and L. M. Venanzi, *Discuss. Faraday Soc.* **26**, 131 (1958); D. M. Adams, J. Chatt, and B. L. Shaw, *J. Chem. Soc.* p. 2047 (1960).

257. D. M. Adams, J. Chatt, J. Gerratt, and A. D. Westland, *J. Chem. Soc.* p. 734 (1964).

258. M. J. Church and M. J. Mays, *J. Chem. Soc.*, A p. 3074 (1968).

259. J. Chatt, L. A. Duncanson, and L. M. Venanzi, *J. Chem. Soc.* p. 3203 (1958): see also, F. H. Allen, A. Pidcock, and C. R. Waterhouse, *J. Chem. Soc.*, A p. 2087 (1970).

260. R. Layton, D. W. Sink, and J. R. Durig, *J. Inorg. Nucl. Chem.* **28**, 1965 (1966).

261. J. Chatt and R. G. Wilkins, *J. Chem. Soc.* p. 70 (1953).

262. G. E. Coates and C. Parkin, *J. Chem. Soc.* p. 421 (1963).

263. A. Pidcock, *Chem. Commun.* p. 92 (1968); R. G. Goodfellow, *ibid.* p. 114.

264. R. L. Keiter, *Diss. Abstr.* **28**, 4455B (1968); S. O. Grim and R. L. Keiter, *Inorg. Chim. Acta* **4**, 56 (1970).

265. R. J. Goodfellow, J. G. Evans, P. L. Goggin, and D. A. Duddell, *J. Chem. Soc.*, A p. 1604 (1968).

266. J. Chatt and R. G. Wilkins, *J. Chem. Soc.* p. 2532 (1951).

267. J. Chatt and R. G. Wilkins, *J. Chem. Soc.*, p. 273 (1952).

268. J. Chatt and R. G. Wilkins, *J. Chem. Soc.* p. 4300 (1952).

268a. P. Haake and R. M. Pfeiffer, *J. Amer. Chem. Soc.* **92**, 4996 (1970).

269. J. M. Jenkins and B. L. Shaw, *J. Chem. Soc.*, A p. 770 (1966).

270. N. Bartlett and P. R. Rao, *Proc. Chem. Soc.*, *London* p. 393 (1964).

271. F. S. Clements and E. V. Nutt (INCO (Mond) Ltd.), Brit. Pat., 879,074 (1961); *Chem. Abstr.* **56**, 8298c (1962).

272. R. van Helden and T. Jonkhoff (Shell Oil Co.), U.S. Pat., 3,210,152 (1965); *Chem. Abstr.* **63**, 15895a (1965).

273. H. Schäfer, U. Wiese, K. Rincke, and K. Brendel, *Angew. Chem., Int. Ed. Engl.* **6**, 253 (1967).

274. K. Brodersen, G. Thiele, and H. G. Schnering, *Z. Anorg. Allg. Chem.* **337**, 120 (1965).

275. U. Wiese, H. Schäfer, H. G. Schnering, C. Brendel, and R. Rincke, *Angew. Chem., Int. Ed. Engl.* **9**, 158 (1970).

276. W. van Bronswyk and R. Nyholm, *J. Chem. Soc.*, A p. 2084 (1968).

277. W. E. Bell, U. Merten, and M. Tagami, *J. Phys. Chem.* **65**, 510 (1961).

278. J. R. Soulen and W. H. Chappell, *J. Phys. Chem.* **69**, 3669 (1965).

279. G. Paiaro, N. Netto, A. Musco, and R. Palumbo, *Ric. Sic., Parte 2: Sez. A* **8**, 1441 (1965).

280. A. A. Grinberg, M. I. Gel'fman, and N. V. Kiseleva, *Russ. J. Inorg. Chem.* **12**, 620 (1967).

281. M. S. Kharasch, R. C. Seyler, and F. R. Mayo, *J. Amer. Chem. Soc.* **60**, 882 (1938).

282. H. Dietl, H. Reinheimer, J. Moffat, and P. M. Maitlis, *J. Amer. Chem. Soc.* **92**, 2276 (1970).

283. W. Kitching and C. J. Moore, *Inorg. Nucl. Chem. Lett.* **4**, 691 (1968); M. Kubota, B. A. Denechaud, P. M. McKinney, T. E. Needham, and G. O. Spessard, *J. Catal.* **18**, 119 (1970).

283a. G. Beech, G. Marr, and S. J. Ashcroft, *J. Chem. Soc., A* p. 2903 (1970).

284. J. M. Jenkins and J. G. Verkade, *Inorg. Syn.* **11**, 108 (1968).

285. J. Chatt and L. M. Venanzi, *J. Chem. Soc.* p. 2445 (1957).

286. D. M. Adams, P. J. Chandler, and R. G. Churchill, *J. Chem. Soc., A* p. 1272 (1967).

287. J. R. Durig, R. Layton, D. W. Sink, and B. R. Mitchell, *Spectrochim. Acta* **21**, 1367 (1965).

288. D. M. Adams and P. J. Chandler, *Chem. Commun.* p. 69 (1966); *J. Chem. Soc., A* p. 588 (1969).

289. D. M. Adams, "Metal-Ligand and Related Vibrations." Arnold, London, 1967.

289a. "Spectroscopic Properties of Inorganic and Organometallic Compounds" *Chemical Society (London). Specialist Periodical Reports* **1**, 172 (1968); **2**, 289 (1969).

290. G. Thiele and K. Brodersen, *Fortschr. Chem. Forsch.* **10**, 631 (1968).

291. R. G. Brown, J. M. Davidson, and C. Triggs, *Amer. Chem. Soc., Div. Petrol. Chem. Prepr.* **14**, B23 (1969).

292. B. O. Field and C. J. Hardy, *J. Chem. Soc.* p. 4428 (1964).

293. S. M. Morehouse, A. R. Powell, J. P. Heffer, T. A. Stephenson, and G. Wilkinson, *Chem. Ind. (London)* p. 544 (1964).

294. T. A. Stephenson, S. M. Morehouse, A. R. Powell, J. P. Heffer, and G. Wilkinson, *J. Chem. Soc.* p. 3632 (1965).

295. E. A. Hausman, J. R. Grasso, and G. R. Pond (Engelhard Industries Inc.), Fr. Pat., 1,403,398 (1965); *Chem. Abstr.* **63**, 12717h (1965).

295a. A. C. Skapski and M. L. Smart, *Chem. Commun.* p. 658 (1970).

296. R. H. Holm and F. A. Cotton, *J. Phys. Chem.* **65**, 321 (1961); see also A. N. Knyazeva, E. A. Shugam, and L. M. Shkol'nikova, *Zh. Strukt. Khim.* **11**, 938 (1970).

297. R. F. Schramm and B. B. Wayland, *Chem. Commun.* p. 898 (1968).

298. B. B. Wayland and R. F. Schramm, *Inorg. Chem.* **8**, 971 (1969).

299. F. Basolo, J. L. Burmeister, and A. J. Poe, *J. Amer. Chem. Soc.* **85**, 1700 (1963).

300. J. L. Burmeister and F. Basolo, *Inorg. Chem.* **3**, 1587 (1964).

301. A. Sabatini and I. Bertini, *Inorg. Chem.* **4**, 1665 (1965); I. Bertini and A. Sabatini, *Inorg. Chem.* **5**, 1025 (1966).

302. J. Burmeister, R. L. Hassel, and R. J. Phelan, *Chem. Commun.* p. 679 (1970); D. W. Meek, P. E. Nipcon, and V. I. Meek, *J. Amer. Chem. Soc.* **92**, 5351 (1970); G. Beran and G. J. Palenik, *Chem. Commun.* p. 1354 (1970).

303. A. Mawby and G. E. Pringle, *Chem. Commun.* p. 385 (1970).

304. M. A. Bennett and P. A. Longstaff, *J. Amer. Chem. Soc.* **91**, 6266 (1969).

305. F. A. Cotton, B. G. DeBoer, M. D. La Prade, J. R. Pipal, and D. A. Ucko, *J. Amer. Chem. Soc.* **92**, 2926 (1970).

306. G. Thiele and P. Woditsch, *Angew. Chem.* **81**, 706 (1969).

307. N. S. Garif'yanov, I. I. Kalinichenko, I. V. Ovchinnikov, and Z. F. Mant'yanova, *Dokl. Akad. Nauk SSSR* **176**, 328 (1967).

308. A. S. Foust and R. H. Soderberg, *J. Amer. Chem. Soc.* **89**, 5507 (1967).

309. L. F. Warren and M. F. Hawthorne, *J. Amer. Chem. Soc.* **90**, 4823 (1968).

310. N. Bartlett and P. R. Rao, *Abstr., 154th Amer. Chem. Soc.* paper K15 (1967).

311. H. Henkel and R. Hoppe, *Z. Anorg. Allg. Chem.* **359**, 160 (1968).

312. K. Ito, D. Nakamura, Y. Kurita, K. Ito, and M. Kubo, *J. Amer. Chem. Soc.* **83**, 4526 (1961).

313. R. S. Nyholm, *Proc. Chem. Soc., London* p. 273 (1961).
314. F. Cariati and R. Ugo, *Chim. Ind. (Milan)* **48**, 1288 (1966).
315. R. G. Vranka, L. F. Dahl, P. Chini, and J. Chatt, *J. Amer. Chem. Soc.* **91**, 1574 (1969); J. Chatt and P. Chini, *J. Chem. Soc., A* 1538 (1970); P. Chini and G. Longoni, *J. Chem Soc. A* 1542 (1970).
316. W. Manchot and J. König, *Chem. Ber.*, **59** Part II, 883 (1926).
317. R. J. Irving and E. A. Magnusson, *J. Chem. Soc.* p. 2283 (1958).
318. E. O. Fischer and A. Vogler, *J. Organometal. Chem.* **3**, 161 (1965).
319. A. Treiber, *Tetrahedron Lett.* p. 2831 (1966).
320. W. Schnabel and E. Kober, *J. Organometal. Chem.* **19**, 455 (1969).
321. A. D. Gel'man and E. Meilakh, *Dokl. Akad. Nauk SSSR* **36**, 188 (1942).
322. J. V. Kingston and G. R. Scollary, *Chem. Commun.* p. 455 (1969).
323. Z. Burianec and J. Burianova, *Collect. Czech. Chem. Commun.* **28**, 2138 (1963).
324. H. C. Clark, K. R. Dixon, and W. J. Jacobs, *Chem. Commun.* p. 93 (1968); *J. Amer. Chem. Soc.* **90**, 2259 (1968).
325. J. V. Kingston and G. R. Scollary, *Chem. Commun.* p. 362 (1970).
326. D. Medema, R. van Helden, and C. F. Kohll, *Abstr., Inorg. Chim. Acta Symp., 1968* p. E3 (1968); *Inorg. Chim. Acta* **3**, 255 (1969).
327. J. Powell and B. L. Shaw, *J. Chem. Soc., A* p. 1839 (1967).
328. W. T. Dent, R. Long, and A. J. Wilkinson, *J. Chem. Soc.* p. 1585 (1964).
329. A. B. Fasman, V. A. Golodov, and D. V. Sokolskii, *Zh. Fiz. Khim.* **38**, 1545 (1964).
330. V. A. Golodov, G. G. Kutyukov, A. B. Fasman, and D. V. Sokolskii, *Russ. J. Inorg. Chem.* **9**, 1257 (1964).
331. A. B. Fasman, G. G. Kutyukov, and D. V. Sokolskii, *Dokl. Akad. Nauk SSSR (English Transl.)* **158**, 958 (1964).
332. G. G. Kutyukov, A. B. Fasman, A. E. Lyuts, Yu. A. Kushnikov, V. F. Vozdvizhenskii, and V. A. Golodov, *Zh. Fiz. Khim.* **40**, 1468 (1966).
333. I. V. Fedoseev and V. I. Spitsyn, *Dokl. Akad. Nauk SSSR* **174**, 371 (1967).
334. V. F. Vozdvizhenskii, Yu. A. Kushnikov, G. G. Kutyukov, and A. B. Fasman, *Russ. J. Inorg. Chem.* **12**, 798 (1967).
335. G. G. Kutyukov, A. B. Fasman, V. F. Vozdvizhenskii, and Yu. A. Kushnikov, *Zh. Neorg. Khim.* **13**, 1542 (1968).
336. V. I. Spitsyn, I. V. Znamenskii, and I. V. Fedoseev, *Dokl. Akad. Nauk SSSR* **181**, 617 (1968).
337. E. O. Fischer and H. Werner, *Chem. Ber.* **93**, 2075 (1960).
338. D. Jones, G. W. Parshall, L. Pratt, and G. Wilkinson, *Tetrahedron Lett.* p. 48 (1961).
339. E. O. Fischer and H. Werner (Badische Aniline u. Soda-Fabrik AG), Ger. Pat., 1,132,923 (1962); *Chem. Abstr.* **58**, 1494a (1963).
340. G. F. Svatos and E. E. Flagg, *Inorg. Chem.* **4**, 422 (1965).
341. E. O. Fischer and H. Werner, *Chem. Ber.* **95**, 695 (1962).
341a. E. M. Badley, J. Chatt, R. L. Richards, and G. A. Sim, *Chem. Commun.* p. 1323 (1969); B. Crociani, T. Boschi, and U. Belluco, *Inorg. Chem.* **9**, 2021 (1970); F. Bonati, G. Minghetti, T. Boschi, and B. Crociani, *J. Organometal. Chem.* **25**, 255 (1970); A. Burke, A. L. Balch, and J. H. Enemark, *J. Amer. Chem. Soc.* **92**, 2555 (1970).
341b. Y. Yamamoto and H. Yamazaki, *Bull. Chem. Soc. Jap.* **43**, 2653 (1970); T. Kajimoto, H. Takahashi, and J. Tsuji, *J. Organometal. Chem.* **23**, 275 (1970).
341c. M. H. Chisholm and H. C. Clark, *Chem. Commun.* p. 763 (1970).
342. W. Beck and E. Schuierer, *Chem. Ber.* **98**, 298 (1965).
343. P. W. R. Corfield and H. M. M. Shearer, *Acta Crystallogr.* **20**, 502 (1966).
344. P. W. R. Corfield and H. M. M. Shearer, *Acta Crystallogr.* **21**, 957 (1966).

345. M. I. Bruce, D. A. Harbourne, F. Waugh, and F. G. A. Stone, *J. Chem. Soc., A* p. 356 (1968).
346. J. Chatt and B. L. Shaw, *J. Chem. Soc.* p. 1718 (1960).
347. J. Chatt and B. L. Shaw, *J. Chem. Soc.* p. 4020 (1959).
348. P. Rigo, C. Pecile, and A. Turco, *Inorg. Chem.* **6**, 1636 (1967).
349. D. M. Roundhill and H. B. Jonassen, *Chem. Commun.* p. 1233 (1968); J. H. Nelson, H. B. Jonassen, and D. M. Roundhill, *Inorg. Chem.* **8**, 2591 (1969).
350. J. P. Collman and J. W. Kang, *J. Amer. Chem. Soc.* **89**, 844 (1967).
351. F. Glockling and K. A. Hooton, *J. Chem. Soc., A* p. 1066 (1967).
352. G. R. Davies, R. H. B. Mais, and P. G. Owston, *J. Chem. Soc., A* p. 1750 (1967).
353. W. P. Spofford, P. D. Carfagna, and E. L. Amma, *Inorg. Chem.* **6**, 1553 (1967).
354. M. R. Churchill and T. A. O'Brien, *Chem. Commun.* p. 992 (1967); *J. Chem. Soc., A* p. 266 (1970).
355. M. R. Churchill and T. A. O'Brien, *J. Chem. Soc. A*, p. 2970 (1968).
356. W. J. Pope and S. J. Peachey, *J. Chem. Soc.* p. 571 (1909).
357. R. E. Rundle and J. H. Sturdivant, *J. Amer. Chem. Soc.* **69**, 1561 (1947).
358. J. Chatt and B. L. Shaw, *J. Chem. Soc.* p. 705 (1959).
359. M. L. H. Green, "Organometallic Compounds," 3rd ed., Vol. 2, p. 223. Methuen London, 1968.
360. J. R. Moss and B. L. Shaw, *J. Chem. Soc., A* p. 1793 (1966).
361. K. Matsuzaki and T. Yasukawa, *J. Phys. Chem.* **71**, 1160 (1967).
362. K. Matsuzaki and T. Yasukawa, *Chem. Commun.* p. 1460 (1968).
363. J. Chatt and B. L. Shaw, *Chem. Ind. (London)* p. 675 (1959).
364. P. G. Owston and J. M. Rowe, *J. Chem. Soc.* p. 3411 (1963).
365. J. Chatt, L. M. Vallarino, and L. M. Venanzi, *J. Chem. Soc.* p. 3413 (1957).
366. H. Yamazaki and N. Hagihara, *Bull. Chem. Soc. Jap.* **37**, 907 (1964).
367. D. W. McBride, E. Dudeck, and F. G. A. Stone, *J. Chem. Soc.* p. 1752 (1964).
368. H. Yamazaki, T. Nishido, Y. Matsumoto, S. Sumida, and N. Hagihara, *J. Organometal. Chem.* **6**, 86 (1966).
369. I. A. Ustyniuk, T. I. Voevodskaia, N. A. Zharikova, and N. A. Ustyniuk, *Dokl. Akad. Nauk SSSR* **181**, 372 (1968).
370. Y. Yamamoto, A. Yamazaki, and N. Hagihara, *Bull. Chem. Soc. Jap.* **41**, 532 (1968).
370a. J. Thomson and M. C. Baird, *Can. J. Chem.* **48**, 3442 (1970).
370b. R. J. Cross and R. Wardle, *J. Organometal. Chem.* **23**, C4 (1970).
370c. G. Yugupsky, W. Mowat, A. Shortland, and G. Wilkinson, *Chem. Commun.* p. 1369 (1970); M. R. Collier, M. F. Lappert, and M. M. Truelock, *J. Organometal. Chem.* **25**, C36 (1970).
371. F. A. Cotton and J. A. McCleverty, *Inorg. Chem.* **4**, 490 (1965).
372. H. C. Clark and J. H. Tsai, *J. Organometal. Chem.* **7**, 515 (1967).
373. H. A. Skinner, *Advan. Organometal. Chem.* **2**, 49 (1964).
374. S. J. Ashcroft and C. T. Mortimer, *J. Chem. Soc., A* p. 930 (1967).
375. P. L. Goggin and R. J. Goodfellow, *J. Chem. Soc., A* p. 1462 (1966).
376. G. Calvin and G. E. Coates, *Chem. Ind. (London)* p. 160 (1958).
377. A. J. Rest, D. T. Rosevear, and F. G. A. Stone, *J. Chem. Soc., A* p. 66 (1967).
378. F. J. Hopton, A. J. Rest, D. T. Rosevear, and F. G. A. Stone, *J. Chem. Soc., A* p. 1326 (1966).
379. I. I. Moiseev and M. N. Vargaftik, *Dokl. Akad. Nauk SSSR (English Transl.)* **166**, 81 (1966).
380. A. Kasahara and T. Izumi, *Bull. Chem. Soc. Jap.* **42**, 1765 (1969).
380a. R. J. Cross and R. Wardle, *J. Chem. Soc., A* p. 840 (1970).

381. R. F. Heck, *J. Amer. Chem. Soc.* **90**, 5518 (1968).
382. R. F. Heck, *J. Amer. Chem. Soc.* **90**, 5526 (1968).
383. R. F. Heck, *J. Amer. Chem. Soc.* **90**, 5531 (1968).
384. R. F. Heck, *J. Amer. Cheml Soc.* **90**, 5535 (1968).
385. R. F. Heck, *J. Amer. Chem. Soc.* **90**, 5538 (1968).
386. R. F. Heck, *J. Amer. Chem. Soc.* **90**, 5542 (1968).
387. R. F. Heck, J. *Amer. Chem. Soc.* **90**, 5546 (1968).
388. P. M. Henry, *Tetrahedron Lett.* p. 2285 (1968).
389. G. Wilke and G. Herrmann, *Angew. Chem., Int. Ed. Engl.* **5**, 581 (1966).
390. T. Saito, Y. Uchida, and A. Misono, *J. Amer. Chem. Soc.* **88**, 5198 (1966); A. Yamamoto, K. Morifuji, S. Ikeda, T. Saito, Y. Uchida, and A. Misono, *ibid.* **87**, 4652 (1965).
391. A. Yamamoto and S. Ikeda, *J. Amer. Chem. Soc.* **89**, 5989 (1967).
392. T. Saito, M. Araki, Y. Uchida, and A. Misono, *J. Phys. Chem.* **81**, 2370 (1967); S. Castellano and H. Günther, *ibid.* p. 2368.
393. M. Tsutsui and H. Zeiss, *J. Amer. Chem. Soc.* **81**, 6090 (1959).
394. D. O. Cowan, N. G. Krieghoff, and G. Donnay, *Acta Crystallogr., Sect. B* **24**, 287 (1968); M. N. Hoechstetter and C. H. Brubaker, *Inorg. Chem.* **8**, 400 (1969).
395. J. D. Ruddick and B. L. Shaw, *Chem. Commun.* p. 1135 (1967).
396. J. D. Ruddick and B. L. Shaw, *J. Chem. Soc., A* pp. 2964 and 2969 (1969).
397. D. R. Coulson, *Chem. Commun.* p. 1530 (1968).
398. P. Fitton, J. E. McKeon, and B. C. Ream, *Chem. Commun.* p. 370 (1969).
399. M. Green, R. B. L. Osborn, A. J. Rest, and F. G. A. Stone, *J. Chem. Soc., A* p. 2525 (1968).
400. M. Baird, G. Hartwell, R. Mason, A. M. Rae, and G. Wilkinson, *Chem. Commun.* p. 92 (1967); R. Mason and A. M. Rae, *J. Chem. Soc., A* p. 1767 (1970)
401. G. Wilke and H. Schott, *Angew. Chem., Int. Ed. Engl.* **5**, 583 (1966).
402. K. A. Hofmann and J. von Narbutt, *Chem. Ber.* **41**, 1625 (1908).
403. J. Chatt, L. M. Vallarino, and L. M. Venanzi, *J. Chem. Soc.* p. 2496 (1957).
404. H. Reinheimer, H. Dietl, J. Moffat, D. Wolff, and P. M. Maitlis, *J. Amer. Chem. Soc.* **90**, 5321 (1968).
405. B. Crociani, P. Uguagliati, T. Boschi, and U. Belluco, *J. Chem. Soc., A* p. 2869 (1968).
406. W. A. Whitla, private communication (1970).
407. C. Pannatoni, G. Bombieri, E. Forsellini, B. Crociani, and U. Belluco, *Chem. Commun.* p. 187 (1969).
408. J. K. Stille and R. A. Morgan, *J. Amer. Chem. Soc.* **88**, 5135 (1966).
409. G. Paiaro, A. de Renzi, and R. Palumbo, *Chem. Commun.* p. 1150 (1967).
410. C. B. Anderson and B. J. Burreson, *J. Organometal. Chem.* **7**, 181 (1967); *Chem. Ind. (London)* p. 620 (1967).
411. D. A. White, *Organometal. Chem. Rev.* **3**, 497 (1968).
412. J. Tsuji and H. Takahashi, *J. Amer. Chem. Soc.* **87**, 3275 (1965).
413. J. Tsuji and H. Takahashi (Toyo Rayon Co. Ltd.), Jap. Pat. 23,180/67; *Chem. Abstr.* **69**, 52305x (1968).
414. H. Takahashi and J. Tsuji, *J. Amer. Chem. Soc.* **90**, 2387 (1968).
415. B. F. G. Johnson, J. Lewis, and M. S. Subramanian, *J. Chem. Soc., A* p. 1993 (1968).
416. R. Palumbo, A. de Renzi, A. Panunzi, and G. Paiaro, *J. Amer. Chem. Soc.* **91**, 3874 (1969); A. Panunzi, A. de Renzi, R. Palumbo, and G. Paiaro, *ibid.* p. 3879.
417. M. Tada, Y. Kuroda, and T. Sato, *Tetrahedron Lett.* p. 2871 (1969).
418. C. R. Kistner, J. H. Hutchinson, J. R. Doyle, and J. C. Storlie, *Inorg. Chem.* **2**, 1255 (1963); J. R. Doyle, J. H. Hutchinson, N. C. Baenziger, and L. W. Tresselt, *J. Amer. Chem. Soc.* **83**, 2768 (1961).

419. M. Avram, E. Sliam, and C. D. Nenitzescu, *Justus Liebigs Ann. Chem.* **636**, 184 (1960).
420. J. K. Stille, R. A. Morgan, D. D. Whitehurst, and J. R. Doyle, *J. Amer. Chem. Soc.* **87**, 3282 (1965).
421. R. G. Schultz, *J. Organometal. Chem.* **6**, 435 (1966).
421a. R. N. Haszeldine, R. V. Parish, and D. W. Robbins, *J. Organometal. Chem.* **23**, C33 (1970).
422. W. Kitching, *Organometal. Chem. Rev.* **3**, 35 and 61 (1968).
423. A. C. Cope, J. M. Kliegman, and E. C. Friedrich, *J. Amer. Chem. Soc.* **89**, 287 (1967).
424. A. Kasahara, K. Tanaka, and T. Izumi, *Bull. Chem. Soc. Jap.* **42**, 1702 (1969).
425. T. Yukawa and S. Tsutsumi, *Inorg. Chem.* **7**, 1458 (1968).
426. K. Matsumoto, Y. Odaira, and S. Tsutsumi, *Chem. Commun.* p. 832 (1968).
427. J. Ashley-Smith, J. Clemens, M. Green, and F. G. A. Stone, *J. Organometal. Chem.* **17**, P23 (1969).
428. R. G. Miller, D. R. Fahey, and D. P. Kuhlmann, *J. Amer. Chem. Soc.* **90**, 6248 (1968).
429. A. C. Cope and R. W. Siekman, *J. Amer. Chem. Soc.* **87**, 3272 (1965).
430. A. C. Cope and E. C. Friedrich, *J. Amer. Chem. Soc.* **90**, 909 (1968).
431. A. Kasahara, *Bull. Chem. Soc. Jap.* **41**, 1272 (1968).
432. A. L. Balch and D. Petridis, *Inorg. Chem.* **8**, 2247 (1969).
432a. G. E. Hartwell, R. V. Lawrence, and M. J. Smas, *Chem. Commun.* p. 912 (1970).
433. H. Takahashi and J. Tsuji, *J. Organometal. Chem.* **10**, 511 (1967).
434. P. M. Maitlis and F. G. A. Stone, *Chem. Ind. (London)* p. 1865 (1962).
435. C. R. Kistner, D. A. Drew, J. R. Doyle, and G. W. Rausch, *Inorg. Chem.* **6**, 2036 (1967).
436. G. Booth and J. Chatt, *Proc. Chem. Soc., London* p. 67 (1961).
437. G. Booth and J. Chatt, *J. Chem. Soc., A* p. 634 (1966).
438. D. M. Adams and G. Booth, *J. Chem. Soc.* p. 1112 (1962).
438a. H. C. Clark and R. J. Puddephat, *Chem. Commun.* p. 92 (1970); *Inorg. Chem.* **10**, 18 (1971).
439. J. Chatt, R. S. Coffey, A. Gough, and D. T. Thompson, *J. Chem. Soc., A* p. 190 (1968).
440. D. A. Harbourne, D. T. Rosevear, and F. G. A. Stone, *Inorg. Nucl. Chem. Lett.* **2**, 247 (1966).
441. H. C. Clark and W. S. Tsang, *J. Amer. Chem. Soc.* **89**, 529 (1967).
442. J. Cooke, W. R. Cullen, M. Green, and F. G. A. Stone, *Chem. Commun.* p. 170 (1968).
442a. A. J. Deeming, B. F. G. Johnson, and J. Lewis, *Chem. Commun.* p. 598 (1970).
443. E. Lodewijk and D. Wright, *J. Chem. Soc., A* p. 119 (1968).
444. G. W. Parshall, *J. Amer. Chem. Soc.* **87**, 2133 (1965).
445. C. D. Cook and G. S. Jauhal, *J. Amer. Chem. Soc.* **90**, 1464 (1968).
446. A. G. Swallow and M. R. Truter, *Proc. Roy. Soc., Ser. A.* **254**, 205 (1960).
447. A. C. Hazell and M. R. Truter, *Proc. Roy. Soc., Ser. A* **254**, 218 (1960).
448. N. A. Bailey, R. D. Gillard, M. Keeton, R. Mason, and D. R. Russell, *Chem. Commun.* p. 396 (1966).
449. D. T. Rosevear and F. G. A. Stone, *J. Chem. Soc.* p. 5275 (1965); J. R. Phillips, D. T. Rosevear, and F. G. A. Stone, *J. Organometal. Chem.* **2**, 455 (1964).
450. U. Belluco, U. Croatto, P. Uguagliati, and R. Pietropaolo, *Inorg. Chem.* **6**, 718 (1967).
451. U. Belluco, M. Giustiniani, and M. Graziani, *J. Amer. Chem. Soc.* **89**, 6494 (1967).
452. M. A. Bennett, J. Chatt, G. J. Erskine, J. Lewis, R. F. Long, and R. S. Nyholm, *J. Chem. Soc., A* p. 501 (1967); M. A. Bennett, G. J. Erskine, and R. S. Nyholm, *ibid.* p. 1260.
453. W. Keim, *J. Organometal. Chem.* **14**, 179 (1968).

453a. A. J. Cheney, B. E. Mann, B. L. Shaw, and R. M. Slade, *Chem. Commun.* p. 1176 (1970).
454. R. van Helden and G. Verberg, *Rec. Trav. Chim. Pays-Bas* **84**, 1263 (1965).
455. J. Müller and P. Göser, *Angew. Chem., Int. Ed. Engl.* **6**, 364 (1967).
456. W. Keim, *J. Organometal. Chem.* **8**, P25 (1967).
457. M. Green, R. N. Haszeldine, and J. Lindley, *J. Organometal. Chem.* **6**, 107 (1966).
457a. D. A. White and G. W. Parshall, *Inorg. Chem.* **9**, 2358 (1970).
458. B. Bogdanovic, H. Bönnemann, and G. Wilke, *Angew. Chem., Int. Ed. Engl.* **5**, 582 (1966).
459. C. Kowala and J. M. Swan, *Aust. J. Chem.* **19**, 547 (1966).
460. H. Gilman and L. A. Woods, *J. Amer. Chem. Soc.* **70**, 550 (1948).
461. D. Gibson, J. Lewis, and C. Oldham, *J. Chem. Soc., A* p. 72 (1967).
462. M. Green and R. I. Hancock, *J. Chem. Soc., A* p. 2054 (1967).
463. B. F. G. Johnson, J. Lewis, and D. A. White, *J. Amer. Chem. Soc.* **91**, 5186 (1969); *J. Chem. Soc., A* p. 1738 (1970).
464. D. M. Adams, J. Chatt, and R. G. Guy, *Proc. Chem. Soc., London* p. 179 (1960); D. M. Adams, J. Chatt, R. Guy, and N. Sheppard, *J. Chem. Soc.* p. 738 (1961).
465. S. E. Binns, R. H. Cragg, R. D. Gillard, B. T. Heaton, and M. F. Pilbrow, *J. Chem. Soc., A* p. 1227 (1969).
465a. E. Vedejs and M. F. Salomon, *J. Amer. Chem. Soc.*, **92**, 6965 (1970).
466. D. R. Coulson, *J. Amer. Chem. Soc.* **91**, 201 (1969).
467. H. Reinheimer, J. Moffat, and P. M. Maitlis, *J. Amer. Chem. Soc.* **92**, 2285 (1970).
468. R. F. Heck, *J. Amer. Chem. Soc.* **90**, 313 (1968).
469. C. C. Hunt and J. R. Doyle, *Inorg. Nucl. Chem. Lett.* **2**, 283 (1966).
470. M. Graziani, L. Busetto, M. Guistiniani, M. Nicolini, and A. Palazzi, *Ric. Sci.* **37**, 632 (1967).
471. G. Faraone, V. Ricevuto, R. Romeo, and M. Trozzi, *Gazz. Chim. Ital.* **98**, 480 (1968).
472. R. Pietropaolo, S. Sergi, and G. Gaetano, *Ric. Sci.* **38**, 195 (1968).
473. J. Chatt, L. A. Duncanson, and B. L. Shaw, *Proc. Chem. Soc., London* p. 343 (1957).
474. J. Chatt, L. A. Duncanson, and B. L. Shaw, *Chem. Ind. (London)* p. 859 (1958).
475. E. H. Brooks and F. Glockling, *Chem. Commun.* p. 510 (1966).
476. E. H. Brooks and F. Glockling, *J. Chem. Soc., A* p. 1030 (1967).
477. E. H. Brooks and F. Glockling, *J. Chem. Soc., A* p. 1241 (1966).
478. R. Eisenberg and J. A. Ibers, *Inorg. Chem.* **4**, 773 (1965).
479. J. M. Jenkins and B. L. Shaw, *Proc. Chem. Soc., London* p. 279 (1963).
480. E. H. Brooks, R. J. Cross, and F. Glockling, *Inorg. Chim. Acta* **2**, 17 (1968).
481. M. L. H. Green, H. Munakata, and T. Saito, *Chem. Commun.* pp. 208 and 1287 (1969); p. 881 (1970).
482. K. Jonas and G. Wilke, *Angew. Chem., Int. Ed. Engl.* **8**, 519 (1969).
482a. C. A. Tolman, *J. Amer. Chem. Soc.* **92**, 6777 and 6785 (1970).
483. W. C. Zeise, *Ann. Phys. Chem.* [2] **9**, 632 (1827); **21**, 497 (1831).
484. J. Chatt and L. A. Duncanson, *J. Chem. Soc.*, p. 2939 (1953).
485. M. J. S. Dewar, *Bull. Soc. Chim. Fr.* **18**, C79 (1951).
485a. Y. Takahashi, T. Ito, S. Sakai, and Y. Ishii, *Chem. Commun.* p. 1065 (1970).
486. O. S. Mills and B. W. Shaw, *J. Organometal. Chem.* **11**, 595 (1968).
487. R. B. King, M. I. Bruce, J. R. Phillips, and F. G. A. Stone, *Inorg. Chem.* **5**, 684 (1966).
488. J. Chatt, L. A. Duncanson, and R. G. Guy, *Chem. Ind. (London)* p. 430 (1959).
489. J. Chatt, R. G. Guy, and L. A. Duncanson, *J. Chem. Soc.* p. 827 (1961).
490. J. Chatt, R. G. Guy, L. A. Duncanson, and D. T. Thompson, *J. Chem. Soc.* p. 5170 (1963).

491. J. Chatt, G. A. Rowe, and A. A. Williams, *Proc. Chem. Soc., London* p. 208 (1957).
492. G. R. Davies, W. Hewertson, R. H. B. Mais, P. G. Owston, and C. G. Patel, *J. Chem. Soc., A* p. 1873 (1970).
493. J. O. Glanville, J. M. Stewart, and S. O. Grim, *J. Organometal. Chem.* **7**, P9 (1967).
494. G. Wilke and G. Herrmann, *Angew. Chem., Int. Ed. Engl.* **1**, 549 (1962).
495. T. Hosokawa, I. Moritani, and S. Nishioka, *Tetrahedron Lett.* p. 3833 (1969).
496. E. Kuljian and H. Frye, *Z. Naturforsch. B* **20**, 204 (1965).
497. R. Hüttel and M. Bechter, *Angew. Chem.* **71**, 456 (1959).
498. R. Hüttel and J. Kratzer, *Angew. Chem.* **71**, 456 (1959).
499. R. Hüttel, H. Dietl, and H. Christ, *Chem. Ber.* **97**, 2037 (1964).
500. S. V. Pestrikov and I. I. Moiseev, *Izv. Akad. Nauk SSSR (English Transl.)* p. 324 (1965).
501. S. V. Pestrikov, I. I. Moiseev, and T. M. Romanova, *Russ. J. Inorg. Chem.* **10**, 1199 (1965).
502. S. V. Pestrikov, I. I. Moiseev, and B. A. Tsvilikhovskaya, *Russ. J. Inorg. Chem.* **11**, 930 (1966).
503. P. M. Henry, *J. Amer. Chem. Soc.* **88**, 1595 (1966).
504. G. F. Pregaglia, M. Donati, and F. Conti, *Chem. Ind. (London)* p. 1923 (1966).
505. G. F. Pregaglia, M. Donati, and F. Conti, *Chim. Ind. (Milan)* **49**, 1277 (1967).
506. M. Donati and F. Conti, *Tetrahedron Lett.* p. 1219 (1966).
507. A. D. Ketley, L. P. Fisher, A. J. Berlin, C. R. Morgan, E. H. Gorman, and T. R. Steadman, *Inorg. Chem.* **6**, 657 (1967).
508. D. M. Roundhill and G. Wilkinson, *J. Chem. Soc., A* p. 506 (1968).
509. E. O. Greaves and P. M. Maitlis, *J. Organometal. Chem.* **6**, 104 (1966).
510. J. Chatt, B. L. Shaw, and A. A. Williams, *J. Chem. Soc.* p. 3269 (1962).
511. G. Wilke, E. W. Müller, M. Kroener, P. Heimbach, and H. Breil (Studiengesellschaft Kohle m.b.H.), Ger. Pat., 1,191,375 (1965); *Chem. Abstr.* **63**, 7045a (1965).
512. R. Cramer, *Inorg. Chem.* **4**, 445 (1965).
513. H. B. Jonassen and J. E. Field, *J. Amer. Chem. Soc.* **79**, 1275 (1957).
514. J. P. Birk, J. Halpern, and A. L. Pickard, *Inorg. Chem.* **7**, 2672 (1968).
515. S. Cenini, R. Ugo, F. Bonati, and G. LaMonica, *Inorg. Nucl. Chem. Lett.* **3**, 191 (1967).
516. G. N. Schrauzer, *J. Amer. Chem. Soc.* **81**, 5310 (1959).
517. G. N. Schrauzer, *Advan. Organometal. Chem.* **2**, 1 (1964).
518. R. Cramer, *Inorg. Chem.* **1**, 722 (1962).
519. S. F. A. Kettle and L. E. Orgel, *Chem. Ind. (London)* p. 49 (1960); A. R. Luxmore and M. R. Truter, *Proc. Chem. Soc., London* p. 466 (1961).
520. H. W. Quinn and D. N. Glew, *Can. J. Chem.* **40**, 1103 (1962).
521. M. Orchin and P. Schmidt, *Coord. Chem. Rev.* **3**, 345 (1968).
522. H. P. Fritz and D. Sellmann, *Z. Naturforsch. B* **22**, 610 (1967).
523. B. F. G. Johnson, C. E. Holloway, G. Hulley, and J. Lewis, *Chem. Commun.* p. 1143 (1967).
524. C. E. Holloway, G. Hulley, B. F. G. Johnson, and J. Lewis, *J. Chem. Soc., A* p. 53 (1969); p. 1653 (1970).
525. T. Kinusaga, M. Nakamura, H. Yamada, and A. Saika, *Inorg. Chem.* **7**, 2649 (1968).
526. G. B. Bokhii and G. A. Kukina, *J. Struct. Chem. (USSR)* **6**, 670 (1965).
527. J. A. Wunderlich and D. P. Mellor, *Acta Crystallogr.* **7**, 130 (1954); **8**, 57 (1955).
528. P. R. H. Alderman, P. G. Owston, and J. M. Rowe, *Acta Crystallogr.* **13**, 149 (1960).
529. M. Black, R. H. B. Mais, and P. G. Owston, *Acta Crystallogr., Sect. B* **25**, 1753 (1969); see J. A. Jarvis, B. T. Kilbourn, and P. G. Owston, *Acta Crystallogr., Sect. B* **26**, 876 (1970).
530. W. C. Hamilton, *Acta Crystallogr., Sect. A* **15**, Suppl., S172 (1969).

531. M. L. Maddox, S. L. Stafford, and H. D. Kaesz, *Advan. Organometal. Chem.* **3**, 1 (1965).
532. H. W. Quinn, J. S. McIntyre, and D. J. Peterson, *Can. J. Chem.* **43**, 2896 (1965).
533. C. D. Cook and K. Y. Wan, *J. Amer. Chem. Soc.* **92**, 2595 (1970).
534. M. J. Grogan and K. Nakamoto, *J. Amer. Chem. Soc.* **90**, 918 (1968).
535. M. J. Grogan and K. Nakamoto, *Inorg. Chim. Acta* **1**, 228 (1967); M. J. Grogan, *Diss. Abstr.* **28**, 1381B (1967).
536. H. P. Fritz and C. G. Kreiter, *Chem. Ber.* **96**, 1672 (1963).
537. D. B. Powell and N. Sheppard, *Spectrochim. Acta.* **13**, 69 (1958).
538. H. B. Jonassen and W. B. Kirsch, *J. Amer. Chem. Soc.* **79**, 1279 (1957).
539. D. B. Powell and N. Sheppard, *J. Chem. Soc.* p. 2519 (1960).
540. J. Pradilla-Sorzano and J. P. Fackler, *J. Mol. Spectrosc.* **22**, 80 (1967).
541. R. G. Denning, F. R. Hartley, and L. M. Venanzi, *J. Chem. Soc., A* pp. 324 and 328 (1967).
542. R. G. Denning and L. M. Venanzi, *J. Chem. Soc., A* p. 336 (1967).
543. F. R. Hartley and L. M. Venanzi, *J. Chem. Soc., A* p. 333 (1967).
544. R. Cramer, *J. Amer. Chem. Soc.* **89**, 4621 (1967).
545. I. I. Moiseev, A. P. Belov, V. A. Igoshin, and Ya. K. Syrkin, *Dokl. Akad. Nauk SSSR* (*English Transl.*) **174**, 256 (1967).
546. W. Dreissig and H. Dietrich, *Acta Crystallogr., Sect. B* **24**, 108 (1968).
547. C. D. Cook, C. H. Koo, S. C. Nyburg, and M. T. Shiomi, *Chem. Commun.* p. 426 (1967).
547a. J. K. Stalick and J. A. Ibers, *J. Amer. Chem. Soc.* **92**, 5333 (1970).
548. C. Pannatoni, G. Bombieri, U. Belluco, and W. H. Baddley, *J. Amer. Chem. Soc.* **90**, 798 (1968); G. Bombieri, E. Forsellini, C. Pannatoni, R. Graziani, and G. Bandoli, *J. Chem. Soc., A* p. 1313 (1970); C. Pannatoni, R. Graziani, G. Bandoli, D. A. Clemente and U. Belluco, *J. Chem. Soc., A* p. 371 (1970).
549. J. A. McGinnety and J. A. Ibers, *Chem. Commun.* p. 235 (1968).
549a. G. Hulley, B. F. G. Johnson, and J. Lewis, *J. Chem. Soc., A* p. 1732 (1970).
550. R. Cramer, *J. Amer. Chem. Soc.* **86**, 217 (1964).
551. R. Cramer, *J. Amer. Chem. Soc.* **89**, 5377 (1967).
552. R. Cramer, J. B. Kline, and J. D. Roberts, *J. Amer. Chem. Soc.* **91**, 2519 (1969); R. Cramer, *Abstr., Inorg. Chim. Acta Symp., 1969* p. 11 (1969).
553. H. Dietrich and R. Uttech, *Naturwissenschaften* **50**, 613 (1963).
554. H. Dietrich and H. Schmidt, *Naturwissenschaften* **52**, 301 (1965).
555. H. Dierks and H. Dietrich, *Z. Kristallogr., Kristallgeometrie, Kristallphys., Kristall-chem.* **122**, 1 (1965).
556. M. R. Churchill and T. A. O'Brien, *Inorg. Chem.* **6**, 1386 (1967).
557. R. Mason and G. B. Robertson, *J. Chem. Soc., A* p. 492 (1969).
558. G. W. Parshall and F. N. Jones, *J. Amer. Chem. Soc.* **87**, 5356 (1965).
559. C. E. Moore, "Atomic Energy Levels", *Nat. Bur. Stand.* (*U.S.*), *Circ.* **467** (1952).
560. H. Howard and G. W. King, *Can. J. Chem.* **37**, 700 (1959).
561. A. C. Blizzard and D. P. Santry, *J. Amer. Chem. Soc.* **90**, 5749 (1968).
562. C. A. Coulson and E. A. Stewart, *in* "Chemistry of the Alkenes" (S. Patai, ed.), pp. 129 and 139. Wiley (Interscience), New York, 1964.
563. I. Leden and J. Chatt, *J. Chem. Soc.* p. 2936 (1955).
564. D. G. McMane and D. S. Martin, *Inorg. Chem.* **7**, 1169 (1968).
565. T. P. Cheeseman, A. L. Odell, and H. A. Raethel, *Chem. Commun.* p. 1496 (1968).
566. J. Chatt and L. M. Venanzi, *J. Chem. Soc.* p. 2351 (1957).
567. W. H. Clement, *J. Organometal. Chem.* **10**, P19 (1967).

568. J. Chatt and R. G. Wilkins, *Nature (London)* **165**, 859 (1950).
569. J. Chatt and R. G. Wilkins, *J. Chem. Soc.* p. 2622 (1952).
570. J. F. Harrod, *Inorg. Chem.* **4**, 429 (1965).
571. S. Takahashi, T. Shibano, and N. Hagihara, *Tetrahedron Lett.* p. 2451 (1967).
572. S. Takahashi, H. Yamazaki, and N. Hagihara, *Mem. Inst. Sci. Ind. Res., Osaka Univ.* **25**, 125 (1968).
573. S. Takahashi, H. Yamazaki, and N. Hagihara, *Bull. Chem. Soc. Jap.* **41**, 254 (1968).
574. S. Takahashi, T. Shibano, and N. Hagihara, *Bull. Chem. Soc. Jap.* **41**, 454 (1968).
575. S. Takahashi, T. Shibano, and N. Hagihara, *Kogyo Kagaku Zasshi* **72**, 1798 (1969).
576. S. Takahashi, T. Shibano, and N. Hagihara, *Chem. Commun.* p. 161 (1969).
576a. A. Panunzi, A. de Renzi, and G. Paiaro, *J. Amer. Chem. Soc.* **92**, 3488 (1970).
577. A. Aguilo, *Advan. Organometal. Chem.* **5**, 321 (1967).
578. F. C. Phillips, *Amer. Chem. J.* **16**, 255 (1894).
579. J. S. Anderson, *J. Chem. Soc.* p. 971 (1934); p. 1042 (1936).
580. J. Halpern and M. Pribanic, *J. Amer. Chem. Soc.* **90**, 5943 (1968).
581. J. Smidt, W. Hafner, R. Jira, R. Sieber, T. Sedlmeier, and A. Sabel, *Angew. Chem., Int. Ed. Engl.* **1**, 80 (1962); J. Smidt, *Chem. Ind. (London)* p. 54 (1962).
582. I. I. Moiseev, M. N. Vargfatik, and Ya. K. Syrkin, *Dokl. Akad. Nauk SSSR (English Transl.)* **133**, 801 (1960).
583. E. W. Stern and M. L. Spector, *Proc. Chem. Soc., London* p. 370 (1961).
584. A. D. Ketley and L. P. Fisher, *J. Organometal. Chem.* **13**, 243 (1968).
585. J. Tsuji, M. Morikawa, and J. Kiji, *J. Amer. Chem. Soc.* **86**, 4851 (1964).
586. T. Saegusa, T. Tsuda, and K. Nishijima, *Tetrahedron Lett.* p. 4255 (1967).
587. H. Okada and H. Hashimoto, *Kogyo Kagaku Zasshi* **70**, 2152 (1967).
588. A. D. Ketley and J. A. Braatz, *J. Organometal. Chem.* **9**, P5 (1967).
589. A. D. Ketley, J. A. Braatz, J. Craig, and R. Cole, *Amer. Chem. Soc., Div. Petrol. Chem., Prepr.* **14**, B142 (1969); *Chem. Commun.* p. 1117 (1970).
590. B. L. Shaw, *Chem. Commun.* p. 464 (1968).
591. I. Moritani and Y. Fujiwara, *Tetrahedron Lett.* p. 1119 (1967).
592. Y. Fujiwara, I. Moritani, M. Matsuda, and S. Teranishi, *Tetrahedron Lett.* p. 633 (1968).
593. Y. Fujiwara, I. Moritani, and M. Matsuda, *Tetrahedron* **24**, 4819 (1968).
594. I. Moritani, Y. Fujiwara, and S. Teranishi, *Amer. Chem. Soc., Div. Petrol. Chem., Prepr.* **14**, B172 (1969).
595. Y. Fujiwara, I. Moritani, S. Danno, R. Asano, and S. Teranishi, *J. Amer. Chem. Soc.* **91**, 7166 (1969).
596. S. Danno, I. Moritani, and Y. Fujiwara, *Tetrahedron* **25**, 4809 (1969).
597. Y. Fujiwara, I. Moritani, R. Asano, H. Tanaka, and S. Teranishi, *Tetrahedron* **25**, 4815 (1969).
598. S. Danno, I. Moritani, and Y. Fujiwara, *Tetrahedron* **25**, 4819 (1969).
599. R. Hüttel, J. Kratzer, and M. Bechter, *Chem. Ber.* **94**, 766 (1961).
600. A. D. Ketley and J. Braatz, *Chem. Commun.* p. 169 (1968).
601. D. Morelli, R. Ugo, F. Conti, and M. Donati, *Chem. Commun.* p. 801 (1967).
602. D. M. Barlex, R. D. W. Kemmitt, and G. W. Littlecott, *Chem. Commun.* p. 613 (1969).
602a. P. B. Tripathy and D. M. Roundhill, *J. Amer. Chem. Soc.* **92**, 3825 (1970).
602b. B. E. Mann, B. L. Shaw, and N. I. Tucker, *Chem. Commun.* p. 1333 (1970).
603. B. Bogdanovic, M. Kröner, and G. Wilke, *Justus Liebigs Ann. Chem.* **699**, 1 (1966).
604. R. J. Alexander, N. C. Baenziger, C. Carpenter, and J. R. Doyle, *J. Amer. Chem. Soc.* **82**, 535 (1960).
605. T. Hosokawa and I. Moritani, *Tetrahedron Lett.* p. 3021 (1969).

606. P. Mushak and M. A. Battiste, *Chem. Commun.* p. 1146 (1969).
607. A. S. Gow and H. Heinemann, *J. Phys. Chem.* **64**, 1574 (1960).
608. J. T. van Gemert and P. R. Wilkinson, *J. Phys. Chem.* **68**, 645 (1964).
609. K. Kawamoto, T. Imanaka, and S. Teranishi, *Bull. Chem. Soc. Jap.* **42**, 2688 (1969).
610. W. J. Bland and R. D. W. Kemmitt, *Nature (London)* **211**, 963 (1966).
611. L. Malatesta, G. Santarella, L. M. Vallarino, and F. Zingales, *Angew. Chem.* **72**, 34 (1960).
612. A. T. Blomquist and P. M. Maitlis, *J. Amer. Chem. Soc.* **84**, 2329 (1962).
613. H. H. Freedman, *J. Amer. Chem. Soc.* **83**, 2194 and 2195 (1961).
614. P. Chini, F. Canziani, and A. Quarta, *Abstr., Inorg. Chim. Acta Symp., 1968* p. D8 (1968).
615. R. Criegee and G. Schröder, *Justus Liebigs Ann. Chem.* **623**, 1 (1959); *Angew. Chem.* **71**, 70 (1959).
616. P. V. Balakrishnan and P. M. Maitlis, *Chem. Comm.* p. 1303 (1968); *J. Chem. Soc., A* in press (1971).
617. G. B. McCauley, *Diss. Abstr.* **28**, 86B (1967).
618. L. J. Guggenberger, *Chem. Commun.* p. 512 (1968).
619. G. E. Batley and J. C. Bailar, *Inorg. Nucl. Chem. Lett.* **4**, 577 (1968).
620. N. C. Baenziger, R. C. Medrud, and J. R. Doyle, *Acta Crystallogr.* **18**, 237 (1965).
621. J. A. Ibers and R. G. Snyder, *Acta Crystallogr.* **15**, 923 (1962).
622. J. H. van den Hende and W. C. Baird, *J. Amer. Chem. Soc.* **85**, 1009 (1963).
623. M. D. Glick and L. F. Dahl, *J. Organometal. Chem.* **3**, 200 (1965).
624. J. D. Dunitz, H. C. Mez, O. S. Mills, and H. M. M. Shearer, *Helv. Chim. Acta* **45**, 647 (1962).
625. P. M. Maitlis, D. F. Pollock, M. L. Games, and W. J. Pryde, *Can. J. Chem.* **43**, 470 (1965).
626. P. M. Maitlis, *Advan. Organometal. Chem.* **4**, 95 (1966).
627. M. F. Hawthorne, D. C. Young, T. D. Andrews, D. V. Howe, R. L. Pilling, A. D. Pitts, M. Reintjes, L. F. Warren, and P. A. Wegner, *J. Amer. Chem. Soc.* **90**, 879 (1968).
628. P. M. Maitlis and K. W. Eberius, *in* "Nonbenzenoid Aromatics and Related Topics" (J. Snyder, ed.), Vol. 2. Academic Press, New York, 1971.
629. F. A. Cotton, "Chemical Applications of Group Theory," p. 180. Wiley, New York, 1963.
630. H. Frye and G. B. McCauley, *Inorg. Nucl. Chem. Lett.* **4**, 347 (1968).
631. M. S. Kabaloui, *Diss. Abstr.* **28**, 4454B (1968).
632. G. Wilkinson (Ethyl Corp.), U.S. Pat. 3,105,084 (1963); *Chem. Abstr.* **61**, 7046e (1964).
633. E. W. Abel, M. A. Bennett, and G. Wilkinson, *J. Chem. Soc.* p. 3178 (1959).
634. S. D. Robinson and B. L. Shaw, *J. Chem. Soc.* p. 5002 (1964).
635. H. Frye, E. Kuljian and J. Viebrock, *Z. Naturforsch. B* **20**, 269 (1965).
636. J. C. Trebellas, J. R. Olechowski, and H. B. Jonassen, *J. Organometal. Chem.* **6**, 412 (1966).
637. H. Dietl and P. M. Maitlis, *Chem. Commun.* p. 759 (1967).
638. E. E. Van Tamelen and D. Carty, *J. Amer. Chem. Soc.* **89**, 3922 (1967).
639. B. L. Shaw and G. Shaw, *J. Chem. Soc., A* p. 602 (1969).
639a. H. A. Tayim and A. Vassilian, *Chem. Commun.* p. 630 (1970).
640. K. L. Kaiser and P. M. Maitlis, *Chem. Commun.* p. 942 (1970).
641. M. Avram, E. Avram, G. D. Mateescu, I. G. Dinulescu, F. Chiraleu, and C. D. Nenitzescu, *Chem. Ber.* **102**, 3996 (1969).
642. M. Avram, I. G. Dinulescu, G. D. Mateescu, and C. D. Nenitzescu, *Rev. Roum. Chim.* **14**, 1191 (1969).

643. J. Tsuji, S. Hosaka, J. Kiji, and T. Susuki, *Bull. Chem. Soc. Jap.* **39**, 141 (1966).
644. P. Heimbach, *Angew. Chem. Int. Ed. Engl.* **7**, 727 (1968).
645. H. Frye, E. Kuljian, and J. Viebrock, *Inorg. Nucl. Chem. Lett.* **2**, 119 (1966).
646. H. Frye and D. Chinn, *Inorg. Nucl. Chem. Lett.* **5**, 613 (1969).
647. E. Vedejs, *J. Amer. Chem. Soc.* **90**, 4751 (1968).
648. J. Powell and B. L. Shaw, *J. Chem. Soc., A* p. 583 (1968).
649. B. L. Shaw, *Chem. Ind.* (*London*) p. 1190 (1962).
650. S. D. Robinson and B. L. Shaw, *J. Chem. Soc.* p. 4806 (1963).
651. J. Lukas and P. A. Kramer, *Abstr., Inorg. Chim. Acta Symp., 1969* p. 14 (1969).
652. R. Hüttel and H. J. Neugebauer, *Tetrahedron Lett.* p. 3541 (1964).
653. L. M. Vallarino and G. Santarella, *Gazz. Chim. Ital.* **94**, 252 (1964).
654. M. Avram, I. G. Dinulescu, G. D. Mateescu, E. Avram, and C. D. Nenitzescu, *Rev. Roum. Chim.* **14**, 1181 (1969).
655. E. Müller, K. Munk, P. Ziemek, and M. Sauerbier, *Justus Liebigs Ann. Chem.* **713**, 40 (1968).
656. G. Henrici-Olivé and S. Olivé, *Angew. Chem., Int. Ed. Engl.* **6**, 873 (1967).
657. G. Wittig and P. Fritze, *Justus Liebigs Ann. Chem.* **712**, 79 (1968).
658. S. Otsuka and M. Rossi, *J. Chem. Soc., A* p. 2631 (1968).
659. J. Browning, D. J. Cook, C. S. Cundy, M. Green, and F. G. A. Stone, *Chem. Commun.* p. 929 (1968).
660. J. P. Durand, F. Dawans, and P. Teyssie, *J. Polym. Sci., Part B* **5**, 785 (1967).
661. J. P. Durand and P. Teyssie. *J. Polym. Sci., Part B* **6**, 299 (1968).
662. P. J. Hendra and D. B. Powell, *Spectrochim. Acta* **17**, 909 (1961).
663. H. P. Fritz and H. J. Keller, *Chem. Ber.* **95**, 158 (1962).
664. H. P. Fritz and D. Sellmann, *Z. Naturforsch. B* **22**, 20 (1967).
665. W. Partenheimer, *Diss. Abstr.* **29**, 524B (1968).
666. H. P. Fritz, *Z. Naturforsch. B* **16**, 415 (1961).
667. P. M. Maitlis and F. G. A. Stone, *Proc. Chem. Soc., London* p. 330 (1962).
668. R. C. Cookson and D. W. Jones, *J. Chem. Soc.* p. 1881 (1965).
668a. T. Hosokawa and I. Moritani, *Chem. Commun.* p. 905 (1970).
669. D. F. Pollock and P. M. Maitlis, *Can. J. Chem.* **44**, 2673 (1966).
670. V. R. Sandel and H. H. Freedman, *J. Amer. Chem. Soc.* **90**, 2059 (1968).
671. P. M. Maitlis, A. Efraty, and M. L. Games, *J. Organometal. Chem.* **2**, 284 (1964).
672. P. M. Maitlis, A. Efraty, and M. L. Games, *J. Amer. Chem. Soc.* **87**, 719 (1965).
673. R. Criegee and P. Ludwig, *Chem. Ber.* **94**, 2038 (1961); R. B. King, *Inorg. Chem.* **2**, 530 (1963).
674. R. Criegee, F. Foerg, H. A. Brune, and D. Schoenleber, *Chem. Ber.* **97**, 3461 (1964).
675. W. E. Oberhansli and L. F. Dahl, *Inorg. Chem.* **4**, 150 (1965).
676. P. A. Wegner and M. F. Hawthorne, *Chem. Commun.* p. 861 (1966).
677. A. Panunzi, A. de Renzi, and G. Paiaro, *Inorg. Chim. Acta* **1**, 475 (1967).
678. J. K. Stille and D. B. Fox, *Inorg. Nucl. Chem. Lett.* **5**, 157 (1969).
679. H. H. Freedman and D. R. Petersen, *J. Amer. Chem. Soc.* **84**, 2837 (1962); G. S. Pawley, W. N. Lipscomb, and H. H. Freedman, *ibid.* **86**, 4725 (1964).
680. R. Criegee, *Angew. Chem., Int. Ed. Engl.* **1**, 519 (1962).
681. P. M. Maitlis and M. L. Games, *J. Amer. Chem. Soc.* **85**, 1887 (1963).
682. R. C. Cookson and D. W. Jones, *Proc. Chem. Soc., London* p. 115 (1963).
683. H. H. Freedman, G. A. Doorakian, and V. R. Sandel, *J. Amer. Chem. Soc.* **87**, 3019 (1965).
684. P. M. Maitlis and M. L. Games, *Chem. Ind.* (*London*) p. 1624 (1963).
685. P. M. Maitlis and A. Efraty, *J. Organometal. Chem.* **4**, 172 (1965).
686. P. M. Mailis and A. Efraty, *J. Organometal. Chem.* **4**, 175 (1965).

687. D. F. Pollock and P. M. Maitlis, *J. Organometal. Chem.* **26**, 407 (1971).
688. G. N. Schrauzer and H. E. Thyret, *J. Amer. Chem. Soc.* **82**, 6420 (1960); *Z. Naturforsch.* **B 16**, 353 (1961); **17**, 73 (1962).
689. S. D. Robinson and B. L. Shaw, *Tetrahedron Lett.* p. 1301 (1964); G. Winkhaus and H. Singer, *Chem. Ber.* **99**, 3610 (1966).
690. H. C. Volger and H. Hogeveen, *Rec. Trav. Chim. Pays-Bas* **86**, 1066 (1967).
691. K. Vrieze and H. C. Volger, *J. Organometal. Chem.* **11**, P17 (1968).
692. H. C. Volger, K. Vrieze, and A. P. Praat, *J. Organometal. Chem.* **14**, 429 (1968).
693. L. Cattalini, R. Ugo, and A. Orio, *J. Amer. Chem. Soc.* **90**, 4800 (1968).
694. J. Lewis and A. W. Parkins, *J. Chem. Soc., A* p. 1150 (1967).
695. J. Powell and B. L. Shaw, *J. Chem. Soc., A* p. 597 (1968).
696. L. Porri, A. Lionetti, G. Allegra, and A. Immirzi, *Chem. Commun.* p. 336 (1965).
697. L. Porri and A. Lionetti, *J. Organometal. Chem.* **6**, 422 (1966); A. Immirzi and G. Allegra, *Acta Crystallogr., Sect. B* **25**, 120 (1969).
698. J. E. Mahler and R. Pettit, *J. Amer. Chem. Soc.* **85**, 3955 (1963); J. E. Mahler, O. H. Gibson, and R. Pettit, *ibid.* p. 3959.
699. K. G. Ihrman and T. H. Coffield (Ethyl Corp.), U.S. Pat., 3,117,148 (1964); *Chem. Abstr.* **60**, 15914e (1964).
700. G. F. Emerson and R. Pettit, *Advan. Organometal. Chem.* **1**, 1 (1964).
701. W. W. Prichard, U.S. Pat., 2,600,571 (1952) *Chem. Abstr.* **46**, P10188f (1952).
702. P. E. Slade and H. B. Jonassen, *J. Amer. Chem. Soc.* **79**, 1277 (1957).
703. W. R. McClellan, H. H. Hoehn, A. N. Cripps, E. L. Muetterties, and B. W. Howk, *J. Amer. Chem. Soc.* **83**, 1601 (1961).
704. R. Hüttel and H. Christ, *Chem. Ber.* **96**, 3101 (1963).
705. R. Hüttel and H. Christ, *Chem. Ber.* **97**, 1439 (1964).
706. H. C. Volger, *Amer. Chem. Soc., Div. Petrol. Chem., Prepr.* **14**, B72 (1969); *Rec. Trav. Chim. Pays-Bas* **87**, 225 (1969).
707. R. Hüttel and H. Dietl, *Chem. Ber.* **98**, 1753 (1965).
708. M. Donati and F. Conti, *Inorg. Nucl. Chem. Lett.* **2**, 343 (1966).
709. J. Tsuji, S. Imamura, and J. Kiji, *J. Amer. Chem. Soc.* **86**, 4491 (1964).
710. J. Tsuji and S. Imamura, *Bull. Chem. Soc. Jap.* **40**, 197 (1967).
711. G. W. Parshall and G. Wilkinson, *Chem. Ind. (London)* p. 261 (1962).
712. G. W. Parshall and G. Wilkinson, *Inorg. Chem.* **1**, 896 (1962).
713. A. Kasahara, K. Tanaka, and K. Asamiya, *Bull. Chem. Soc. Jap.* **40**, 351 (1967).
714. R. Hüttel and H. Schmid, *Chem. Ber.* **101**, 252 (1968).
715. R. W. Howsam and F. J. McQuillin, *Tetrahedron Lett.* p. 3667 (1968); K. Dunne and F. J. McQuillin, *J. Chem. Soc., C* p. 2196 and 2200 (1970).
716. I. T. Harrison, E. Kimura, E. Bohme, and J. H. Fried, *Tetrahedron Lett.* p. 1589 (1969).
717. Y. Tezuka, T. Ogura, and S. Kawaguchi, *Bull. Chem. Soc. Jap.* **42**, 443 (1969).
718. S. Okeya, T. Ogura, and S. Kawaguchi, *Kogyo Kagaku Zasshi* **72**, 1656 (1969).
719. I. I. Moiseev, A. P. Belov, and Ya. K. Syrkin, *Izv. Akad. Nauk SSSR (English Transl.)* p. 1395 (1963).
720. A. P. Belov, M. N. Vargaftik, and I. I. Moiseev, *Izv. Akad. Nauk SSSR (English Transl.)* p. 1465 (1964).
721. I. I. Moiseev, A. P. Belov, and G. Yu. Pek, *Russ. J. Inorg. Chem.* **10**, 180 (1965).
722. A. P. Belov, G. Yu, Pek, and I. I. Moiseev, *Izv. Akad. Nauk SSSR (English Transl.)* p. 2170 (1965).
723. I. I. Moiseev and A. P. Belov, U.S.S.R. Pat., 175,506 (1965); *Chem. Abstr.* **64**, 5140b (1966).
724. J. Lukas, S. Coren, and J. E. Blom, *Chem. Commun.* p. 1303 (1969).

725. J. R. Doyle, P. E. Slade, and H. B. Jonassen, *Inorg. Syn.* **6**, 216 (1960).

726. J. M. Rowe and D. A. White, *J. Chem. Soc., A* p. 1451 (1967).

727. A. Kasahara and T. Izumi, *Bull. Chem. Soc. Jap.* **41**, 516 (1968).

728. B. L. Shaw and N. Sheppard, *Chem. Ind. (London)* p. 517 (1961).

729. R. Hüttel and H. Dietl (Badische Anilin u. Soda-Fabrik AG.), Ger. Pat., 1,214,232 (1966); *Chem. Abstr.* **64**, 19683b (1966).

730. R. F. Heck, *J. Amer. Chem. Soc.* **90**, 317 (1968).

730a. J. Lukas, P. W. N. M. van Leeuwen, H. C. Volger, and P. Kramer, *Chem. Commun.* p. 799 (1970).

731. P. M. Maitlis and M. L. Games, *Can. J. Chem.* **42**, 183 (1964).

732. A. Efraty and P. M. Maitlis, *Tetrahedron Lett.* p. 4025 (1966).

733. R. G. Schultz, *Tetrahedron Lett.* p. 301 (1964); *Tetrahedron* **20**, 2809 (1964).

734. R. G. Schultz (Monsanto Co.), U.S. Pat., 3,369,035 (1968); *Chem. Abstr.* **68**, 95986s (1968).

735. M. S. Lupin and B. L. Shaw, *Tetrahedron Lett.* p. 15 (1964).

736. M. S. Lupin, J. Powell, and B. L. Shaw, *J. Chem. Soc., A* p. 1687 (1966).

737. R. P. Hughes and J. Powell, *J. Organometal. Chem.* **20**, P17 (1969).

737a. T. Okamoto, *Chem. Commun.* p. 1126 (1970); T. Okamoto, Y. Sakakibara, and S. Kunichika, *Bull. Chem. Soc. Jap.* **43**, 2658 (1970).

738. S. R. Stevens and G. D. Shier, *J. Organometal. Chem.* **21**, 495 (1970).

739. A. D. Walsh, *Trans. Faraday Soc.* **45**, 179 (1949).

740. A. D. Ketley and J. A. Braatz, *Chem. Commun.* p. 959 (1968).

741. J. Tsuji, M. Morikawa, and J. Kiji, *Tetrahedron Lett.* p. 817 (1965).

741a. D. B. Brown, *J. Organometal. Chem.* **24**, 787 (1970).

742. J. Shono, T. Yoshimura, Y. Matsumura, and R. Oda, *J. Org. Chem.* **33**, 876 (1968).

743. R. Noyori and H. Takaya, *Chem. Commun.* p. 525 (1969).

744. P. Mushak and M. A. Battiste, *J. Organometal. Chem.* **17**, P46 (1969).

744a. F. J. Weigert, R. L. Baird, and J. R. Shapley, *J. Amer. Chem. Soc.* **92**, 6630 (1970).

745. J. Smidt and W. Hafner, *Angew. Chem.* **71**, 284 (1959).

746. I. I. Moiseev, E. A. Fedorovskaya, and Ya. K. Syrkin, *Russ. J. Inorg. Chem.* **4**, 1218 (1959).

747. W. Hafner, H. Prigge, and J. Smidt, *Justus Liebigs Ann. Chem.* **693**, 109 (1966).

748. W. H. Urry and M. B. Sullivan, *Amer. Chem. Soc., Div. Petrol. Chem., Prepr.* **14**, B131 (1969).

748a. R. Pietropaolo, P. Uguagliati, T. Boschi, B. Crociani, and U. Belluco, *J. Catal.* **18**, 358 (1970).

749. A. Bright, B. L. Shaw, and G. Shaw, *Amer. Chem. Soc., Div. Petrol. Chem., Prepr.* **14**, B81 (1969).

750. I. A. Zakharova and I. I. Moiseev, *Izv. Akad. Nauk SSSR (English Transl.)* p. 1826 (1964).

751. W. T. Dent and R. Long (ICI Ltd.), Brit. Pat., 1,082,248 (1967); *Chem. Abstr.* **68**, 114748j (1968).

752. J. K. Nicholson, J. Powell, and B. L. Shaw, *Chem. Commun.* p. 174 (1966).

753. M. Sakakibara, Y. Takahashi, S. Sakai, and Y. Ishii, *Chem. Commun.* p. 396 (1969).

754. T. Krück and M. Noack, *Chem. Ber.* **97**, 1693 (1964).

755. E. O. Fischer and A. Maasböl, *Angew. Chem., Int. Ed. Engl.* **3**, 580 (1964); O. S. Mills and A. D. Redhouse, *ibid.* **4**, 1082 (1965).

756. Shell Internationale Research Maatschappij, Neth. Pat., 6,503,362 (1966); *Chem. Abstr.* **66**, 65634s (1967); additional to Neth. Pat., 6,408,476 (1965); *Chem. Abstr.* **63**, 499c (1965).

757. I. I. Moiseev, M. N. Vargaftik and Ya. K. Syrkin, *Izv. Akad. Nauk SSSR* (*English Transl.*) p. 728 (1964).
758. J. Tsuji and N. Iwamoto, *Chem. Commun.* p. 828 (1966).
759. E. O. Fischer and G. Burger, *Z. Naturforsch.* B **16**, 702 (1961).
760. G. Wilke (Studiengesellschaft Kohle m.b.H.), Belg. Pat., 631,172 (1963); *Chem. Abstr.* **61**, 690d (1964).
761. N. M. Klimenko, I. I. Moiseev, and Ya. K. Syrkin, *Izv. Akad. Nauk SSSR* p. 1355 (1961).
762. T. G. Hewitt and J. J. De Boer, *Chem. Commun.* p. 1413 (1968).
762a. G. Raper and W. S. McDonald, *Chem. Commun.* p. 655 (1970).
762b. W. S. McDonald, B. E. Mann, G. Raper, B. L. Shaw, and G. Shaw, *Chem. Commun.* p. 1254 (1969).
763. S. O'Brien, *J. Chem. Soc., A* p. 9 (1970).
764. R. C. Ramey and G. L. Statton, *J. Amer. Chem. Soc.* **88**, 4387 (1966).
765. S. D. Robinson and B. L. Shaw, *J. Organometal. Chem.* **3**, 367 (1965).
766. E. O. Fischer and H. Werner, *Tetrahedron Lett.* p. 17 (1961).
767. D. L. Tibbetts and T. L. Brown, *J. Amer. Chem. Soc.* **91**, 1108 (1969).
768. H. Bönnemann, B. Bogdanovic, and G. Wilke, *Angew. Chem., Int. Ed. Engl.* **6**, 804 (1967).
769. J. K. Becconsall and S. O'Brien, *J. Organometal. Chem.* **9**, P27 (1967).
770. M. S. Lupin, J. Powell, and B. L. Shaw, *J. Chem. Soc., A* p. 1410 (1966).
771. H. P. Fritz, *Chem. Ber.* **94**, 1217 (1961); see also D. M. Adams and A. Squire, *J. Chem. Soc., A* p. 1808 (1970).
772. K. Shobatake and K. Nakamoto, *J. Amer. Chem. Soc.* **92**, 3339 (1970).
773. S. P. Gubin, A. Z. Rubezhov, B. L. Winch, and A. N. Nesmeyanov, *Tetrahedron Lett.* p. 2881 (1964).
774. B. F. Hegarty and W. Kitching, *J. Organometal. Chem.* **6**, 578 (1966).
775. F. R. Hartley, *J. Organometal. Chem.* **21**, 227 (1970).
776. S. P. Gubin and L. I. Denisovich, *Izv. Akad. Nauk SSSR* p. 149 (1966).
777. M. S. Lupin and M. Cais, *J. Chem. Soc., A* p. 3095 (1968).
778. S. F. A. Kettle and R. Mason, *J. Organometal. Chem.* **5**, 573 (1966).
779. A. Veillard, *Chem. Commun.* pp. 1022 and 1427 (1969).
780. I. H. Hillier and R. M. Canadine, *Discuss. Faraday Soc.* **47**, 27 (1969).
781. H. C. Dehm and J. C. W. Chien, *J. Amer. Chem. Soc.* **82**, 4429 (1960).
782. J. E. Nordlander and J. D. Roberts, *J. Amer. Chem. Soc.* **81**, 1769 (1959); G. Whitesides, J. E. Nordlander, and J. D. Roberts, *Discuss. Faraday Soc.* **34**, 185 (1962).
783. J. Powell, S. D. Robinson, and B. L. Shaw, *Chem. Commun.* p. 78 (1965).
784. F. A. Cotton, J. W. Faller, and A. Musco, *Inorg. Chem.* **6**, 179 (1967).
785. K. Vrieze, C. Maclean, P. Cossee, and C. W. Hilbers, *Rec. Trav. Chim. Pay-Bas* **85**, 1077 (1966).
786. K. Vrieze, P. Cossee, A. P. Praat, and C. W. Hilbers, *J. Organometal. Chem.* **11**, 353 (1968).
787. K. Vrieze, A. P. Praat, and P. Cossee, *J. Organometal. Chem.* **12**, 533 (1968).
788. B. L. Shaw and E. Singleton, *J. Chem. Soc., A* p. 1683 (1967).
789. B. L. Shaw and E. Singleton, *J. Chem. Soc., A* p. 1972 (1967).
790. G. L. Statton and K. C. Ramey, *J. Amer. Chem. Soc.* **88**, 1327 (1966).
791. D. L. Tibbetts and T. L. Brown, *J. Amer. Chem. Soc.* **92**, 3031 (1970).
792. D. Walter and G. Wilke, *Angew. Chem., Int. Ed. Engl.* **5**, 897 (1966).
793. G. Paiaro and A. Musco, *Tetrahedron Lett.* p. 1583 (1965).

793a. P. W. N. M. van Leeuwen, A. P. Praat, and M. van Diepen, *J. Organometal. Chem.* **24**, C31 (1970).

793b. C. W. Alexander, W. R. Jackson, and R. Spratt, *J. Amer. Chem. Soc.* **92**, 4990 (1970).

794. C. W. Fong and W. Kitching, *Aust. J. Chem.* **22**, 477 (1969).

795. R. Hüttel and B. König, *Amer. Chem. Soc., Div. Petrol. Chem., Prepr.* **14**, B35 (1969).

796. P. W. M. N. van Leeuwen, K. Vrieze, and A. P. Praat, *J. Organometal. Chem.* **20**, 219 (1969).

797. P. W. M. N. van Leeuwen and A. P. Praat, *J. Organometal. Chem.* **21**, 501 (1970).

798. K. Vrieze, P. Cossee, C. W. Hilbers, and A. P. Praat, *Rec. Trav. Chim. Pays-Bas* **86**, 769 (1967).

799. J. C. W. Chien and H. C. Dehm, *Chem. Ind.* (*London*) p. 745 (1961).

800. F. DeCandia, G. Maglio, A. Musco, and G. Paiaro, *Inorg. Chim. Acta* **2**, 233 (1968).

801. P. Ganis, G. Maglio, A. Musco, and A. L. Segre, *Inorg. Chim. Acta* **3**, 266 (1969).

802. J. W. Faller, M. J. Incorvia, and M. E. Thomsen, *J. Amer. Chem. Soc.* **91**, 518 (1969).

803. J. W. Faller and M. J. Incorvia, *J. Organometal. Chem.* **19**, P13 (1969).

804. P. Corradini, G. Maglio, A. Musco, and G. Paiaro, *Chem. Commun.* p. 618 (1966).

805. J. W. Faller and M. E. Thomsen, *J. Amer. Chem. Soc.* **91**, 6871 (1969); J. W. Faller, M. E. Thomsen, and M. J. Mattina, *J. Amer. Chem. Soc.* in press (1971); G. Maglio, A. Musco, and R. Palumbo, *Inorg. Chim. Acta* **4**, 153 (1970).

806. J. Powell, *J. Amer. Chem. Soc.* **91**, 4311 (1969).

806a. P. W. N. M. van Leeuwen and A. P. Praat, *Rec. Trav. Chim. Pays-Bas* **89**, 321 (1970).

807. K. Vrieze and H. C. Volger, *J. Organometal. Chem.* **9**, 537 (1967).

808. D. N. Lawson, J. A. Osborn, and G. Wilkinson, *J. Chem. Soc., A* p. 1733 (1966).

809. K. C. Ramey, D. C. Lini, and W. B. Wise, *J. Amer. Chem. Soc.* **90**, 4275 (1968).

810. H. C. Volger and K. Vrieze, *J. Organometal. Chem.* **13**, 479 (1968).

811. J. W. Faller and M. J. Incorvia, *Inorg. Chem.* **7**, 840 (1968).

812. E. J. Corey, L. S. Hegedus, and M. F. Semmelhack, *J. Amer. Chem. Soc.* **90**, 2417 (1968).

813. E. J. Corey, M. F. Semmelhack, and L. S. Hegedus, *J. Amer. Chem. Soc.* **90**, 2416 (1968).

814. R. J. Goodfellow and L. M. Venanzi, *J. Chem. Soc., A* p. 784 (1966).

815. B. L. Shaw, *Proc. Chem. Soc., London* p. 247 (1960).

816. R. Riemschneider, E. Hoerner, and F. Herzel, *Monatsh. Chem.* **92**, 777 (1961).

817. A. N. Nesmeyanov, A. Z. Rubezhov, S. P. Gubin, and Z. B. Mitroshina, *Izv. Akad. Nauk SSSR* p. 739 (1966).

818. M. Sakakibara, Y. Takahashi, S. Sakai, and Y. Ishii, *Inorg. Nucl. Chem. Lett.* **5**, 427 (1969).

819. W. Keim, *Angew. Chem., Int. Ed. Engl.* **7**, 879 (1968).

820. S. P. Gubin, A. Z. Rubezhov, L. I. Denisovich, and A. N. Nesmeyanov, *Izv. Akad. Nauk SSSR* p. 1680 (1966).

821. A. N. Nesmeyanov, A. Z. Rubezhov, L. A. Leites, and S. P. Gubin, *J. Organometal. Chem.* **12**, 187 (1968).

822. C. Grindrod, M.Sc. Thesis, McMaster University, Hamilton, Ontario (1966).

823. A. N. Nesmeyanov, A. Z. Rubezhov, and S. P. Gubin, *Izv. Akad. Nauk SSSR* p. 194 (1966).

824. A. N. Nesmeyanov, S. P. Gubin, and A. Z. Rubezhov, *J. Organometal. Chem.* **16**, 163 (1969).

825. B. C. Benson, W. R. Jackson, K. K. Joshi, and D. T. Thompson, *Chem. Commun.* p. 1506 (1968).

826. L. M. Zaitsev, A. P. Belov, M. N. Vargaftik, and I. I. Moiseev, *Russ. J. Inorg. Chem.* **12**, 203 (1967).

827. M. Donati and F. Conti, *Tetrahedron Lett.* p. 4953 (1966).

828. J. Tsuji, H. Takahashi, and M. Morikawa, *Tetrahedron Lett.* p. 4387 (1965).

829. J. Tsuji, H. Takahashi, and M. Morikawa, *Kogyo Kagaku Zasshi* **69**, 920 (1966).

830. Y. Takahashi, S. Sakai, and Y. Ishii, *Chem. Commun.* p. 1092 (1967); Y. Takahashi, K. Tsukiyama, S. Sakai, and Y. Ishii, *Tetrahedron Lett.* p. 1913 (1970).

831. H. Christ and R. Hüttel, *Angew. Chem.* **75**, 921 (1963).

832. R. Hüttel and P. Kochs, *Chem. Ber.* **101**, 1043 (1968).

833. T. A. Schenach and F. F. Caserio, *J. Organometal. Chem.* **18**, P17 (1969).

834. D. Medema and R. van Helden, *Amer. Chem. Soc., Div. Petrol. Chem., Prepr.* **14**, B92 (1969).

834a. A. P. Belov, I. I. Moiseev, N. G. Satsko, and Ya. K. Syrkin, *Izv. Akad. Nauk SSSR* p. 2573 (1969).

834b. H. C. Volger, K. Vrieze, J. W. F. M. Lemmers, A. P. Praat, and P. W. N. M. van Leeuwen, *Inorg. Chim. Acta*, **4**, 435 (1970).

835. J. Tsuji, J. Kiji, and M. Morikawa, *Tetrahedron Lett.* p. 1811 (1963).

836. J. Tsuji, J. Kiji, and S. Hosaka, *Tetrahedron Lett.* p. 605 (1964).

837. J. Tsuji and T. Susuki, *Tetrahedron Lett.* p. 3027 (1965).

838. J. Tsuji and S. Hosaka, *J. Amer. Chem. Soc.* **87**, 4075 (1965).

839. T. Susuki and J. Tsuji, *Bull. Chem. Soc. Jap.* **41**, 1954 (1968).

840. R. Long and G. H. Whitfield, *J. Chem. Soc.* p. 1852 (1964).

841. Shell Internationale Research Maatschappij N. V., Brit. Pat., 1,080,867 (1965); *Chem. Abstr.* **63**, 499c (1965).

842. Y. Takahashi, S. Sakai, and Y. Ishii, *J. Organometal. Chem.* **16**, 177 (1969).

843. L. F. Fieser and M. Fieser, "Advanced Organic Chemistry," p. 486. Reinhold, New York, 1961.

844. G. D. Shier, *Amer. Chem. Soc., Div. Petrol. Chem., Prepr.* **14**, B123 (1969).

845. K. Vrieze, H. C. Volger, M. Gronert, and A. P. Praat, *J. Organometal. Chem.* **16**, P19 (1969); **21**, 467 (1970).

846. T. Kashiwagi, N. Yasuoka, N. Kasai, and M. Kukudo, *Chem. Commun.* p. 317 (1969).

847. P. Racanelli, G. Pantini, A. Immirzi, G. Allegra, and L. Porri, *Chem. Commun.* p. 361 (1969).

848. W. T. Dent, R. Long, and G. H. Whitfield, *J. Chem. Soc.* p. 1588 (1964).

849. R. Long and G. H. Whitfield (ICI Ltd.), Brit. Pat., 987,274 (1965); *Chem. Abstr.* **62**, 16065h (1965).

850. I. L. Mador and J. A. Scheber (National Distillers and Chem. Corp.), Fr. Pat., 1,419,758 (1965); *Chem. Abstr.* **65**, 13553d (1966).

851. E. J. Smutny, *J. Amer. Chem. Soc.* **89**, 6793 (1967).

852. R. K. Armstrong, *J. Org. Chem.* **31**, 618 (1966).

853. F. D. Mango and I. Dvoretsky, *J. Amer. Chem. Soc.* **88**, 1654 (1966).

854. H. Werner and J. H. Richards, *J. Amer. Chem. Soc.* **90**, 4976 (1968).

855. H. D. Kaesz, R. B. King, and F. G. A. Stone, *Z. Naturforsch. B* **15**, 682 (1960).

856. E. J. Corey and M. F. Semmelhack, *J. Amer. Chem. Soc.* **89**, 2755 (1967).

857. Studiengesellschaft Kohle m.b.H., Neth. Pat., 6,409,179 (1965); *Chem. Abstr.* **63**, 5770h (1965).

858. G. Allegra, F. Lo Giudice, G. Natta, U. Giannini, G. Fagherazzi, and P. Pino, *Chem. Commun.* p. 1263 (1967).

859. E. W. Gowling and S. F. A. Kettle, *Inorg. Chem.* **3**, 604 (1964).

860. D. L. Weaver and R. L. Tuggle, *J. Amer. Chem. Soc.* **91**, 6506 (1969); see also M. D. Rausch, R. M. Tuggle, and D. L. Weaver, *J. Amer. Chem. Soc.* **92**, 4981 (1970).

861. K. Ofele, *J. Organometal. Chem.* **22**, C9 (1970).

862. M. Dubeck and A. H. Filbey, *J. Amer. Chem. Soc.* **83**, 1257 (1961); cf. K. W. Barnett, F. D. Mango, and C. A. Reilly, *ibid.* **91**, 3387 (1969).

863. E. O. Fischer and C. Palm, *Chem. Ber.* **91**, 1725 (1958).

864. D. W. McBride, R. L. Pruett, E. Pitcher, and F. G. A. Stone, *J. Amer. Chem. Soc.* **84**, 497 (1962).

865. G. Wilkinson, *Progr. Inorg. Chem.* **1**, 17 (1959); personal communication (1968).

866. E. O. Fischer and H. Schuster-Woldan, *Chem. Ber.* **100**, 705 (1967).

867. M. Rosenblum, "Chemistry of the Iron Group Metallocenes," p. 13. Wiley, New York, 1965.

868. E. O. Fischer and H. Wawersik, *J. Organometal. Chem.* **5**, 559 (1966).

869. H. J. Keller and H. Wawersik, *J. Organometal. Chem.* **8**, 185 (1967).

870. H. P. Fritz and K. E. Schwartzhans, *J. Organometal. Chem.* **5**, 181 (1966).

871. E. O. Fischer and A. Vogler, *Z. Naturforsch. B* **18**, 771 (1963).

872. E. O. Fischer and H. Schuster-Woldan, *Z. Naturforsch. B* **19**, 766 (1964).

873. E. O. Fischer and R. Jira, *Z. Naturforsch. B* **9**, 618 (1954); E. O. Fischer, O. Beckert, W. Hafner, and H. O. Stahl, *ibid.* **10**, 598 (1955); T. S. Piper, F. A. Cotton, and G. Wilkinson, *J. Inorg. Nucl. Chem.* **1**, 165 (1955).

874. A. P. Cox, L. F. Thomas, and J. Sheridan, *Nature (London)* **181**, 1157 (1958).

875. E. O. Fischer, H. Schuster-Woldan, and K. Bittler, *Z. Naturforsch. B* **18**, 429 (1963).

876. R. D. Fischer, A. Vogler, and K. Noack, *J. Organometal. Chem.* **7**, 135 (1967); K. Noack, *ibid.* p. 151; A. R. Manning, *J. Chem. Soc., A* p. 1319 (1968); p. 1498 (1969); J. G. Bullitt, F. A. Cotton, and T. J. Marks, *J. Amer. Chem. Soc.* **92**, 2155 (1970).

877. J. Smidt and R. Jira, *Angew. Chem.* **71**, 651 (1959).

878. S. D. Robinson and B. L. Shaw, *J. Chem. Soc.* p. 1529 (1965).

879. V. A. Semion, A. Z. Rubezhov, Yu. T. Struchkov, and S. P. Gubin, *Zh. Strukt. Khim.* **10**, 151 (1969).

880. E. O. Fischer and H. H. Lindner, *J. Organometal. Chem.* **1**, 307 (1964); **2**, 229 (1964).

881. H. H. Lindner and E. O. Fischer, *J. Organometal. Chem.* **12**, P18 (1968).

882. E. A. Hall and E. L. Amma, *Chem. Commun.* p. 622 (1968); R. W. Turner and E. L. Amma, *J. Amer. Chem. Soc.* **88**, 1877 and 3243 (1966).

883. F. A. Cotton, W. A. Dollase, and J. S. Wood, *J. Amer. Chem. Soc.* **85**, 1543 (1963).

884. P. J. Wheatley, *Perspect. Struct. Chem.* **1**, 1 (1967).

885. R. E. Davis and R. Pettit, *J. Amer. Chem. Soc.* **92**, 716 (1970).

886. J. W. Kang, R. F. Childs, and P. M. Maitlis, *J. Amer. Chem. Soc.* **92**, 720 (1970).

Author Index

Numbers in parentheses are reference numbers and indicate that an author's work is referred to although his name is not cited in the text. Numbers in italics show the page on which the complete reference is listed.

A

Abel, E. W., 151(633), *282*
Adams, D. M., 41(256, 257), 46, 47, 86(438), 97(464), 191(464), 205, *272, 273, 277, 278, 286*
Addison, C. C., 4(11), 33(11), 47, 52(11), *265*
Aguilo, A., 134(577), *281*
Ahrland, S., 35, *270*
Albano, V., 7(24), 15(24, 79), 16, *265, 267*
Alderman, P. R. H., 113, *279*
Alexander, C. W., 214, *287*
Alexander, R. J., 104, 151(604), *281*
Allegra, G., 28(157, 158), 38(157, 158), 39(157, 158), 174(696, 697), 249(847), 251(858), 259, 260, *269, 284, 288*
Allen, A. D., 17, 21(124), 22(124), 122, 133(94), *267, 268*
Allen, F. H., 42(259), *272*
Alyea, E. C., 6(16), 50(16), *265*
Amma, E. L., 63, 261(882), *275, 289*
Anderson, C. B., 76, 77(410), 98(410), *276*
Anderson, J. S., 134, 136, *281*
Andrews, T. D., 148(627), 163(627), *282*
Angoletta, M., 10(34), 14(34, 60), 17(34), 59(34, 60), *266*
Araki, M., 71(392), *276*
Arloth, W., 21(122), 22(122), *268*
Armstrong, R. K., 250(852), *288*
Asamiya, K., 157(713), 181(713), 228(713), *284*
Asano, R., 137(595, 597), *281*
Ascher, E., 4(9), 51, *265*
Ashcroft, S. J., 45(283a), 69, *273, 275*
Ashley-Smith, J., 81, 251(427), *277*
Avram, E., 151(641), 155(654), *282, 283*
Avram, M., 77(419), 151(419, 641, 642), 155, *277, 282, 283*

B

Baddley, W. H., 119(548), 123(548), 129(548), *280*
Badley, E. M., 60(341a), *274*
Baenziger, N. C., 38(226), 39(226, 247, 248), 46, 77(418), 103(418), 112, 136(418), 140(604), 146(620), 148(226), 151(604), *271, 272, 276, 281, 282*
Bailar, J. C., 10(36), 146(619), 159(619), *266, 282*
Bailey, N. A., 34(176), 38(176), 39(176, 254), 88(448), 97(448), 123(448), 191(448), *270, 272, 277*
Baird, M. C., 21(120), 22(120), 23(120), 24(120), 28(120), 68(370a), 72(120), 73(400), 86(120), 213, 224(120), 255(370a), *268, 275, 276*
Baird, R. L., 193(744a), *285*
Baird, W. C., 146(622), 173(622), *282*
Balakrishnan, P. V., 145(616), 154(616), 160(616), 256, *282*
Balch, A. L., 60(341a), 82(432), *274, 277*
Bamford, C. H., 24(136), *268*
Bandoli, G., 119(548), 123(548), 129(548), *280*
Barlex, D. M., 139, *281*
Barnett, K. W., 254(862), *289*
Bartlett, N., 4(8, 10), 38(246), 43(8, 10, 246, 270), 50(270), 52, *265, 271, 272, 273*
Basch, H., 31(166), *269*
Basolo, F., 9(30), 11(43), 12(30), 14(30), 15(81), 16(30, 82, 83), 17, 18(82, 96), 35, 36(43, 193), 37, 49, 51(43), 103, *266, 267, 270, 273*
Basso Ricci, G. M., 7(25), 15(25), 54(25), *265*
Batley, G. E., 146(619), 159(619), *282*

291

Dewar, M. J. S., 106(485), *278*
Dewhirst, K. C., 17(93), 230(93), 245(93), 247(93), 250(93), *267*
Dietl, H., 45(282), 74(282, 404), 80(282, 404), 99(282), 108(282, 404), 109(499), 111(282, 404), 151(637), 159(237), 160 (637), 161(637), 180(499, 707), 185, 203, 205(499), 235(707), 236(707), 238(499), *272, 276, 279, 282, 284, 285*
Dietrich, H., 119(546), 123(546, 553, 554, 555), 146(555), 200(553), 204(553), *280*
Di Luzio, J. W., 18(100), *267*
Dinulescu, I. G., 151(641, 642), 155(654), *282, 283*
Dixon, K. R., 57(324), *274*
Dobinson, G. C., 10(41), *266*
Dolcetti, G., 33(173), 34(173, 186), 39(173), *269, 270*
Dollase, W. A., 261(883), *289*
Donati, M., 109(504, 505), 110(506), 112 (504, 506), 137, 138(601), 151(505, 506), 153(506), 180(601), 181(708), 184(601), 186, 229, 233, *279, 281, 284, 288*
Donnay, G., 72(394), *276*
Doorakian, G. A., 170(683), *283*
Doyle, J. R., 38(226), 39(226), 77(418, 420), 84(435), 102(469), 103, 136(418), 140 (604), 146(620), 148(226), 151(420, 604), 184, 257(469), *271, 276, 277, 278, 281, 282, 285*
Dreissig, W., 119(546), 123(546), *280*
Drew, D. A., 84(435), *277*
Drinkard, W. C., 21(115), 105(115), *268*
Duax, W. L., 38(236), 39(236), *271*
Dubeck, M., 254(862), *289*
Dubini, M., 28, 112, *269*
Duddell, D. A., 42(265), *272*
Dudeck, E., 68(367), 255(367), *275*
Duncanson, L. A., 41(256), 42(259), 103, 106, 108 (488, 489, 490), 111(484), 114 (484), 120 (488, 489, 490), 131(484, 489, 490), *272, 278*
Dunitz, J. D., 148(624), *282*
Dunne, K., 181(715), 183(715), 228(715), 233, 239(715), *284*
Durand, J. P., 159(660, 661), 162(660), *283*
Durig, J. R., 42(260), 46(287), *272, 273*
Duval, C., 4(6), 52(6), *265*
Dvoretsky, I., 251(853), *288*

E

Eaborn, C., 21(125), 22(125), *268*
Eastess, J. W., 14(68), *266*
Eastmond, G. C., 24(136), *268*
Eaton, D. R., 19(108), 21(115), 105(115), *268*
Eatough, D. J., 33(169a), 35(189), *269, 270*
Eberius, K. W., 149(628), 160(628), 168 (628), 171(628), 174(628), *282*
Efraty, A., 163(671, 672), 165(672), 171(672, 685, 686), 188(672, 732), 228(672), 241 (672, 732), 257(672), 258(672), *283, 285*
Eisenberg, R., 104(478), *278*
Emerson, G. F., 174, *284*
Enemark, J. H., 60(341a), *274*
Ercoli, R., 7, 14(19), 55(19), *265*
Eremenko, N. K., 7(21), 54(21), *265*
Erskine, G. J., 40(196), 91(196, 452), 123 (196), *270, 277*
Ettore, R., 33(173), 34, 39(173), *269, 270*
Evans, J. G., 42(265), *272*

F

Fackler, J. P., 114(540), *280*
Fagherazzi, G., 251(858), *288*
Fahey, D. R., 81(428), 96(428), *277*
Faller, J. W., 209(784), 210(784), 214(784), 221, 222, 223(805), 224(811), *286, 287*
Faraone, G., 103(471), *278*
Fasman, A. B., 58, *274*
Fedorovskaya, E. A., 193(746), 200(746), *285*
Fedoseev, I. V., 58(333, 336), *274*
Fernelius, W. C., 14(69), *267*
Ferraris, G., 38(238), 39(238), *271*
Field, B. O., 47, *273*
Field, J. E., 111(513), 114(513), 133(513), *279*
Fieser, L. F., 246(843), *288*
Fieser, M., 246(843), *288*
Filbey, A. H., 254(862), *289*
Fischer, E. O., 12(46, 47), 15(46, 47), 56 (318), 59, 151, 185, 195(755), 197, 200 (759), 204(766), 254, 255, 256, 259, *266, 274, 285, 286, 289*

Johnson, M. P., 21(117), 72(117), 86(117), 197(117), *268*
Jonas, K., 105(482), *278*
Jonassen, H. B., 63(349), 111(513), 114 (513, 538), 121(349), 133(513, 538), 151 (636), 152(636), 154(636), 159(636), 176, 184, *275, 279, 280, 282, 284, 285*
Jones, D., 59(338), 185(338), *274*
Jones, D. W., 162(668), 169(668), 170(682), *283*
Jones, F. N., 124(558), *280*
Jonkhoff, T., 43(272), *272*
Joshi, K. K., 232(825), *287*

K

Kaesz, H. D., 114(531), 251(885), *280, 288*
Kaiser, K. L., 151(640), *282*
Kajimoto, T., 38(223), 221(223), *271*
Kakudo, M., 7(23), 14(23), 23(23), 24(140), 38(23, 220), 39(220), 72(23), 73(23), 181(220), 200(220), *265, 268, 271*
Kalaloui, M. S., 151(631), *282*
Kalinichenko, I. I., 50, *273*
Kang, J. W., 63(350), 261(886), *275, 289*
Kangas, L. F., 18(96), *267*
Kapoor, P. N., 21(125), 22(125), *268*
Kasahara, A., 70(380), 79, 82(431), 98, 157 (713), 181(713), 185(727), 228(713), *275, 277, 284, 285*
Kasai, N., 7(23), 14(23), 23(23), 24(140), 38(23, 220), 39(220), 72(23), 73(23), 181(220), 200(220), 249(846), *265, 268, 271, 288*
Kashiwagi, T., 7(23), 14(23), 23(23), 24 (140), 38(23), 72(23), 73(23), 249(846), *265, 268, 288*
Kaska, W. C., 27(150), *269*
Kawaguchi, S., 38(220), 39(220), 181(220, 717, 718), 200(220), *271, 284*
Kawamoto, K., 143, *282*
Keating, T., 12(50), 94(50), *266*
Keeton, M., 88(448), 97(448), 123(448), 191 (448), *277*
Keim, W., 13(54), 17(93), 92, 94, 96(453, 456), 98(456), 211(54), 225(54), 226(54), 230, 245(93), 247(93), 248(54), 250(54, 93), 251(54), *266, 267, 277, 278, 287*

Keiter, R. L., 42(264), *272*
Keller, H. J., 30(164), 159(663), 255(869), *269, 283, 289*
Kemmitt, R. D. W., 21(121, 131), 22(121, 131, 134), 23(121, 134, 135), 72(121, 135), 139(602), 143, *268, 281, 282*
Kennard, C. H. L., 39(249), *272*
Ketley, A. D., 110, 135, 136, 138, 141, 178, 179(600), 181, 183, 191, 192(589), 193 (589), 226, 239(740), 240, *279, 281, 285*
Kettle, S. F. A., 112(519), 206, 252, *279, 286, 288*
Kharasch, M. S., 45, 107, 109, 133(281), *272*
Kiji, J., 135(585), 151(643), 168(643), 181 (709), 191(741), 242(709, 835, 836), 250 (835), *281, 282, 284, 285, 288*
Kilbourn, B. T., 38(219, 239), 39(219, 239), 113(529), 123(529), 201, 203(219), *271, 279*
Kimura, E., 181(716), *284*
King, G. W., 126(560), 127, *280*
King, R. B., 108, 251(855), *278, 288*
Kingston, J. V., 57, 105, *274*
Kinoshita, Y., 38(234), 39(234), *271*
Kinusaga, T., 113(525), *279*
Kirsch, W. B., 114(538), 133(538), *280*
Kiseleva, N. V., 45(280), *272*
Kistner, C. R., 77(418), 84, 103(418), 136 (418), *276, 277*
Kitano, Y., 38(223), 221(223), *271*
Kitching, W., 45(283), 78(422), 205(774), 215, 223(794), *273, 277, 286, 287*
Klanberg, F., 29, *269*
Kliegman, J. M., 79, 98(423), *277*
Klimenko, N. M., 197, *286*
Kline, J. B., 122(552), *280*
Knoth, W. H., 20(111), *268*
Knyazeva, A. N., 49(296), *273*
Kober, E., 56, *274*
Kochs, P., 235(832), *288*
Kohll, C. F., 57(326), 135(326), 239(326), 241(326), 242(326), 243(326), 244(326), 245(326), 246(326), 248(326), 250(326), *274*
König, B., 216, 225, *287*
König, J., 55(316), *274*
Koo, C. H., 119(547), *280*
Kowala, C., 96(459), *278*
Kramer, P., 187(730a), *285*
Kramer, P. A., 153, *283*

O

P

Subject Index

Palladium complexes are indexed under the name of the ligand, other complexes under the metal.

A

Acetaldehyde, formation of, 134
Acetals, formation of, 134, 135
Acetylene complexes, 108, 110, 119–122, 133
  bonding in, 123–130
  electrophilic attack on, 129, 138
  nucleophilic attack on, 129
Acetylenes, reactions with PdCl$_2$, 79, 80, 108
Acetylides, 61
Acidity, of coordinated water, 131, 195
Acids and bases, hard and soft, 35
Acyl complexes, 85, 99
Aldehydes (and ketones), from olefin oxidation, 177
Alkyl complexes
  decomposition of, 70
  insertion reactions of, 99, 100
  preparation of, 69, 72, 136
  reactions with acetate, 101
    with acids, 96
    with bases, 97, 103
    with cyanide, 97
    with halogens, 100
    with hydrides, 98
    with hydrogen, 98
    with phosphines, 95, 99
  stereochemistry of, 70
Alkynyl, complexes, 61
Allene
  π-complexes of, 249
  reactions with π-allylic complexes, 190, 248, 249
    with Pd-alkyls, 190
    with PdCl$_2$, 188
Allylation reactions, 234
π-Allylic complexes
  asymmetric, 207–227
    exchange in, 211–221
    NMR, 210, 219, 222, 223, 225

π-Allylic complexes—*cont.*
  bonding, 175, 205–207
  cationic, 213, 242, 249
  electronic spectra, 205
  five-coordinate, 211–213, 224, 225, 251
π-Allylic palladium acetates, 200, 223, 228, 249
  reactions with allenes, 248, 249
    with 1,5-COD, 249
    with 1,3-dienes, 245, 247
    with Hg, 231
  reduction of, 239
π-Allylic palladium acetylacetonates, 190, 202, 228, 242, 245
π-Allylic palladium carboxylates, 228, 245, 249
π-Allylic palladium cyclopentadienyls, 201, 228, 230, 257
π-Allylic palladium halides
  adducts with arsines, 219–221
    with cyclohexanone oxime, 221
    with PdCl$_2$, 229
    with phosphines, 208–216, 225, 227
  as catalysts, 250
  carbonylation to unsaturated acids, 242, 243
  decomposition
    to alkanes, 239
    to allylic halides, 233, 238, 241
    to benzofulvenes, 233
    to 1,4-dichloro-2-butene, 243
    to dienes, 227, 233, 236, 247
    of EtOOCCHN$_2$ by, 250
    to glyoxals, 237
    to ketones, 238
    to olefins, 233, 235, 236
    to trienes, 247
    to tetraenes, 247
    to unsaturated aldehydes, 237